Birds of the UNITED ARAB EMIRATES

Dedicated to

All Our Feathered Friends

Published with the assistance of:

هيئة البيئة - أبوظبي
Environment Agency - ABU DHABI

Abu Dhabi Global
Environmental Data Initiative

The Environment Agency - Abu Dhabi (EAD) is charged with conservation and protection of the environment, flora and fauna of the Emirate of Abu Dhabi, part of the United Arab Emirates, and, through the Abu Dhabi Global Environmental Data Initiative (AGEDI), launched in association with UNEP, is responsible for the collection and management of environmental data. Through a Memorandum of Understanding with the Emirates Bird Records Committee (EBRC), it also maintains the national database of UAE bird records.

www.ead.ae www.agedi.ae

HELM FIELD GUIDES

Birds of the UNITED ARAB EMIRATES

Simon Aspinall and Richard Porter

Illustrated by
John Gale, Mike Langman and Brian Small

CHRISTOPHER HELM
LONDON

HELM
Bloomsbury Publishing Plc
50 Bedford Square, London, WC1B 3DP, UK

BLOOMSBURY, HELM and the Helm logo are trademarks of Bloomsbury Publishing Plc

First published in the United Kingdom in 2011

A catalogue record for this book is available from the British Library

Library of Congress Cataloguing-in-Publication data has been applied for

ISBN: PB: 978-1-4729-8257-5

Typeset in the United Kingdom by Julie Dando, Fluke Art
Printed and bound in Great Britain

MIX
Paper from responsible sources
FSC® C013604

To find out more about our authors and books visit www.bloomsbury.com and sign up for our newsletters

CONTENTS

FOREWORD

At the junction of two of the world's great bio-geographical zones, Africa and the Palearctic, covering much of Eurasia, the United Arab Emirates is internationally recognised for having one of the most diverse bird populations in the world. Several million birds, of over three hundred species, pass through the country or winter here each year. The UAE national checklist, compiled by the Emirates Bird Records Committee (EBRC), a partner of the Environment Agency – Abu Dhabi (EAD), numbers nearly 450 species – an impressive total for such a small state.

The study and conservation of birds has long been of interest, not just for birdwatchers, both resident and visiting, but also for the UAE's government. Today, EAD undertakes a variety of conservation initiatives to monitor and protect Abu Dhabi's bird population including an annual census of waterfowl and waders, regular counts of important seabird colonies and management of a network of protected areas.

Other EAD research includes a Satellite Tracking Programme, currently involving studies of Greater Flamingo, Osprey, Sooty Falcon and Macqueen's Bustard, which is providing valuable data on migration routes and stopover sites. EAD also works closely with a number of partners to promote awareness about the UAE's birdlife, including Dubai Municipality, Fujairah Municipality and the Emirates Bird Records Committee.

Internationally, the UAE, through EAD, has also taken numerous initiatives, one of which was the initiation with Britain of discussions that led to the signing of a new multilateral Memorandum of Understanding on the protection of migrant raptors throughout Eurasia and Africa.

Much of this is still work-in-progress, of course – a lot more has yet to be done. Thanks to studies undertaken by both resident and visiting birdwatchers over many decades, it is now possible to present a coherent and well-informed picture of the country's birdlife, as is shown in the text that follows.

Connoisseurs of the region's birds will notice that this book is based on the recently published (2010) second edition of *Birds of the Middle East* by the same authors, with some minor updating and a re-drawing of the maps to show the status of species in the UAE. EAD was proud to be a sponsor of that earlier book. We now present this field guide for those who wish to learn more, as dedicated scientists, as experienced birdwatchers or simply as people with an interest in the great variety of species to be found within the UAE. It is our hope that this guide will further increase awareness of this treasured part of our natural heritage that we are determined to cherish and protect.

I congratulate the authors for their detailed study of the UAE's birdlife and I take particular pleasure in acknowledging the contribution made by Simon Aspinall over the last two decades to studies of the UAE's environment, including its birds. His books on the subject, including this one, provide the foundations on which others will be able to build for many, many years to come.

Razan Khalifa Al Mubarak
Secretary General
Environment Agency – Abu Dhabi

This map is not an authority on international borders

Map of the United Arab Emirates

ACKNOWLEDGEMENTS

In preparing this field guide we have received help and encouragement from a great many people. We would particularly like to thank the following individuals who made invaluable contributions: Mike Blair, Peter Hellyer, Mike Jennings, Dick Newell, Colin Richardson and the current UAE bird recorder, Tommy Pedersen. Mohammed Al Bowardi and Majid Al Mansouri were the driving force behind production of a field guide exclusive to the United Arab Emirates, with Razan Al Mubarak seeing the project through to completion.

The preparation of the maps and status sections was greatly assisted by advice from Gary Brown (Kuwait), Jamie Buchan (Qatar), Hanne & Jens Eriksen (Oman), Mike Jennings (Arabia), Howard King (Bahrain), Mike Pope (Kuwait) and David Stanton (Yemen), while problematic identification issues were discussed with Oscar Campbell, Pete Clement, Alan Dean, Philippe Dubois, Dick Forsman, Steve James, Nick Moran, Tommy Pedersen, Roger Riddington, Abdulrahman Al-Sirhan, Andy Stoddart and the guide's three artists. We would particularly like to thank Dick Newell who amended the relevant text and drafted the table for the ever-problematic large 'white-headed' gulls.

Many other friends and colleagues helped in providing records, other information or support, some over many years, and all being thanked accordingly (with sincere apologies to anyone inadvertently omitted): Abdul Hakim Mohamed Abdi, Ahmed Al Ali, Mohammed Ali, Mohammed Almazrouei, Jack & Sylvia Aspinall, Tom & Teri Bailey, John Bannon, Ingrid Barcelo, Mark Beech, Benno Boer, Paul Bourdin, Ian Boustead, Phil Brett, Gary Brown, Roger Brownsword, Peter Castell, Chris Clark, Dave Clark, Graham Clarkson, Olivier Combreau, Marie-Ann D'Aloia, Ian Dawson, Dave Diskin, Chris & Lucy Drew, Mike Evans, Creena Forrest, Michael Gallagher, Drew Gardner, Colin & Joy Glendenning, Derek Gliddon, Lynda Graham, Ian Harrison, Erik Hirschfeld, Nissar Ahmed Hoath, Jenny Hollingworth, Jacqui Hookham, Mike & Mary Hooper, Dick Hornby, Judith Howlett, Kevin Hyland, Carol James, Nigel Jarrett, Salim Javed, Alan Jones, Reza Khan, Fred Launay, Robert Llewellyn-Smith, Simon Lloyd, Ron Loughland, Ron May, Martin McGill, Abdulaziz Al Midfa, Oystein Mortensen, John Norton, Andy Owen, Veryan Pappin, Alan Pimbley, Julia Porter, Toni Potts, Nora Powell, Trevor Poyser, Rob Quested, Kara Rawden, Len Reaney, Huw Roberts, Alec Rollo, Dave Roshier, Omar al-Saghier, Jaime Samour, Dave Sargeant, Clive Saunders, Derek Scott, Dave Sheldon, Rob Sheldon, Mohammed Shobrak, Nigel Simpson, David Stanton, Ahmed Saeed Suleiman, Nadeem Taleb, Graham Talbot, Christophe Tourenq, Neil Tovey, Simon Tull, Andrew Twyman, Maarten Verhage, Andrew Ward, Ollie Wardman, Rob Western, Deryk Wilby and Mike Wood. Thanks are also due to BirdLife International, the Darwin Initiative, the Wetland Trust and also the Natural History Museum (Tring) where we have examined skins over many years.

The three artists, John Gale, Mike Langman and Brian Small, are particularly thanked by the authors for their ornithological advice and consummate professionalism in the production of the plates and for their tolerance and good humour in the face of last-minute additions and changes. The authors should like to thank Nigel Redman at A & C Black Publishers; Julie Dando and Marc Dando at Fluke Art; and Richard's co-authors of the first edition of the Middle East field guide, Per Schiermacker-Hansen and the late Steen Christensen.

Finally, and most importantly, the Environment Agency – Abu Dhabi (EAD) is thanked for commissioning this first-ever UAE bird guide, and also for their generous financial sponsorship of the earlier, second edition of the Middle East field guide, which this book is based on.

All that now remains is to wish everyone safe travels and good birding.

Simon Aspinall and Richard Porter

INTRODUCTION

This field guide to the birds of the United Arab Emirates, UAE, is an updated abridgement of the second edition of *Birds of the Middle East* (Porter & Aspinall 2010). Full coverage is given to all species, some 445 in total, known to have occurred in the wild in the UAE up to mid 2011. This total includes a number of naturalised species, with a selection of regularly observed free-flying escapes also being illustrated and described. In addition, certain species not yet recorded in the UAE, but which have been reliably recorded in neighbouring states (albeit mostly as rare visitors or vagrants), and are thus deemed likely to occur at some point in time in the UAE, are also included. Amongst these are some potential 'confusion' species and hence we believe it helpful to depict them. Every year additional species find their way onto the UAE's national checklist, some being long-expected, others coming as complete surprises. Indeed, many new species for the Middle East checklist have come from sightings made in the UAE alone.

About 120 species of bird breed annually in the UAE, the breeding avifauna being predominantly Palearctic but with an obvious Oriental component, especially in the eastern half of the country. Of the 445 species on the national checklist, more than 300 occur solely as migrant visitors, with more than 100 of these being especially rare, so-called 'vagrants'. This exceptional diversity is the result of the country lying on a migratory cross-roads, with western and eastern flyways coming into contact with each other and having no clear divide. Unsurprisingly the UAE is a popular destination for visiting birdwatchers, joining a small but growing number of nationals and resident expatriates also pursuing this most absorbing hobby. Hopefully this field guide will both enable and encourage many more people to enjoy identifying those birds that they encounter, whether in city, park or garden, on the coast, or in the desert or mountains. Additional recommended reading dealing with the UAE and its birds, and also with Middle Eastern birds in general, is provided under 'References and Further Reading' (pages 218–220). Birdwatchers are specifically directed to the UAE recorder's website (www.uaebirding.com), which provides detailed directions to key sites, current access arrangements and up-to-date checklists, as well as a comprehensive gallery of bird photographs, contact details for local guides, an online forum, news items and recent reports.

In the UAE, the Emirates Bird Records Committee (EBRC) maintains the country's national list and records' database, collating all observations submitted and assessing claims of so-called vagrants. The definition of a vagrant varies between countries: in the United Arab Emirates the EBRC uses a minimum of 20 records as the 'cut-off' point above which a species ceases to be considered a vagrant, but as a rare visitor instead. We have accepted the published assessment country by country elsewhere, except where published records permit rarely occurring species to be adjudged vagrants on the basis of, on average (and as also used in the UAE), fewer than one record a year over the preceding 10 or more years (except for irruptive species). Of course, in countries with poor observer coverage, today's vagrant may become tomorrow's regular visitor.

The *Birds of the Middle East* covered the following countries: Bahrain, Cyprus, Iran, Iraq, Israel, Jordan, Kuwait, Lebanon, Oman, Qatar, Saudi Arabia, Syria, Turkey, Palestinian territories (West Bank and Gaza), the United Arab Emirates (UAE) and Yemen, including the Socotra archipelago. We have retained substantially the same distribution maps from that publication in this guide since it is helpful to see the bigger, regional, picture, and one can see at a glance which species occurs where and when (if resident or migrant, and if the latter whether it breeds or not, and so on). To the best of our knowledge, only species accepted onto a checklist by the relevant national body in any country are included in the maps or status text. Readily accessible national avifaunas are listed under 'References and Further Reading' on pages 218–220.

ESCAPES AND INTRODUCED BIRDS

Around the world, birds are trapped and transported across borders for the cagebird trade, many subsequently either escaping or being released. In this respect the Middle East is no exception, and in recent decades several escaped species have established naturalised (self-sustaining) breeding populations in the wild in the region, notably parakeets, mynas and weavers. Any such species deemed naturalised in the UAE at the present day, of which there is a growing number, is included in this book. A selection of others from among a long list of those not yet naturalised (some perhaps unlikely to become so), but which may nonetheless be observed free-flying are also included here. Many are large and long-lived. Regrettably, the establishment of exotic species has sometimes had an adverse impact on the region's indigenous avifauna.

SEQUENCE, TAXONOMY AND NOMENCLATURE

This is a very confusing and contentious area of ornithology, and newcomers to birding will have every reason to be mystified by the various treatments given by different books. There are those taxonomists who prefer to 'lump' species and those who prefer to 'split' them (thus subspecies become species in their own right).

The species order, taxonomy and nomenclature (English and scientific names) followed here is that adopted by the Ornithological Society of the Middle East in its OSME Region list, ORL. This can be viewed at: www.osme.org/orl. The ORL, which has been fundamental to preparation of this guide, primarily follows Dickinson (2003) *The Howard & Moore Complete Checklist of the Birds of the World* (3rd edition), and Gill & Wright (2006) *Birds of the World: Recommended English Names* (the published list of the International Ornithological Congress, IOC).

Only rarely have English names deviated from those proposed by the IOC, in each case involving logical divergence from that proposed, rather than (and despite temptation) any particular personal preference.

We have also given alternative names where relevant (under 'Alt' at the end of the species text), but only those that are still often used or are of help in preventing confusion. The treatment given to alternative names is not comprehensive but does include those still in everyday use. If a problem arises, the scientific name should prevent ambiguity, although these too have been subject to change in recent years.

In vogue with the times, though not without substantial scientific backing, the trend has been toward splitting and recognition of new species. With such a constant state of flux, it is impossible to stay ahead of developments. This field guide is not a taxonomic authority and has simply embraced the latest thinking on what may or may not constitute a species. However, we describe all recognisable taxa known to occur in the region.

Some explanation of the use of binomial and trinomial scientific names is required:

- Where there is universal agreement on full species status a binomial is used in the scientific name, e.g. *Eremophila bilopha* (Temminck's Lark).
- When a taxon is clearly identifiable in the field but where there is still debate over whether full species status is merited we have used a trinomial, e.g. Indian Reed Warbler *Acrocephalus (stentoreus) brunnescens*, with the bracketed name referring to the parent taxon.
- For taxa which are universally recognised as different forms (subspecies) of the same species then the trinomial is used without brackets, e.g. *Buteo buteo vulpinus* Steppe Buzzard.

Waiting in the wings, potential or incipient species include, for example: Indian Reef Heron *Egretta (gularis) schistacea*; Eastern Cattle Egret *Bubulcus (ibis) coromandus* and Indian Golden Oriole *Oriolus (oriolus) kundoo*, amongst several others in the immediate region.

ILLUSTRATIONS AND IDENTIFICATION TEXT

Our aim has been to make the illustrations and species accounts of help to beginner and expert alike. We have concentrated on those features which are important for identification and, accordingly, these are highlighted in the text. For each species, length from bill tip to tail (L) is given in centimetres; for larger birds wingspan (W) is also given. Where the identification of a species does not present a problem, the texts are often brief, whereas more difficult species have necessarily warranted more detailed description, sometimes including biometrics.

There is no need to stress the increasing importance of digital photography in helping in making correct identifications, of tricky species especially. This enables examination and discussion with experts at a later date.

Abbreviations used on the plates ad – adult, juv – juvenile, imm – immature, ♂ – male, ♀ – female.

BIRD TOPOGRAPHY

The illustrations below show the various features of a bird – its topography – used in the identification texts. Knowledge of these, especially the feather tracts, is vital for describing a bird and its plumage. Making field sketches, even if embarrassingly poor, is a good way to learn.

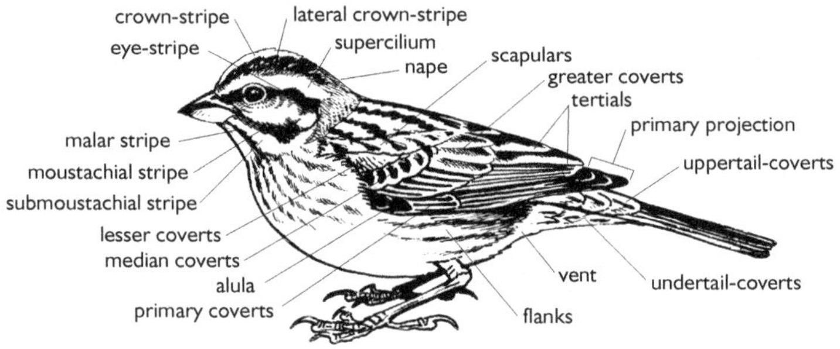

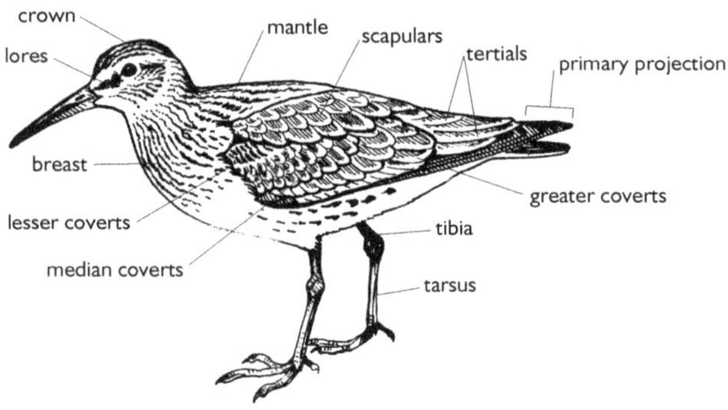

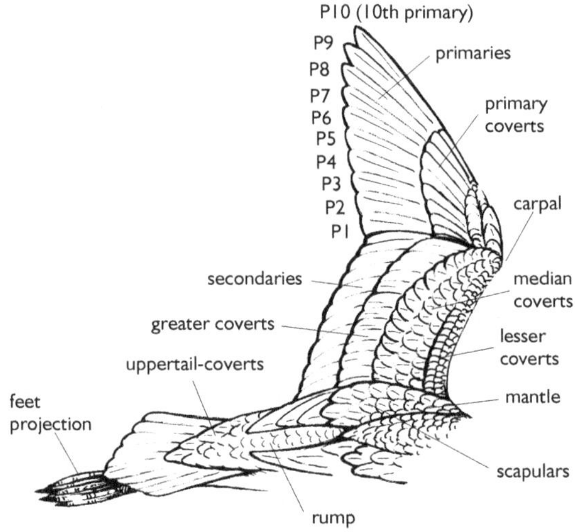

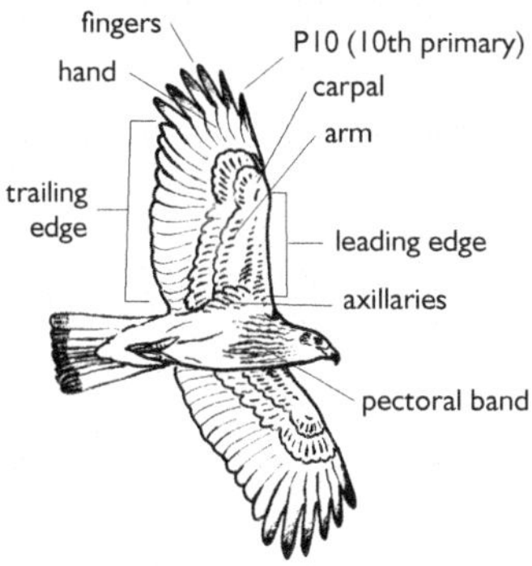

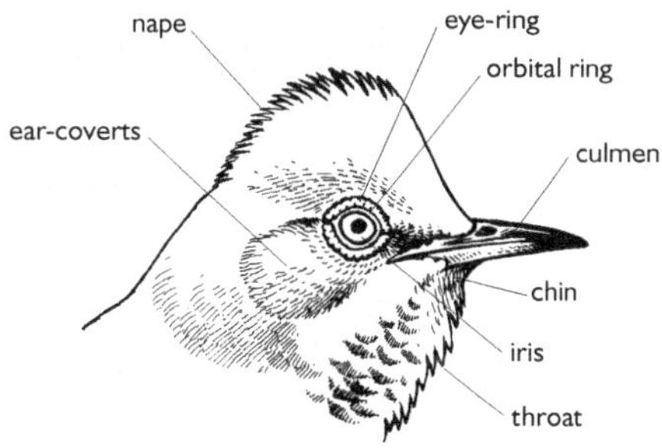

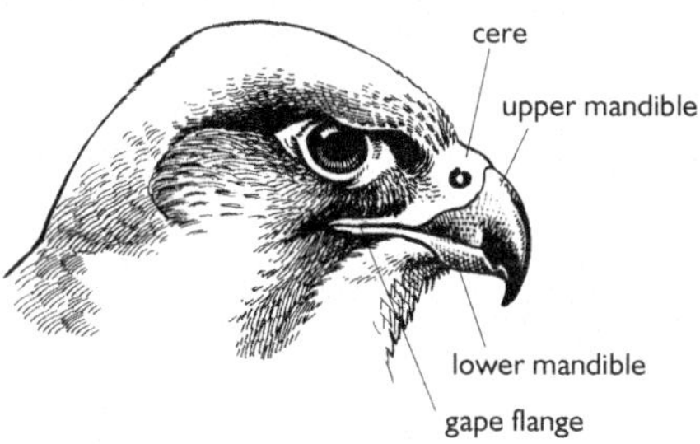

VOICE

Calls and song are only given where a species occurs regularly in the region or where it might be helpful for identification. Song is of course given for all species breeding in the UAE, but many species of migrant also give at least occasional bursts of song or sub-song on passage or in winter, even though moving on to breed. Where this is known to be the case in UAE then description of those songs is also given.

A large number of sound guides are now widely available, and given the limitations of phonetic transcription we strongly recommend that one or other be purchased. (The internet can also be checked for sound recordings). There is no substitute for learning vocalisations, or, if unable to commit them to memory, in comparing recordings. This stated, use of the technology to record calls and songs has strangely somewhat lagged behind digital photography as a means for confirming the identity of a bird. Few of us use sonograms, but the way forward certainly lies in this field. In our region, for example, the calls of the different subspecies of Common Chiffchaff *Phylloscopus collybita* may only be separable by analysis in this way. Indeed, many recently described new species of birds (and even some mammals) have been first noticed as such on the basis of their vocalisations.

There is also, as amply demonstrated by our own species, *Homo sapiens*, the matter of differing regional dialects. Songs and calls of bird species commonly found in western Europe may often sound quite different in the Middle East. Even within the region, strong intraspecific variations can often be noticed.

HABITAT

The habitats described are those occupied in the Middle East and this may help in identification or at least narrow down the list of possible contenders, or direct you where to look for a particular species. Habitats occupied by species visiting the United Arab Emirates in winter and/or on passage are detailed when differing from those utilised in their Middle East breeding grounds.

MAPS AND STATUS

As a ready-reckoner, the status of each species in the UAE appears to the right of the species header, using the codes given below:

RB/rb – Resident breeder
MB/mb – Migrant breeder
PM/pm – Passage migrant
WV/wv – Winter visitor (Dec–Feb at least)
SV/sv – Summer visitor (non-breeding)
V – Vagrant (<20 records)
E/I – Escape or Introduction

Upper and lower case are used to indicate relative abundance of that species, upper case reflecting large numbers, or greater abundance either at that season and/or in comparison with other closely related species. When PM and WV (or pm and wv) share the same case the season when that species is the more common will appear first, unless they are actually equally common at the two seasons.

The regional distribution maps distinguish between resident and migrant breeders and also show the passage and/or winter range for all regularly occurring species. Note that the maps show the current known ranges (and not historical ones). Further details on status in the Middle East, to complement the maps, are given as necessary under 'Note' at end of the species accounts. Also listed here, apart from the UAE as appropriate, are any other countries of the Arabian peninsula, and if relevant sometimes also other Middle Eastern states, to which that species is a known vagrant.

Sometimes broad geographical divisions are used:

NEAR EAST – the eastern Mediterranean countries (excluding Turkey and Cyprus)
SW ARABIA – Yemen and adjacent areas of Saudi Arabia
SE ARABIA – Oman, UAE and adjacent areas of Saudi Arabia (east of the Qatar peninsula).

The breeding, passage and winter distributions of birds are now relatively well known for much of the Middle East, although gaps still exist. Regrettably, it has not been possible at this time, even when appropriate, to map passage and winter ranges separately. In any case, in the south of the region, where the seasons are not clear-cut, autumn migration may continue, for example, until December, while spring migrants may reappear as early as January. In general, however, a species shown by hatch across the entirety of Arabia will generally be a passage migrant absent in winter or present then only in relatively low numbers.

Note that many non-breeding waterfowl, often immature birds, may be found year-round on the coast or at inland wetlands of the UAE, including in particular many different shorebirds, amongst them even arctic breeding species. Don't be surprised to see, for example, Ruddy Turnstone *Arenaria interpres* or Sanderling *Calidris alba* on the foreshore of the Gulf in mid-summer.

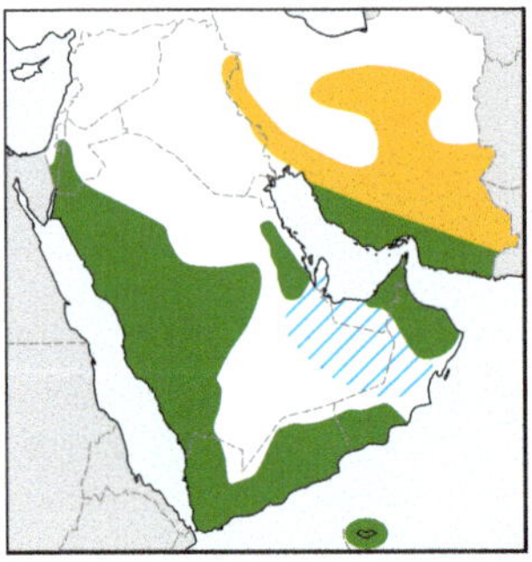

Green – Resident (present all year)

Orange – Migrant breeder (seen in the breeding season and on passage)

Blue hatch – Passage and/or winter visitor

Some species mapped as breeding visitors may actually be resident in some areas, usually more southerly localities, and/or perhaps involving only a small proportion of the population. (Interestingly, some classic Palearctic-African migrants are now found wintering in newly vegetated areas of Arabia.) Moreover, as the Tropic of Cancer traverses Arabia it is perhaps hardly surprising that the breeding season is not always the spring and summer; winter breeding is the norm for some species, including migrant species, which elsewhere otherwise breed conventionally, in summer. The breeding season of yet others, in Dhofar (southern Arabia), including some migrants from Africa, is regulated by the late summer monsoon and occurs in the 'autumn' months.

Extensive use has been made of the excellent maps prepared for the *Atlas of the Breeding Birds of Arabia*, to which many individuals have contributed, and on the country avifaunas listed in 'References and Further Reading' (pages 218–220). The mapping scale and accuracy varies from one country avifauna to another, and at times we have had to extrapolate or interpret to the best of our combined knowledge those maps produced with, say, a 'broader brush' approach.

CODE OF CONDUCT AND SUBMITTING RECORDS

The welfare of birds must come first, particularly of nesting birds. Many ground-nesting species may not incubate by day, merely shade their eggs, and prolonged disturbance may result in the embryos baking alive, or in nest-desertion. Try to minimise time spent near nests or chicks; even if you haven't actually located them it is usually evident from the behaviour of the adult/s that they are nearby. Photographers please take note! Roosting and feeding birds, notably waders, are easily disturbed so please be thoughtful when, for example, trying to get that frame-filling shot of a Crab-plover (*Dromas ardeola*).

We all now recognise the importance of digital photography in helping in making correct identifications of tricky species and also, all-importantly, in convincing national records committees of any rarer sightings you may make. On this point, please do remember to submit your observations to the national recorder – all may be helpful in promoting ecotourism, lobbying for conservation and in bringing added protection to important sites and species. The relevant body in the UAE for submission of your sightings is the EBRC (www.uaebirding.com).

Finally, observe cultural and religious sensitivities in the region. Hospitality shown to visitors is second to none, but be aware of and observe local custom and etiquette. Always accept a cup of tea, unless you are going to miss your flight home, and even if you don't take sugar.

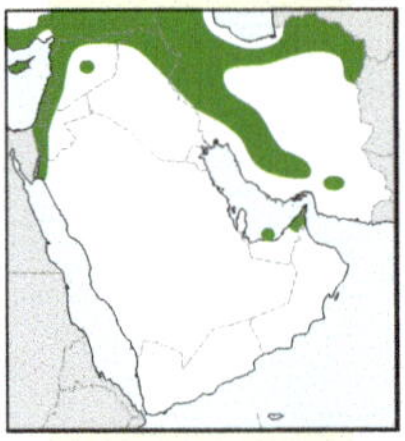

Chukar Partridge *Alectoris chukar* RB; E/I

L: 33. A distinctive plump gamebird; ***grey-brown with striped flanks. Bill and legs red***. Wary, runs fast; usually in small flocks. **Voice** Characteristic accelerating *ka-ka-kaka-kaka-kaka...* followed by a chuckling *chukara-chukar-chukara-chukar*, frequent at dawn and dusk especially. **Habitat** Rocky slopes and hillsides, semi-deserts, agricultural land; from sea-level to 3,000m, occasionally higher. **Note** Native to northern Hajar mountains, but introduced UAE islands.

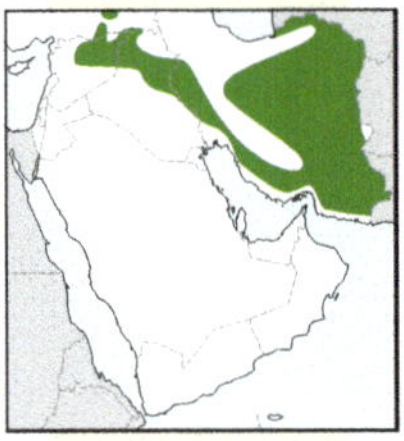

See-see Partridge *Ammoperdix griseogularis* E/I

L: 24. W: 41. Introduced in UAE to Sir Bani Yas island. Size of Sand Partridge; runs fast when disturbed, or flies low with whirr of wings. In flight, chestnut outer tail feathers in both sexes. Male has ***grey head and throat with black forehead and supercilium bordered below by white eye-stripe***. Female and juvenile lack head and flank marks, thus resembling female Sand Partridge closely. **Voice** Far-carrying repeated *who-it* or *wuid,wuid*. **Habitat** Dry, stony or gravelly hillsides, open broken ground, steep rocky banks, avoids tall or thick vegetation.

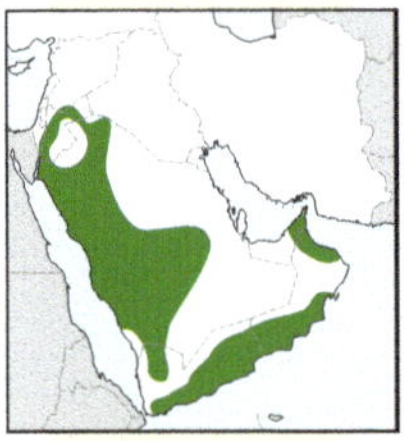

Sand Partridge *Ammoperdix heyi* RB

L: 24. W: 40. Larger than Common Quail, but much smaller than Chukar Partridge. Runs nimbly over rocks; wings whirr noisily in flight. ***Male has white patch on forehead and behind eye, but lacks black head-marks of See-see***. Male of subspecies *intermedia*, occurring Arabia, dark pinky-cinnamon above (fading sandier when worn); flanks heavily banded black and chestnut. Both sexes show chestnut outer tail feathers in flight. Female and immature are uniform sandy-grey, usually inseparable from female See-see in field. Often forms crèches, with large flocks arriving on foot to drink from small pools. **Voice** Commonest call a repeated, metronomic, *qwei, qwei, qwei*; alarm call *wit-wit* in flight. **Habitat** Desolate, arid rocky and stony slopes, wadis and cliffs.

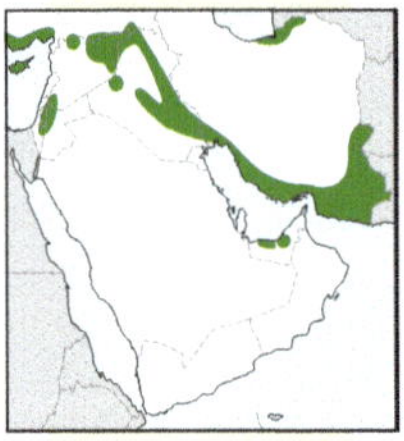

Black Francolin *Francolinus francolinus* E/I

L: 35. W: 53. Large, plump and stub-tailed. More often heard than seen and hard to flush. Male distinctive, ***mostly black with chestnut collar***; white spots on mantle and breast-sides merging into ***white chevrons on flanks; obvious white patch behind eye***. In flight, black- and brown-barred wings and blackish tail. Female ***warm brown*** with paler head and ***arrowhead-shaped feather-centres***, most obvious on lower breast and flanks; ***indistinct chestnut patch on hindneck***. In flight note dark tail with blackish outer tail feathers. **Voice** Male calls most intensely at sunrise and evening, ***loud, penetrating and very distinctive*** with grating character, mostly from a mound or a bush, *gldlri, djjii, djji-djji-djjii, djji-djjii*, first note lower and second and fifth stressed. **Habitat** Densely scrub-covered lowlands and wadis, dunes with scattered vegetation, reedy flood-plains, wetland margins; often not far from water. **Note** Feral breeding UAE.

Grey Francolin *Francolinus pondicerianus* RB; E/I

L: 30. Rather drab, stub-tailed, greyish-brown gamebird separated from slightly larger female Black Francolin by ***much finer mottling and barring on upper- and underparts with chestnut forehead and cheeks***. Lacks chestnut patch on hindneck and has ***pale throat-patch bordered below by black 'U'***. In flight, shows chestnut tail. Not secretive, often feeding in small parties in open. Usually runs from danger but, if pressed, rises with explosive whirr of wings. **Voice** Loud far-carrying series of 9–15 notes: *kik-kjyw-ku, kik-kjyw-ku, kik-kjyw-ku* commonly heard. **Habitat** Scrub, edges of cultivation and semi-desert. **Note** Range expanding naturally following deliberate introduction.

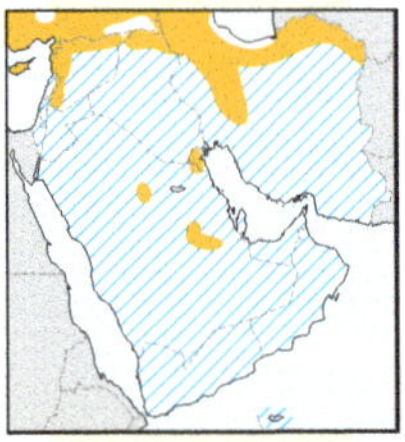

Common Quail *Coturnix coturnix* PM, wv, mb?

L: 17. W: 33. Small gamebird, ***more often heard than seen***, and hard to flush. In flight ***size of Common Starling***; rather pointed, narrow wings bowed, fast shallow wingbeats, ***plain wings and striped back***. Creeps about on ground inconspicuously. Female has paler head pattern and lacks neck-band of male. **Voice** Characteristic sound in open country over much of the region especially in farmland, by day or night, an explosive, rhythmically repeated ***trisyllabic whistle***; **pit, pil-it** (rendered as 'wet my lips'). **Habitat** Grasslands, cereal crops, meadows (sometimes in mountains). **Note** Passage hatched, rare in winter.

ad
See-see Partridge
♀
Chukar Partridge
♂
Sand Partridge
Black Francolin
♀
♂
♀
Grey Francolin
♂
♀
♂
Common Quail
ad

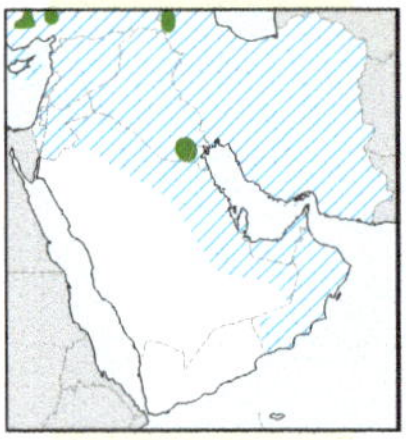

Greylag Goose *Anser anser* wv

L: 83. W: 164. Only the paler eastern subspecies *rubrirostris* has been recorded in the region. A large goose with thick neck, ***heavy, pink bill, pale greyish head and neck, and pink legs***. In flight, shows distinctive ***pale grey forewing***. Juvenile similar to adult but lacks sharply defined transverse lines of upperparts and dark belly marks. From all other geese in the region by combination of size, bill and leg colour and strikingly pale forewing. **Voice** Similar to domestic goose; loud, characteristic *ang-ang-ang* in flight. **Habitat** Grasslands, arable fields, marshes, estuaries; breeds in marshes, reedbeds, boggy thickets, islets. **Note** Passage and winter hatched.

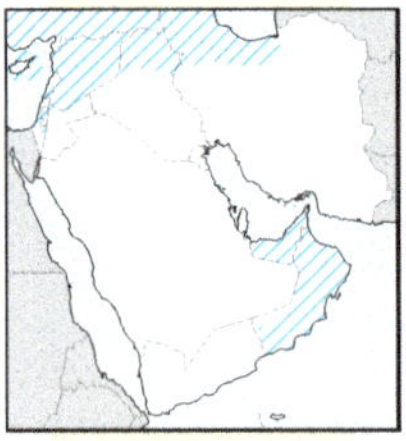

Greater White-fronted Goose *Anser albifrons* wv

L: 72. W: 148. Smaller than Greylag, and normally larger than Lesser White-fronted Goose. Warmer grey-brown than Greylag, with slightly darker head and hindneck; has ***orange legs*** and lacks silvery forewing. Wing pattern similar to that of Lesser White-front. Bill not heavy, but quite long, ***pink with white nail***; no eye-ring visible in field. ***Adult has large white area surrounding base of bill*** (rarely also forecrown) and ***black bars on underparts***. Juvenile browner, lacks white forehead and black bars on underparts; ***forehead and area round bill very dark, bill having dark nail***. **Voice** Musical and much higher pitched than other grey geese except Lesser White-front; repeated disyllabic, sometimes trisyllabic with metallic, laughing quality *kow-lyow* or *lyo-lyck*. **Habitat** Grasslands, marshes, estuaries. **Note** Passage and winter hatched; vagrant Kuwait, Saudi Arabia, Yemen.

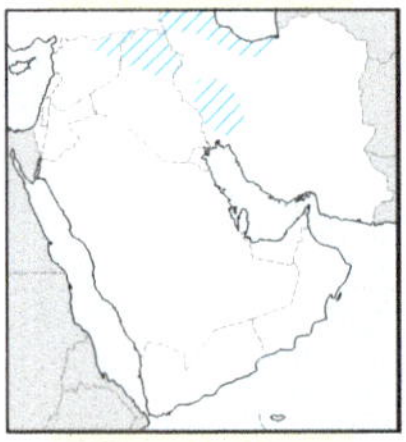

Lesser White-fronted Goose *Anser erythropus* V

L: 59. W: 127. Resembles Greater White-fronted Goose, but smaller (though a few overlap in size), with proportionally ***shorter neck and bill, steep forehead, yellow eye-ring*** and slightly longer wings, often extending past tail. Adult has ***extensive white on forehead and forecrown to above eye***; upperparts dusky-brown with only dull transverse lines; underparts with only a few black blotches. Juvenile similar to juvenile Greater White-fronted Goose, but told by size, head and neck proportions and yellow eye-ring. Compared to Greater White-fronted has faster wingbeats, more compact silhouette and noticeably faster walk. **Voice** Higher pitched, squeakier and faster than Greater White-fronted; a di- or trisyllabic yelp *kow-yow, kyu-yu-yu* or piping *yi-yi-yi*. **Habitat** Grasslands, marshes. **Note** Winter hatched; vagrant Kuwait, Oman, UAE.

Egyptian Goose *Alopochen aegyptiaca* E/I

L: 68. W: 144. Slightly larger than Ruddy Shelduck, with broader wings, but similar wing-pattern in flight. Otherwise ***brownish-buff*** with brownish or greyish upperparts, pale foreneck and face and ***prominent dark eye-patch***; dark collar and belly-patch. Juvenile/immature browner, lacking eye-patch, collar and belly-patch, and with yellowish-grey legs. Told from Ruddy Shelduck by longer legs, dark eye-patch and body colour. **Voice** Deep nasal braying. **Habitat** Freshwater marshes, lakes and rivers with nearby grassland. **Note** Introduced UAE.

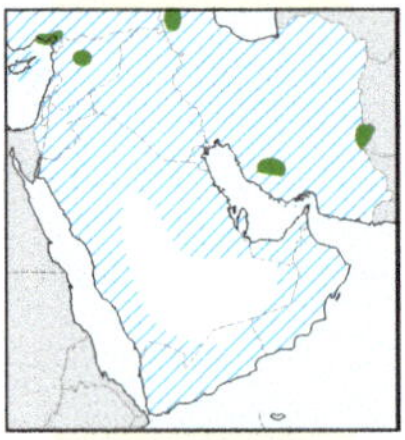

Common Shelduck *Tadorna tadorna* WV

L: 63. W: 122. Large with white body, ***black head with green gloss, broad chestnut band round breast and prominent red bill with fleshy knob in breeding male***. In flight black flight feathers contrast with white upper- and underwing-coverts and black shoulders. Juvenile lacks chestnut band, grey-brown head, hindneck and upperparts, and pale greyish bill. **Voice** Male in breeding season has high whistling *siss-siss-siss*, answered by the female's whinnying *gehehehehehеh*; sometimes nasal *ah-hang*. **Habitat** Sandy and muddy coasts, fresh and saltwater lakes and marshes; nests in burrow, hole in tree or under thick cover. **Note** Passage and winter hatched, but uncommon much of Arabia.

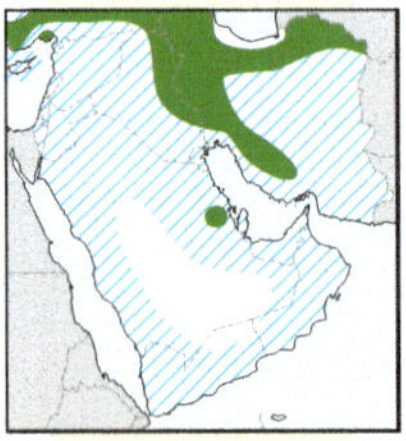

Ruddy Shelduck *Tadorna ferruginea* wv

L: 64. W: 133. Size of Common Shelduck with ***orange-chestnut body*** and paler cinnamon-buff head; ***black flight feathers*** with green speculum, ***striking white forewing and underwing-coverts*** (similar to Egyptian Goose). Female and non-breeding male lack black neck-collar; female also has paler head. Juvenile like adult female, but back browner. From Egyptian Goose by shape, narrower, more pointed wings, head pattern and darker chestnut underparts. **Voice** Nasal, trumpeting and rather penetrating *ang, ang*. **Habitat** Sandy lake shores, river banks, fields, arid steppe; nests in hole or burrow. **Note** Passage and winter hatched, but rare or irregular south Arabia.

Greater White-fronted Goose
ad
juv
ad
Greylag Goose
ad
Lesser White-fronted Goose
ad
juv
Egyptian Goose
ad
juv
♀
♂
juv
Common Shelduck
Ruddy Shelduck
♀
♂
♀

Mute Swan *Cygnus olor* V

L: 153. W: 223. ***Orange bill with black knob*** in adults (larger in male). ***Typically with graceful S curve to long neck, bill often pointing down.*** Juvenile dingy-brown with ***black-based grey bill*** gradually becoming pink and then orange. Plumage grows increasingly white during first winter and spring, sometimes still partly brownish above until second winter. ***In flight, wingbeats produce loud, rhythmic, singing sound*** **vaou-vaou-vaou.** **Voice** Mostly silent. **Habitat** Lakes, marshes, deltas; in winter also sheltered sea-coasts; builds large nest near water. **Note** Winter hatched; vagrant to Arabian Gulf States and Oman.

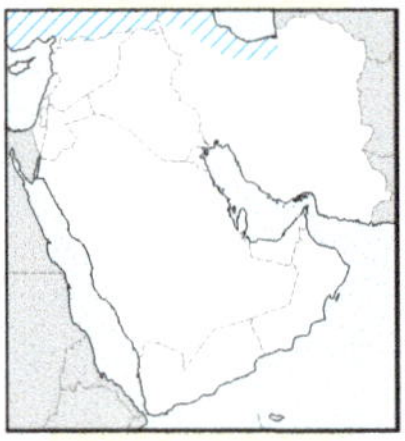

Whooper Swan *Cygnus cygnus* V

L: 153. W: 231. Similar to Bewick's Swan, but larger (size of Mute Swan) with proportionally longer neck and bill; ***black bill with prominent yellow base extending in wedge to below nostril or beyond.*** Flight silent but with powerful wingbeats like Mute Swan. ***Juvenile/immature similar to Bewick's, but note longer neck and wedge-shaped head profile.*** Young birds often with parents during first winter. **Voice** Similar to Bewick's but deeper and stronger and with musical, trumpet-like quality *ahng-ha* or *ko-ko-ko*; often trisyllabic. **Habitat** Tidal waters, lakes, rivers and floodplains, fields. **Note** Winter hatched; vagrant Oman, Qatar, UAE.

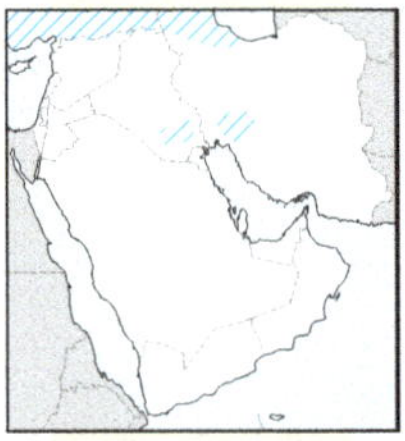

Bewick's Swan *Cygnus (columbianus) bewickii* V

L: 121. W: 195. ***Smallest swan, with shorter neck and bill*** than Mute and Whooper Swan. Similar to larger Whooper (both having straight neck), but ***yellow on bill reduced to base, ending well behind nostril.*** Immature greyer than same age Mute Swan, and ***pink bill lacks black base***, becoming white and later yellow during first winter. From immature Whooper by ***shorter neck and bill and more rounded head.*** In flight wingbeats slightly faster, more goose-like and with no humming sound. **Voice** Similar to Whooper but higher pitched, monosyllabic or disyllabic (Whooper often has three syllables), sometimes recalling distant barking dogs. **Habitat** Inland and coastal wetlands. **Note** Winter hatched; vagrant Oman, Saudi Arabia, UAE.

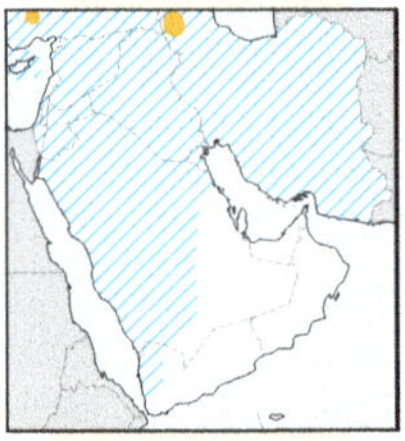

Great White Pelican *Pelecanus onocrotalus* V

L: 140–175. W: 270–330. Large with huge wingspan. Adult ***white with contrasting, solid black flight feathers below,*** but body tinged yellowish-rosy in breeding plumage; (Dalmatian Pelican has greyish underwing, and body appears greyish-white). At close range note ***short, shaggy crest on nape*** (in breeding season), ***dark eye surrounded by naked rosy skin, fleshy-yellow legs and pointed forehead-feathers where meeting culmen***; these latter characters are also useful when separating immature from similar Dalmatian Pelican. Immature Great White Pelican has ***clearly darker grey-brown upperparts than the grey-buff Dalmatian.*** Flight consists of a few slow wingbeats followed by a glide; flocks often fly in regular lines, or circle in formation. **Habitat** Large inland wetlands and shallow coastal lagoons; nests colonially in reeds. **Note** Has bred Kuwait; passage hatched, some winter S. Turkey southwards; vagrant Bahrain, Oman, UAE.

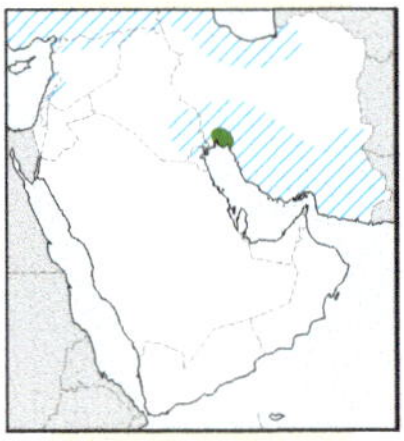

Dalmatian Pelican *Pelecanus crispus* V

L: 150. W: 310–345. Resembles Great White Pelican but told in flight by ***greyish underwing with pale band through centre and greyish-white body*** (in Great White, flight feathers below solidly black, and white body tinged with yellowish-rosy). At close range note ***nape feathers curl upwards*** (drooping in Great White), ***pale eye*** (dark in Great White) ***and grey legs***; also shape of bare skin round eye and of feathers where meeting culmen useful at all ages. Immature dirty-white below, ***pale grey-buff above*** (similar Great White is dark grey-brown above). **Habitat** As Great White Pelican; nests in reeds and trees. **Note** Partial migrant; vagrant Kuwait, Oman, UAE.

Mute Swan
imm
ad
Whooper Swan
juv
ad
juv
ad
Bewick's Swan
ad winter
juv
Great White Pelican
ad winter
juv
Dalmatian Pelican

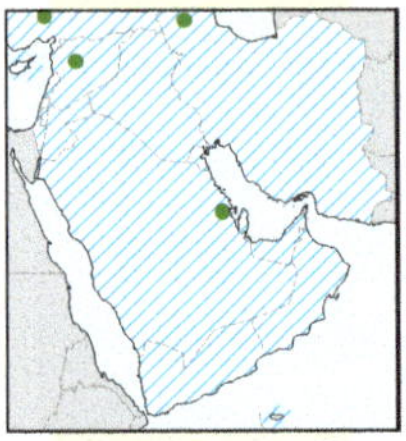

Northern Shoveler *Anas clypeata* PM, WV

L: 51. W: 78. ***Huge, spatulate bill.*** Swims with front end low and bill often dabbling in water. In flight, wings appear set far back. Male unmistakable; in flight, looks black, white and chestnut with ***distinctive blue forewing***. Female and juvenile on water resemble female Mallard and Gadwall, but bill distinctive; in flight, show bluish forewing, somewhat resembling smaller Garganey male. Male in eclipse largely resembles female, but brighter blue forewing. **Voice** Male calls a hollow, double-note *g-dunk, g-dunk*, often in flight, while female simultaneously quacks *pe-ett*. **Habitat** Marshes, lakes, ponds. **Note** Passage and winter hatched.

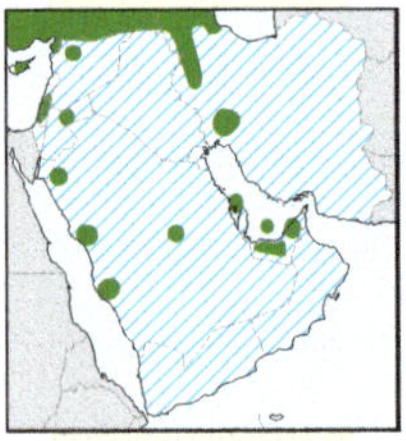

Mallard *Anas platyrhynchos* WV; E/I

L: 56. W: 95. Large dabbling duck. Male has ***grey wings, dark blue speculum distinctly bordered by white, and white underwing-coverts***. Female mottled brown like other female ducks and told by size, bill shape and colour, and wing pattern. Male in eclipse resembles female, but bill greenish-yellow. **Voice** Male has a soft nasal *raehb* and, during courtship, a weak, high-pitched whistle *piu*. Female has familiar deep quacking. **Habitat** Any wetland, including estuaries in winter. **Note** Arabian population feral; passage and winter hatched.

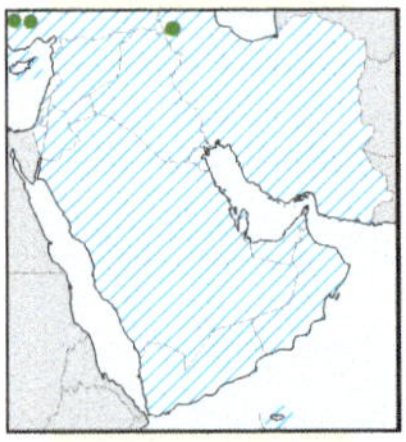

Gadwall *Anas strepera* pm, wv

L: 51. W: 89. Medium-sized, slightly built dabbling duck. Flies with rapid wingbeats and pointed wings like Eurasian Wigeon. ***Male mainly dark grey with black around tail; in flight, white speculum bordered black in front, and white belly***. Female and juvenile resemble larger and heavier female Mallard, but tail grey-brown (not white), and thin dark bill is orange along sides; ***in flight, shows white speculum patch close to body*** (often smaller in male) and ***white belly*** bordered by dark flanks. **Voice** Male has a rasping, low *rrep* call. Female's call resembles Mallard's but higher. **Habitat** Any wetland. **Note** Passage and winter hatched.

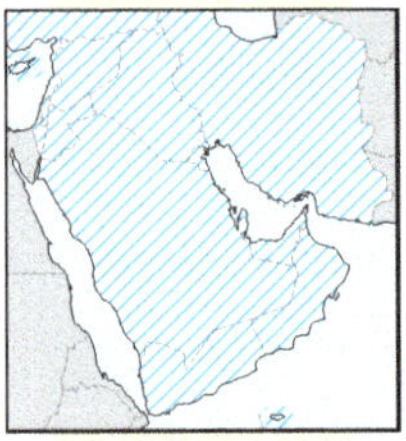

Northern Pintail *Anas acuta* PM, WV

L: 56 (excluding long tail feathers of male). W: 88. Slim and elegant dabbling duck. Male has ***white neck and underparts*** with ***dark head and tail and dark slender bill***. Female recalls other female dabbling ducks, but note ***dark bill, greyer plumage and much slimmer appearance***. In flight both sexes show distinctly ***longer neck and tail***, and slender, more pointed wings than other ducks; also note ***green speculum with conspicuous white border to rear***. Female has distinctive ***white border at rear of secondaries***, upperwing otherwise brownish. **Voice** Male has a low-pitched, weak whistle; female a hoarse quack. **Habitat** Sheltered coasts and estuaries in winter, also shallow inland waters. **Note** Passage and winter hatched.

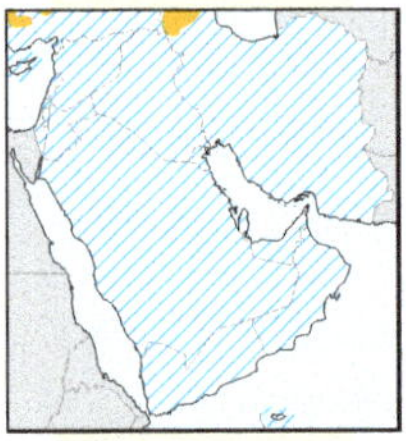

Garganey *Anas querquedula* PM, wv

L: 39. W: 63. Small dabbling duck. ***Male shows long white stripe on head*** and, in flight, ***striking blue-grey forewing***. Female similar to slighter female Eurasian Teal, but ***longer, heavier bill and more contrasting dark and light head-stripes, widening in front into pale patch at bill base***, and whiter throat. In flight, female's forewing is slightly paler and white border along secondaries is distinctive (Eurasian Teal has white wing-bar in middle of wing in front of speculum). Male in eclipse like female, but wing pattern as adult male. Juvenile similar to female. **Voice** Male calls a dry, drawn-out rattled *knerreck*; female a short, sharp quack. **Habitat** Freshwater wetlands. **Note** Passage hatched; some winter Arabia.

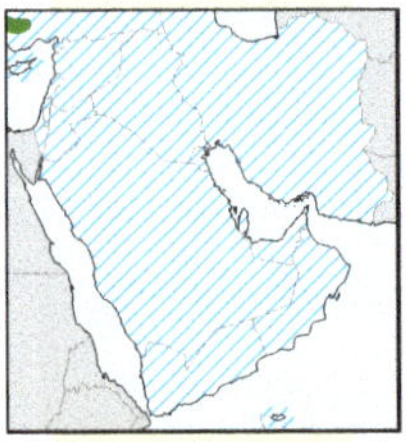

Eurasian Teal *Anas crecca* WV, pm

L: 36. W: 61. Commonest small duck; readily taking to the wing, both sexes showing white belly and ***greyish underwing with light band through middle***. Adult male distinctive; female rather similar to Garganey (which see for differences). Often in fast-flying, tight-knit twisting and turning flocks. **Voice** Male has far-carrying ringing whistle *kreek-kreek*; female a high-pitched, nasal quacking. **Habitat** Wetlands, from saltmarshes to lakes and ditches. **Note** Passage and winter hatched.

Northern Shoveler
Mallard
Gadwall
Northern Pintail
Garganey
Eurasian Teal

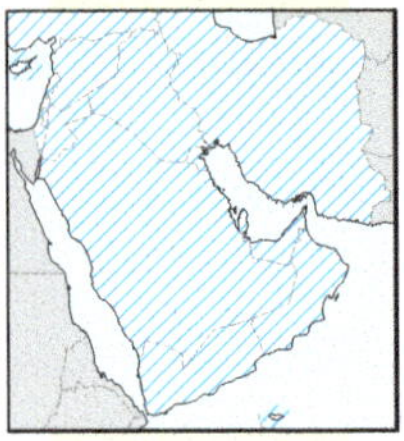

Eurasian Wigeon *Anas penelope* WV, pm

L: 48. W: 80. ***Rather rufous duck*** with short neck, rounded, ***steep forehead*** and ***small blue-grey bill. In all plumages has white belly-patch***, obvious in flight and when grazing. Adult ***male has creamy-yellow forehead***. Adult female rufous to greyish-brown; lacks white forewing. Juvenile resembles adult female (immature male does not usually assume white forewing until second winter). In flight in all plumages shows narrow wings, pointed tail, and contrasting, white belly-patch; ***white forewing contrasting with blackish-green speculum is distinctive in adult male***. Female and juvenile have grey-brown forewing, darker flight feathers and black speculum. **Voice** Male has characteristic clear whistle *wheeooo*, both in flight and on water; female growls in flight. **Habitat** Coastal mudflats, marshes, lakes. **Note** Passage and winter hatched.

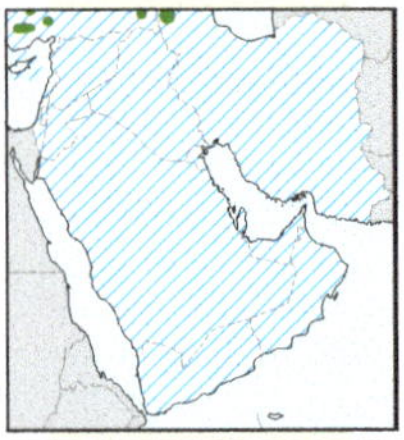

Common Pochard *Aythya ferina* WV, pm

L: 45. W: 80. ***Sloping forehead grading into long bill*** characteristic. Male has ***chestnut head and neck, contrasting with black breast***; black bill has pale blue-grey band in centre. In eclipse resembles female but greyer above. Female has dull brownish head and breast with paler chin and eye-stripe; dark bill becomes paler towards broad black tip. In flight looks longer and plumper than Tufted Duck; greyish-brown wings (palest in male) have ***indistinct pale grey wing-bar***. **Voice** Courting male has low whistle; also hoarse wheezing note; female has rough harsh *krra-krra*. **Habitat** Well-vegetated wetlands. **Note** Passage and winter hatched.

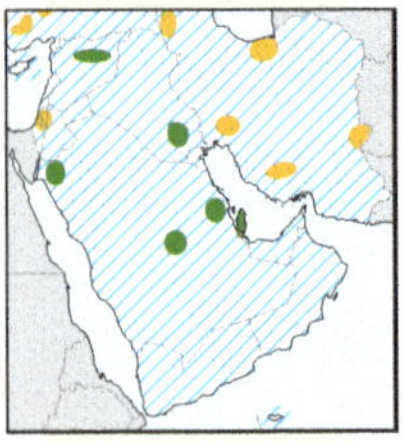

Ferruginous Duck *Aythya nyroca* wv, pm

L: 40. W: 66. Slightly smaller than Tufted Duck, head shape close to Common Pochard's. ***Compared to Tufted Duck has higher crown, flatter sloping forehead and longer, dark grey bill with black nail only. Male rich chestnut-brown with white eye, sharply defined white undertail-coverts***. Female and immature have dark eyes (but white in one-year-old male); they lack whitish at base of bill of some Tufteds; ***female has warmer brown head than Tufted*** (which has yellow eye); pure white and sharply defined white undertail-coverts (sometimes seen in Tufted). ***First-year birds told by head shape, length and pattern of bill and dark eye***. In flight, ***broader, more conspicuous white wing-bar than Tufted Duck***. **Voice** In flight high-pitched *crr-err*. **Habitat** Shallow, well-vegetated wetlands. **Note** Passage and winter hatched, mostly scarce. Occasionally summers UAE.

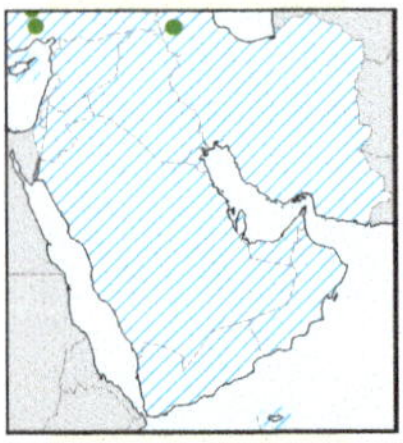

Tufted Duck *Aythya fuligula* pm, wv

L: 42. W: 70. Small diving duck with roundish head but fairly steep forehead; ***crest at nape***, long and drooping in male, minute in female. ***Bill blue-grey*** (darker in female) ***with broad black tip. Male has black upperparts and purple head sheen***. Female and male in eclipse have brownish sides to body; female sometimes with whitish band round base of bill. In summer female has ***darkish back, short crest and broad black tip to bill***. Some females show white undertail-coverts. In rapid flight has conspicuous white wing-bar, almost the full length of wing. **Habitat** Lakes, ponds. **Note** Passage and winter hatched.

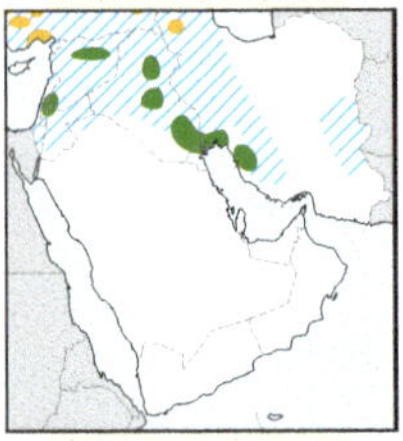

Marbled Duck *Marmaronetta angustirostris* V

L: 41. In flight resembles small female Northern Pintail, having fairly long neck, wings and tail. On the water identified by ***pale plumage, dappled dark and cream*** (marbled) with ***dark oval patch around eye***. Head appears large and rounded with steep forehead and, in adult, bulky crest on lower nape. In flight, wings are fairly pale with slightly paler secondaries but ***no speculum***. Often secretive, hiding in vegetation. **Voice** Low nasal wheezing *jeak* or double whistling note. **Habitat** Well-vegetated lakes. **Note** Winter hatched but mostly rare; vagrant Bahrain, Kuwait, Oman, Qatar, Saudi Arabia, UAE.

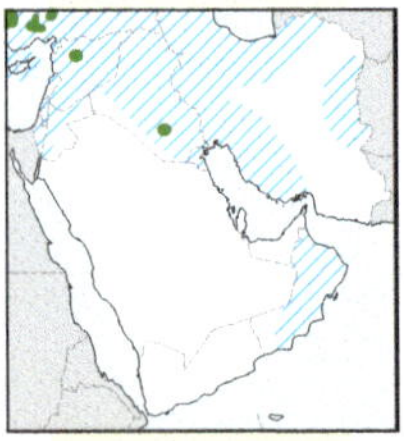

Red-crested Pochard *Netta rufina* V

L: 56. Large diving duck which sits high on water. Male unmistakable ***with large dark-orange head and red bill***; in eclipse resembles female but bill red. Female has pale grey cheeks contrasting with dark crown; dark bill has pink band near tip. In flight shows broad white wing-bar. **Voice** Male has a double *weep-weep*; female a grating *keerr*. **Habitat** Freshwater lakes, estuaries. **Note** Winter hatched, uncommon Oman; vagrant Bahrain, Qatar, Saudi Arabia, UAE.

Eurasian Wigeon
♀
♂
Common Pochard
♀
♂
Ferruginous Duck
♀
♂
Tufted Duck
♀
♂
Marbled Duck
ad
Red-crested Pochard
♀
♂

PLATE 6: MERGANSER, PYGMY GOOSE & GREBES

Red-breasted Merganser *Mergus serrator* V

L: 56. W: 78. Conspicuous ragged wispy crest in both sexes. In fast direct flight, appears elongated with long head and neck, and shows much white on inner wing. Male's greenish-black head, thin red bill, white neck collar and rusty black-spotted breast, easily separate it. Female and juvenile closely resemble female Goosander *M. merganser*, which unrecorded in UAE, but smaller and slimmer-billed, with more brownish (less greyish) upperparts, less contrasting head pattern, and without sharp demarcation between brown head and greyish-buff neck and breast; flanks darkish grey. **Habitat** Chiefly maritime. **Note** Winter hatched, often rare; vagrant Arabian Gulf to Oman.

Cotton Pygmy Goose *Nettapus coromandelianus* V

L: 33. W: 55. Smallest duck to occur in region. Short neck and short, stubby, goose-like bill; round body with, on water, high rear end. Male told by white head, neck and underparts with black cap, eye and band across white breast. In fast flight looks white with green-tinged black wings with conspicuous white trailing edge across full length of wing, very broad on primaries. Female a drab version of male, browner with dark eye-line and greyish flanks; in flight brown wings and narrow white trailing edge to secondaries. **Habitat** Well-vegetated wetlands, lakes. **Note** Rare on passage and winter Oman as hatched; vagrant Bahrain, Qatar, Saudi Arabia, UAE, Yemen (Socotra) and Iran and Iraq. [Alt: Cotton Teal]

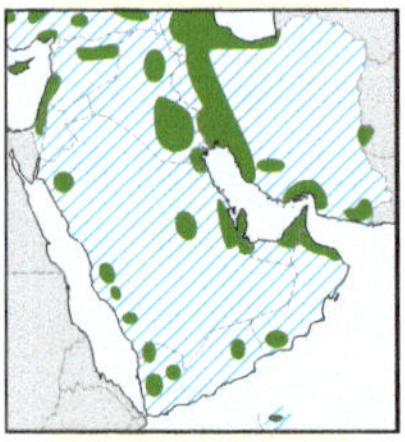

Little Grebe *Tachybaptus ruficollis* RB, wv

L: 27. W: 43. Small size, ***blunt-ended body***, short neck and bill. ***In flight amount of white on secondaries and inner primaries varies, being obvious in the subspecies*** capensis (which occurs in Arabia). Adult has ***bright chestnut throat and cheeks and conspicuous yellow gape-patch***. In winter, adult and young paler brown above, buff below, variably mixed with dull chestnut on foreneck (adult); light gape often reduced or absent. Juvenile has white striped head. Dives with fast jump; when alarmed dives rather than flies. Flight is only for a short distance with rapid wingbeats low over water. **Voice** When breeding more often heard than seen; distinctive ***high-pitched trilling*** recalling whinny of horse. **Habitat** Well-vegetated lakes and pools; also estuaries in winter. **Note** Passage and winter hatched.

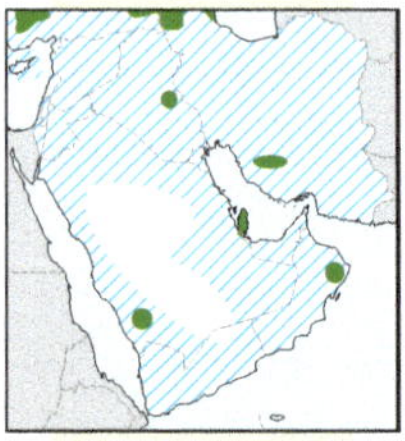

Black-necked Grebe *Podiceps nigricollis* WV, pm

L: 31. W: 58. Small grebe with ***short, uptilted bill, steep forecrown and peaked head***; sometimes looks puff-backed, recalling Little Grebe. Breeding plumage ***mostly black (including foreneck)*** with golden ear-coverts and chestnut flanks. Winter plumage basically black and white. Uptilted bill can be difficult to see at distance. In flight long, white wing-patch extending to inner primaries; also lacks white on inner forewing. **Voice** In breeding area a plaintive whistle *ooo-eep*. **Habitat** Freshwater lakes and pools, also coastal in winter. **Note** Has bred UAE; passage and winter hatched. [Alt: Eared Grebe]

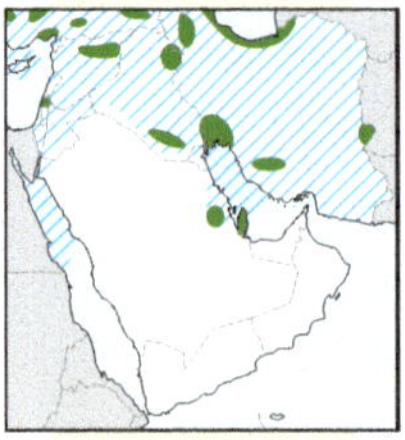

Great Crested Grebe *Podiceps cristatus* V

L: 49. W: 88. ***Large grebe with long, white slender neck with pink, dagger-like bill*** held horizontally. In flight, extended neck and feet are held below line of body; wingbeats rapid, ***large white patch on secondaries, white border to forewing and on shoulders along body***. Easily visible on open water and usually submerges smoothly, without leaping. Breeding adult told by ***black crest and black and chestnut tippets***. Winter plumage lacks tippets, but retains short, dark grey crest; ***black lores and narrow white line over eye*** making eye clearly visible at distance; Juvenile has black-and-white striped head and neck. **Voice** In breeding season a loud, harsh, far-carrying *rah-rah-rah....* **Habitat** Open fresh water, also coastal in winter. **Note** Winter hatched; vagrant Oman, UAE.

Red-breasted Merganser
♀
♂
Cotton Pygmy Goose
♂
♀
♀
♂
Little Grebe
ad winter
ad summer
Black-necked Grebe
ad winter
ad summer
ad winter
Great Crested Grebe
ad summer

Cory's Shearwater *Calonectris (diomedea) borealis* V

L: 46. W: 113. Large shearwater with ***yellowish bill and lack of clear demarcation between grey-brown head and white chin and throat***. Underwing white with narrow brownish margins; may show ill-defined whitish band at base of uppertail. Flight relaxed with fairly slow wingbeats interspersed with long glides, often in arcs well above the waves, on ***characteristically bowed wings***. Scopoli's Shearwater *C. diomedea* breeding in the Mediterranean Sea has been recently split from Cory's Shearwater and they remain very difficult to separate. Cory's is slightly ***heavier, darker on head, with brighter yellow bill and more extensive and solidly demarcated dark tips to primaries below. Scopoli's is slightly smaller with a thinner, duller bill, slightly narrower wings with white on underwing typically extending in fingers on the primaries***. **Habitat** Maritime. **Note** Vagrant UAE (from North Atlantic); reports Iran and Oman do not distinguish between Cory's and Scopoli's.

Wedge-tailed Shearwater *Puffinus pacificus* V

L: 43. W: 100. ***Sooty-brown*** with longish tail, ***the wedge shape being difficult to observe unless tail is spread***. May show paler bar on greater coverts above and ***light edging to all coverts, forming narrow bars. Bill grey with dark tip***. Flight lazy with 3–4 fairly quick wingbeats followed by short glide, wings bowed forwards and downwards; flight erratic when windy, often changing direction and soaring in low arcs. Told from Jouanin's Petrel by larger size, ***broader-based wings, longer grey bill with dark tip, which is held horizontally***, and pale feet; from Sooty Shearwater by longer tail, broader wings, ***all-dark underwings and pale feet***; and from Flesh-footed Shearwater by slightly smaller size, all-dark underwing (pale underside to primaries in Flesh-footed) and thinner bill, which is grey with dark tip (flesh with dark tip in Flesh-footed). **Habitat** Maritime. **Note** Rare in summer Arabian Sea; vagrant UAE.

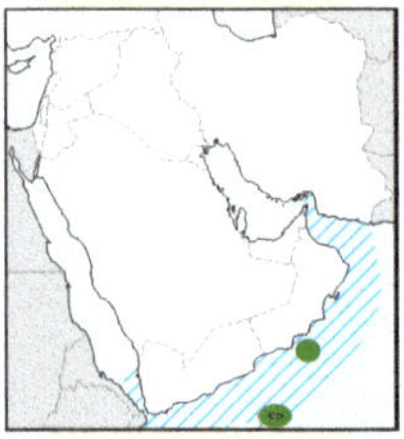

Persian Shearwater *Puffinus persicus* SV, WV

L: 32. W: 70. Small; upperparts brownish (but looks black unless close) extending to about level of eye, often showing ***suggestion of a whitish forehead and supercilium, especially in autumn birds***. Underwing with ***dark axillaries extending in a dark wedge onto most of the secondary coverts and a broad dark trailing edge to whole wing. Legs pink***. Fast wingbeats interspersed with short low glides, though often rising in low arcs in strong winds. At fish shoals patters over sea with outstretched wings. **Habitat** Maritime. **Note** Throughout year in hatched area; rare Arabian Gulf.

Sooty Shearwater *Puffinus griseus* sv

L: 42. W: 100. Dark brown shearwater, similar in size to Wedge-tailed Shearwater but with shorter tail, much narrower wings and ***variable white on underwing***, which can often flash in contrast to dark plumage; feet dark, project beyond tail in flight. Flight fast and direct with fast wingbeats interspersed with glides on stiffly held wings, sometimes arcing high above the waves. **Habitat** Maritime. **Note** Regular in summer as hatched, though rare; vagrant Arabian Gulf, Iran.

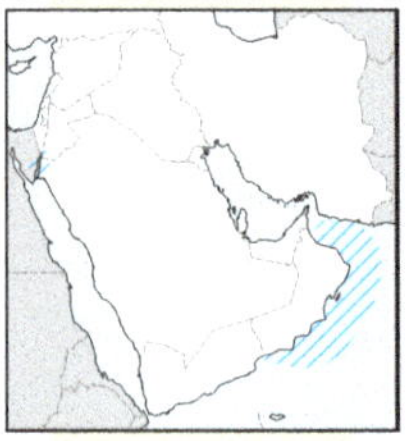

Flesh-footed Shearwater *Puffinus carneipes* V

L: 43. W: 100. Uniformly dark nut-brown often with slightly paler coverts and ***silvery flash on primaries below***, visible even at distance, distinguishes from Wedge-tailed Shearwater; if close, note Flesh-footed's larger, paler bill, slightly larger size, broader wings and ***shorter tail***. Lazy flight, generally more steady than Wedge-tailed with straighter wings: a series of slow, heavy flaps followed by a long glide, rising well above the waves in windy conditions. **Habitat** Maritime. **Note** Regular in hatched area; vagrant Gulf of Aqaba and UAE.

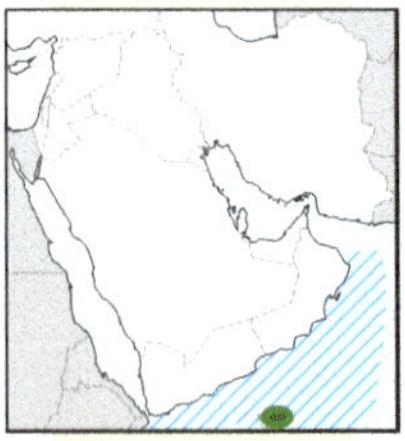

Jouanin's Petrel *Bulweria fallax* V

L: 30. W: 75. Smaller than similar sooty-brown Wedge-tailed Shearwater with less languid flight; ***all-dark stout bill held down at 45° angle***. Tail fairly long, the graduation in tail not visible unless close. In windy conditions will rise 2–5m above the waves in ***long banking arcs, interspersed with spells of 4 or 5 rather leisurely flaps***, which tend to be at the peak of the arcs. Otherwise flies fairly close to the sea, progressing through troughs with a mixture of wingbeats and long glides. **Habitat** Maritime. **Note** Regular in hatched area throughout year; vagrant UAE.

Cory's Shearwater
ad
Wedge-tailed Shearwater
ad
Persian Shearwater
ad
ad
Sooty Shearwater
ad
ad
Flesh-footed Shearwater
Jouanin's Petrel

PLATE 8: STORM PETRELS, TROPICBIRD & BOOBIES

Wilson's Storm Petrel *Oceanites oceanicus* — SV

L: 18. W: 40. Most numerous storm petrel in region. In flight the ***bold white rump extends onto sides of uppertail-coverts, and legs protrude beyond square-ended tail;*** also with fairly conspicuous panel on the upperwing-coverts and ***underwing wholly dark***. ***Webs of feet yellowish***. Flies with series of flaps interspersed with short glides. Will follow ships, unlike most storm petrels. **Habitat** Maritime. **Note** Regular summer visitor in hatched area; rare Arabian Gulf, vagrant to Gulf of Aqaba.

Leach's Storm Petrel *Oceanodroma leucorhoa* — V

L: 20. W: 46. Larger and longer-winged than Wilson's Storm Petrel with ***forked tail***. Sooty-brown, often looking black, with a pale bar across the upperwing-coverts and rather ***dirty white rump***. ***Flight erratic and bounding***, with long, rather angled wings and deep wingbeats. **Habitat** Maritime. **Note** Vagrant Gulf of Aqaba and UAE.

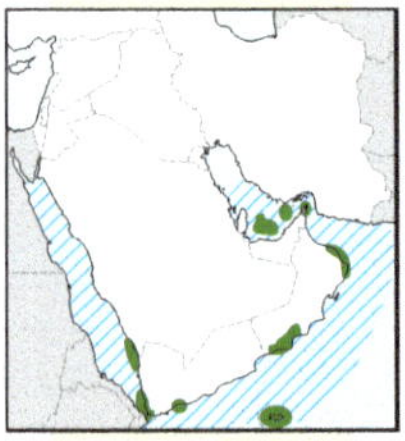

Red-billed Tropicbird *Phaethon aethereus* — MB

L: 48 (plus 50cm tail). W: 105. Unlikely to be confused with any other seabird in the Middle East. Plump-bodied, white with exceptionally ***long, white tail-streamers and conspicuous red bill***; the white plumage is relieved by a black eye-stripe, black outer primaries and narrow black barring on upperparts and coverts. Juvenile (which lacks tail-streamers), has black-tipped tail, yellowish bill and blackish collar. Flight a useful character: direct with ***fast wingbeats and interspersed with glides on horizontally held wings***, usually fairly high. Will settle on sea. **Voice** Shrill, rapid rasping notes. **Habitat** Maritime; nests colonially on rocky mainland or island cliffs, or rocky slopes on islands. **Note** Occurs at sea in hatched area throughout year; vagrant Kuwait.

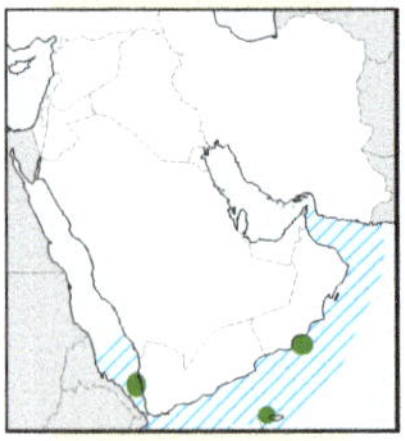

Masked Booby *Sula dactylatra* — V

L: 85. W: 150. Adults and sub-adults distinctive, but juveniles recall Brown Booby. Juvenile Masked Booby has ***brown head and neck separated from paler brown back by white collar***, which broadens with age, while upperparts become mottled with white especially (first) on the scapulars; underparts are white, also ***underwing-coverts white with black line through centre*** and black flight feathers. Juvenile differs from Brown Booby primarily in larger size, variable white mottling on upperparts, white neck-collar and more prominent band on underwing-coverts. See also Red-footed Booby for differences. **Habitat** Maritime. **Note** Occurs at sea in hatched area throughout year; rare Arabian Gulf; vagrant Iran, UAE.

Red-footed Booby *Sula sula* — V

L: 75. W: 100. This vagrant to the Arabian seas has a white and a brown colour morph, which can be can be similar in some plumages to Masked and Brown Boobies. Birds from the Indian Ocean (the most likely to occur) have, in ***adult plumage, irrespective of morph, all-white tails. The adult white morph resembles Masked Booby but is smaller, has white tail, black carpal-patches below***, often yellowish wash to head and ***red feet***, and lacks a black mask. ***Brown morph adult is grey-brown*** with darker back and wings (both above and below) and ***white tail***. Juvenile is rather featureless, all-brown with noticeably dark underwing lacking white or pale on coverts (as shown by juvenile Brown Booby). **Habitat** Maritime. **Note** Vagrant Oman, UAE, from Indian Ocean.

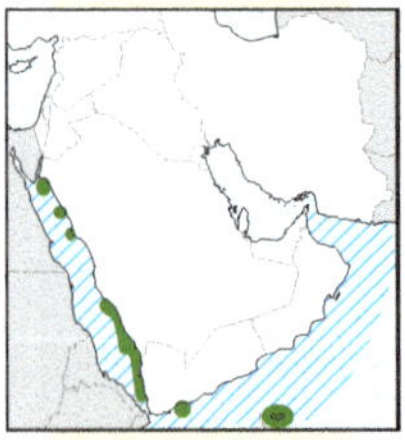

Brown Booby *Sula leucogaster* — SV

L: 70. W: 145. The commonest booby in the Red and Arabian Seas and readily identified in adult plumage by ***uniform chocolate-brown upperparts, head and neck and conspicuously white underbody and underwing-coverts***. The ***pale greenish-yellow bill*** contrasts with the dark head even at a distance. Juvenile plumage similar to adult but underparts buffish-brown (thus less contrast with brown head, neck and upper breast); white underwing-coverts appear at an early age but in some very young birds can look quite brownish. (See Masked Booby for differences.) Flight comprises a series of wingbeats followed by a long glide with the wings held fairly horizontal. Catches fish by diving, with folded wings, often at a shallow angle from a short height above the sea. See Masked Booby for separation of respective juveniles. **Habitat** Maritime. **Note** Occurs at sea in hatched area throughout year; rare Arabian Gulf.

ad
Storm Petrels
not to scale
ad
Wilson's Storm Petrel
Leach's Storm Petrel
imm
ad
ad
Red-billed Tropicbird
juv
pale morph
dark morph
Masked Booby
juv
juv
ad
Red-footed Booby
Brown Booby

PLATE 9: FLAMINGOS, STORKS & CRANES

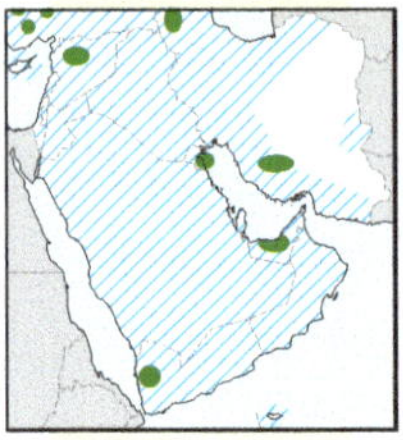

Greater Flamingo *Phoenicopterus roseus* WV, sv, rb

L: 130. W: 155. Very large, long-legged, long-necked, with characteristic bill shape.. The white plumage of adults gradually acquires a pink hue and red wing-coverts. Juveniles, which are half the height of adults, are greyish with brown markings. Most readily told from Lesser Flamingo by larger size (but beware as much size variation in Greater) and ***pink (adult) or greyish (juvenile) bill with black tip*** (all-blackish bill in Lesser Flamingo). **Voice** Cacophony of deep honks and grunts, strongly recalling domestic geese. **Habitat** Coastal lagoons, salt-lakes, mudflats; breeds colonially on mud banks or in shallow water of salt-lakes, building mud-heap nest. **Note** Passage and winter hatched.

Lesser Flamingo *Phoeniconaias minor* V

L: 85. W: 130. ***Smaller than Greater Flamingo and generally deeper and brighter pink with all-blackish bill*** though very close views show dark carmine near tip. In full adult plumage has deep rose-pink on face bordering bill (lacking in Greater) and on the long scapulars (pinkish-white to white in Greater). In flight, shows rose-pink patch across centre of upperwing-coverts, bordered pinkish-white (in Greater all secondary coverts are rose-pink, but can bleach to white). Iris red (pale yellow in Greater). ***Juvenile similar to Greater Flamingo in plumage but blackish bill readily identifies it***. Lesser Flamingo's smaller size can often be difficult to establish in lone birds; also immatures of both species are smaller than adults. ***In flight note shorter neck and legs, faster wingbeats and more uptilted chin*** than seen in Greater Flamingo. **Voice** Goose or shelduck-like but wavering (yodeling); also thin high-pitched notes recalling large gulls on a rubbish tip. **Habitat** As Greater Flamingo. **Note** Regular Yemen; vagrant Iran, Kuwait, Oman, UAE.

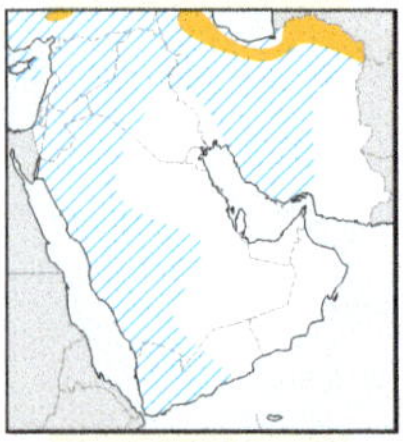

Black Stork *Ciconia nigra* V

L: 95. W: 150. Glossy-black stork with white lower underparts. Told by ***all-black upperparts*** (no white on lower back and rump, as in Abdim's Stork ***C.abdimii*** of S.Arabia) and ***small white axillary-patch on black underwing***. Adult has red bill and legs, whereas browner, less glossy juvenile has greyish-green bill and legs. **Habitat** Lakesides, marshes, riversides and fields on migration. **Note** Passage hatched; occasional in winter S. Arabia; vagrant Bahrain, Kuwait, Oman, Qatar, UAE.

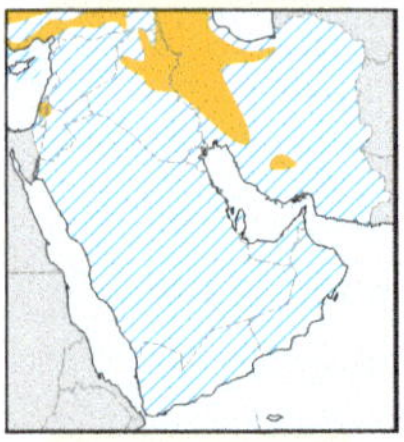

Western White Stork *Ciconia ciconia* pm, wv

L: 100. W: 170. Easily told by ***large size, white plumage with black flight feathers, straight red bill and long red legs***. In flight, the neck is extended and legs protrude beyond tail. Juvenile has duller white plumage and duller red bill and legs. From adult Yellow-billed Stork (an escape in UAE; Plate 100) by straight red bill and all-white tail (black in Yellow-billed). **Voice** Clatters bill at nest; otherwise silent. **Habitat** Wetlands, plains and farmland; nests on buildings and trees. **Note** Passage hatched, some overwinter.

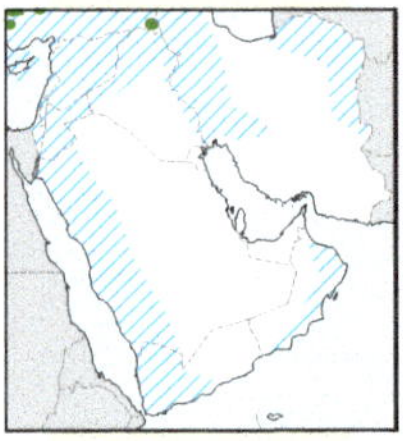

Common Crane *Grus grus* V

L: 115. W: 233. Large and majestic. ***Grey plumage with contrasting black flight feathers, black head and upper neck, and white stripe from eye down side of neck. Looks 'bushy'*** at rear end on ground. Juvenile has brownish head without contrasting head pattern. Adult told from Demoiselle by size and ***absence of black breast***. Gregarious on migration. Neck extended in flight, as in other cranes; powerful wingbeats interspersed with long glides, often soars; flies in 'V' formation. **Voice** Often detected by far-carrying, trumpeting *krrllaa, krrllaa*. **Habitat** Wetlands, fields and steppe. **Note** Passage and winter hatched, but sporadic/local in winter; vagrant Bahrain, Kuwait, Qatar, UAE.

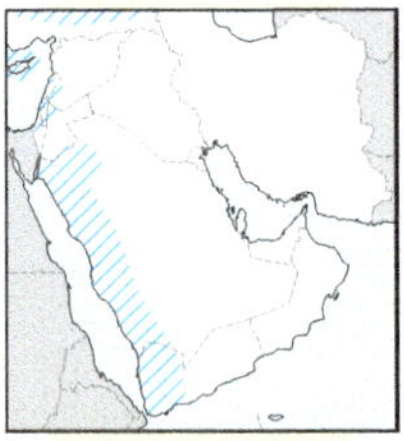

Demoiselle Crane *Anthropoides virgo* V; E/I

L: 95. W: 175. Smaller with shorter neck and bill than Common Crane, but size deceptive without comparison. Pale grey with largely black head and neck with elongated breast feathers hanging down in a narrow black fringe. Juvenile grey on head and neck with short whitish band behind eye. Immature gradually acquires adult plumage, general colouration browner with duller black parts. **Voice** Call higher-pitched than Common Crane. **Habitat** Open plains near water, arable land, wetlands. **Note** Passage hatched; rare Oman; vagrant Iran, Iraq, Kuwait, UAE.

Greater Flamingo
ad
juv
Lesser Flamingo
ad
juv
Black Stork
ad
imm
Western White Stork
ad
imm
Common Crane
juv
ad
Demoiselle Crane
juv
ad

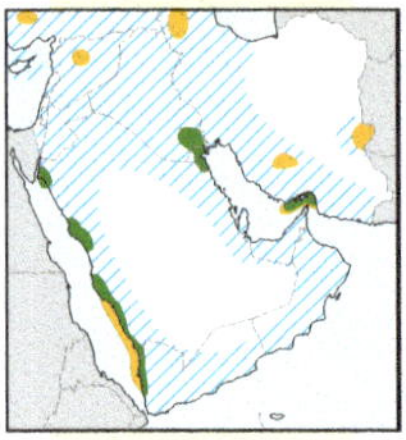

Eurasian Spoonbill *Platalea leucorodia* WV, pm, sv

L: 85. W: 120. Heron-sized with ***all ivory-white plumage, characteristic black spatulate bill with yellow tip and black legs***. Adult has nape plumes and yellowish neck band, which are lost in winter. Immature has dull flesh-coloured bill and legs and black wing-tips. In flight, neck extended and wingbeats fast and shallow, interspersed with short glides, usually in small groups in line-formation. When feeding keeps in close groups, unlike egrets and herons, sweeping head from side to side, with bill submerged. Often asleep by day, with head on back and bill hidden, but ivory-coloured plumage and closeness of birds to each other aids identification. **Habitat** Shallow open water and mudflats; nests colonially in reedbeds, islands and trees. **Note** Passage hatched, winters widely from S. Turkey southwards.

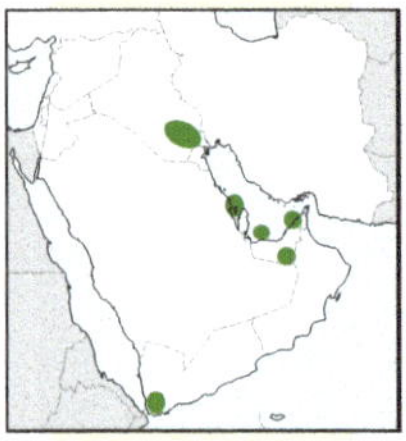

African Sacred Ibis *Threskiornis aethiopicus* E/I

L: 80. W: 112. Unmistakable with black-and-white plumage, ***long black scapulars drooping over rear end and long, decurved bill***. Juvenile has mottled head and neck. In rather heavy flight, ***shows diagnostic black line on rear-edge of wings***; longish neck and legs protruding just beyond tail. **Habitat** Wetlands, cultivated areas, coastal marshes, parks, large gardens; nests colonially in trees. **Note** May breed Iran; feral population in Arabian Gulf; vagrant Kuwait, Oman, Saudi Arabia.

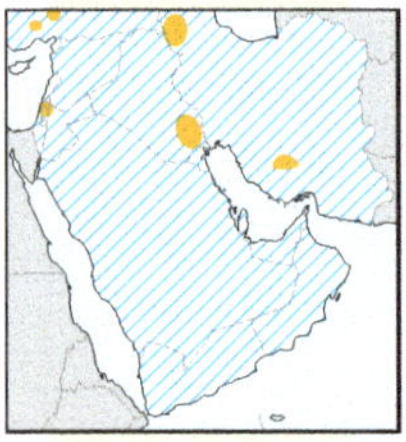

Glossy Ibis *Plegadis falcinellus* WV, pm

L: 65. W: 90. ***Blackish with long, decurved bill*** but close views show adult to have ***deep purple-chestnut plumage***, glossed green on wings. In breeding season has white marks at base of dull pink bill; in winter, bill brownish with fine pale streaks on head and neck. Juvenile much duller. Fast wingbeats in flight often interspersed with long glides; frequently flies in line-formation. **Habitat** Freshwater wetlands and marshes; nests colonially in reedbeds, occasionally in trees. **Note** Passage hatched, some winter.

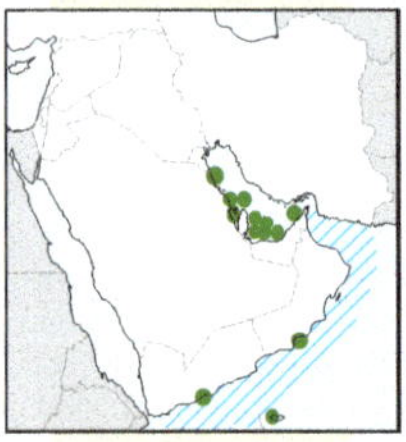

Socotra Cormorant *Phalacrocorax nigrogularis* RB

L: 80. W: 130. Slightly smaller than Great Cormorant, with slimmer head and neck (resembling European Shag *P. aristotelis* in structure, though longer-winged in flight). ***Adult sooty-black with glossy bronze-green wings and back, lacks white face and chin-patch of Great Cormorant and has much slimmer greyish bill. Immature grey-brown above with pale fringes to coverts and scapulars***; breast and belly off-white, sometimes with brownish spotting. Juvenile has less obvious pale fringes to coverts and lacks spotting on breast and belly; best separated from young Great Cormorant by structure, bill shape, pale-fringed coverts and, when present, dark spotting below. Congregates in large flocks in and out of breeding season. **Habitat** Maritime, coastal; nests in large colonies mainly on offshore islands. **Note** Occurs in hatched area outside breeding season.

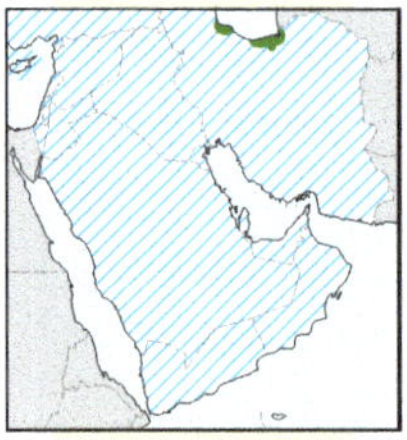

Great Cormorant *Phalacrocorax carbo* PM, WV

L: 90. W: 140. Large; swims low in water and frequently perches with wings outstretched. Breeding birds have ***white on nape and neck and white thigh-patch. In winter retains white patch on chin and throat*** unlike Socotra Cormorant. Juvenile browner than adult with dirty-white underparts. **Habitat** Coasts and inland lakes. **Note** Has bred Bahrain; passage and winter in hatched area, some oversummer UAE.

Eurasian Spoonbill
imm
ad summer
ad
African Sacred Ibis
imm
Glossy Ibis
ad summer
imm
Socotra Cormorant
imm
ad
Great Cormorant
imm
ad summer

PLATE 11: HERONS & BITTERN

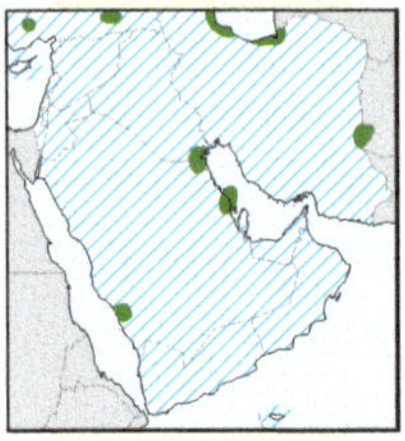

Grey Heron *Ardea cinerea* PM, WV

L: 95. W: 185. Large with long neck and legs and powerful bill. In all plumages has ***grey upperparts and white underparts and neck***. Adult has ***black crest and dark markings down front of neck***. Juvenile has darker grey or browner upperparts and crown with rest of plumage white. Bill yellowish and legs brownish-yellow in adult, but in younger plumages bill browner and legs greyish. Juvenile can superficially resemble Purple Heron, but is much paler with thicker neck, stouter bill and never shows buffish-brown on neck and breast. **Voice** A raucous *waak* in flight. **Habitat** Wetlands including coastal; nests in trees. **Note** Passage and winter in hatched area; some oversummer UAE.

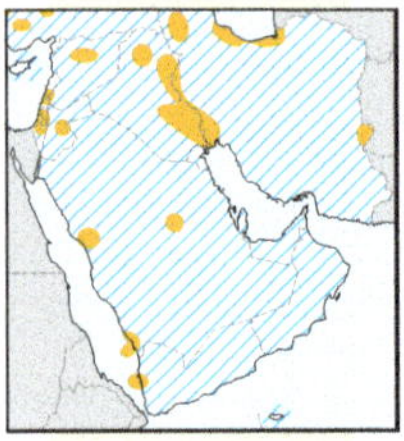

Purple Heron *Ardea purpurea* PM, wv, sv

L: 80. W: 135. Slimmer and much darker than Grey Heron, with ***angular, snake-like neck***. Adult dark grey with black belly, ***chestnut on neck edged by black streak, black crown and line across cheeks***; dull yellow bill. Juvenile has sandy-brown upperparts and hindneck with diffuse dark streaking on neck and blackish crown. ***In flight, has more angular neck (bulging downwards more obviously) than heavier-looking Grey Heron with prominent spread toes and brownish upperwing-coverts***. Mostly seen in or when flushed from vegetation, unlike Grey Heron which feeds in more open areas. **Habitat** Marshes, reedbeds, ditches; nests colonially in reeds or trees. **Note** Mainly summer visitor; passage hatched.

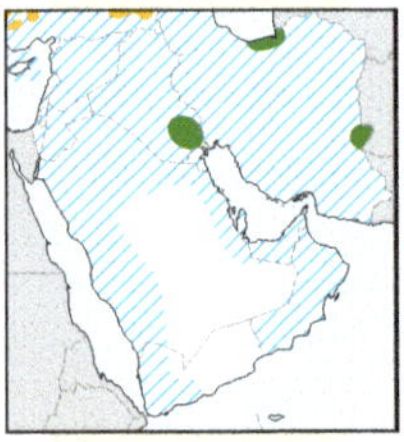

Eurasian Bittern *Botaurus stellaris* wv

L: 75. W: 130. Cryptically patterned brown heron, more often heard than seen. Smaller than Grey Heron with ***stocky neck and entirely dark brown and golden-buff streaked plumage***. In owl-like flight (but with quick wingbeats) neck often extended in front, before being retracted. Juvenile Black-crowned Night Heron is smaller and shows white spots on wings; juvenile Purple Heron lacks streaks on back and has a long, thin neck. **Voice** In breeding season a ***loud boom like blowing across the mouth of an empty bottle,*** **upwhoom**, usually repeated many times; flight call a hoarse *kaau*. **Habitat** Reedbeds; also wetlands with lush vegetation outside breeding season. **Note** Passage and winter hatched, but rare Arabia, including UAE.

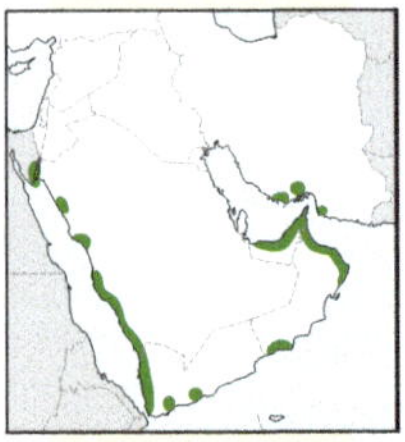

Striated Heron *Butorides striata* RB

L: 43. W: 60. Primarily ***coastal; small, rather dark heron***. Adult identified by black crown with elongated nape plumes, ***bluish-grey upperparts***, ***buff-fringed coverts and greyish neck and underparts***; yellow patch on lores and dark moustachial streak give marked facial pattern. ***Rosy, pinkish or yellowish legs*** extend just beyond tail in flight. ***Immature brownish with white spots on tips of wing-coverts***, brown-and-white streaked upperbreast and yellowish-green legs. May recall young Black-crowned Night Heron, but smaller with dark crown and lacks white spots on mantle. In flight (low with fast wingbeats), all ages show dark upperwings. Solitary, often skulking, adopting crouching position if disturbed; most active at dusk. **Voice** Alarm call *chook-chook-chook*; when flushed a croaking *kweuw*. **Habitat** Rocky or sandy coasts, mangroves, sometimes inland wetlands. **Note** Range expanding in Arabian Gulf.

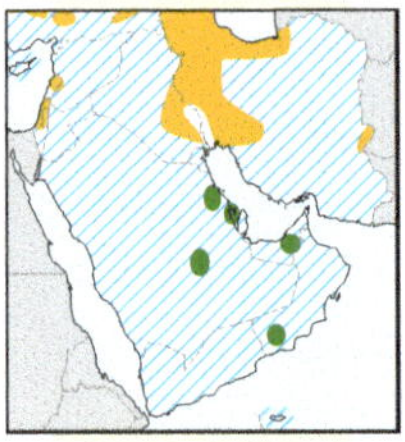

Black-crowned Night Heron *Nycticorax nycticorax* WV, pm, rb

L: 60. W: 110. Stocky, about half the size of Grey Heron; most active at dusk. ***Adult's grey plumage with black back and crown, unmistakable***. ***Brownish juvenile is prominently spotted with white on back and coverts*** and this feature immediately identifies it from other brown heron-types. By second calendar year the spotting is reduced in size and the mantle and scapulars become grey-brown (mirroring the black of the adult). More likely to be flushed than seen in the open during day. **Voice** Harsh deep *kwark*, heard in flight, especially at dusk. **Habitat** Rivers, lakes and marshes with trees and dense vegetation; nests in reeds or trees. **Note** Has bred Kuwait; passage and winter hatched, but few winter in north.

Grey Heron
ad
juv
ad
Purple Heron
juv
Eurasian Bittern
ad
Striated Heron
ad
imm
ad
ad
ad
juv
Black-crowned Night Heron
ad

PLATE 12: BITTERNS, POND HERONS & CATTLE EGRETS

Cinnamon Bittern *Ixobrychus cinnamomeus* V

L: 38. Adult cinnamon-tan above with ***warm cinnamon ear-coverts***; in flight ***wings and tail uniformly cinnamon above*** (Little Bittern shows strong contrast between coverts and flight feathers in both sexes). Darker-crowned female more streaked below than male. Legs yellowish. Immature dark brown, mottled paler above; more darkly streaked below than similar-aged Little Bittern. **Voice** May give croaked *kok* when flushed. **Habitat** Wetlands, dykes and pond margins. **Note** Vagrant UAE.

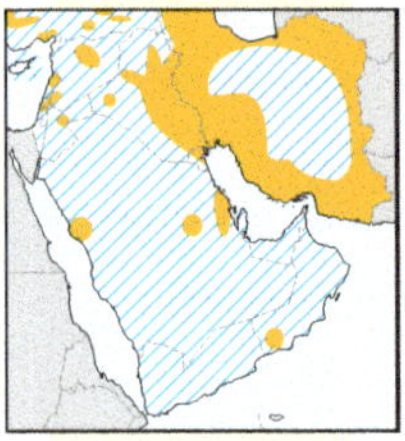

Little Bittern *Ixobrychus minutus* PM

L: 35. W: 55. ***Small; conspicuous pale covert-panels contrasting with black flight feathers and dark back diagnostic***. Female Little Bittern resembles male but duller with a more rufous tinge and buff streaks on upperparts and brownish streaks on underparts. Juvenile more boldly streaked buff and brown with duller, streaked covert-panels. Legs greenish or yellowish. Most often seen in flight, when flushed, usually flying a short distance on rather jerky wingbeats before diving into cover. **Voice** In flight a short *kek* or repeated *kek-kek-kek-kek*; often calls at dusk. Male in breeding season has loud croaking *khok* repeated at intervals of about two seconds. **Habitat** Well-vegetated rivers, ponds and lakes; often nests in loose colonies. **Note** Has bred Bahrain, UAE; passage hatched.

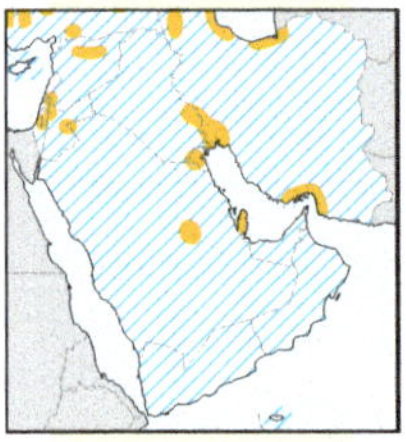

Squacco Heron *Ardeola ralloides* PM, wv

L: 45. W: 85. Small heron, which usually remains in or near lush cover. Adult in breeding plumage is golden-buff with purple sheen on mantle and ***long, black-and-white streaked nape plumes***; bill has a greenish-blue base in summer; nape plumes lost in winter, when neck becomes streaked and bill has yellowish base. Juvenile brownish-buff with streaked neck and upper breast, making it well camouflaged. ***At all ages, white wings revealed in flight, making the bird look predominantly white. Closely resembles Indian Pond Heron (which see)***; care needed to separate in areas where both occur or when vagrant individual suspected. **Voice** Flight note a harsh *kar*. **Habitat** Wetlands, ditches, lakesides, rivers with vegetation; nests colonially in trees, reeds. **Note** Has bred UAE; passage hatched, few winter.

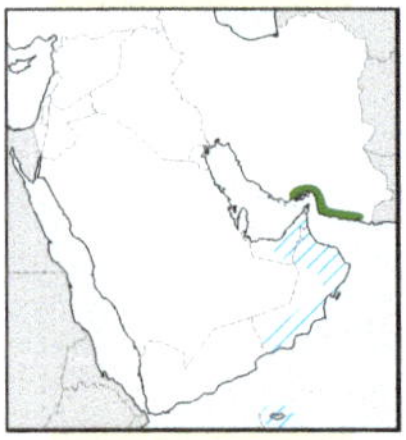

Indian Pond Heron *Ardeola grayii* WV, pm

L: 45. W: 85. ***Very similar to Squacco Heron; often feeds in open. Readily separable in breeding plumage by dark maroon back, unstreaked yellowish-buff to buff-brown crown and hindneck, white nape plumes*** (crown and hindneck streaked blackish, white nape plumes black-edged in Squacco), and pale buff breast (pale ochre in Squacco). In winter, loses head plumes and others are reduced in length; head, neck and upper breast become streaked brownish. ***Winter adult and juvenile told from Squacco by darker, vinous-brown mantle and lack of buff tones***; dark loral line an additional feature (though sometimes absent), the bare skin above it being rectangular in shape (but triangular in Squacco, which lacks dark loral line). Juvenile similar to winter adult but primaries dark-tipped. Legs greenish/yellowish (red in some breeding birds). **Voice** Close to Squacco's. **Habitat** Fresh and saltwater marshes especially with dense vegetation; mangrove mudflats, rivers, ponds. **Note** Some present throughout year in hatched area; vagrant Kuwait.

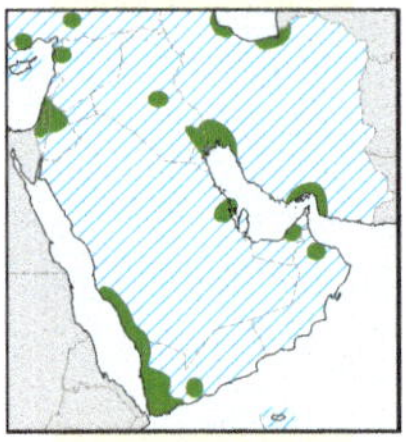

Western Cattle Egret *Bubulcus ibis* PM, WV, rb

L: 50. W: 85. Small, white, grassland-dwelling heron. Separated from Little Egret by stockier build, ***shorter yellow bill, shorter neck and legs, extended 'jowl' under bill*** and faster flight. Also in ***breeding season, orange-buff wash on crown, back and breast, and reddish bill***. Usually seen feeding in small groups, associating with grazing animals, or flying in flocks to and from roost sites. **Voice** A short *ark* and duck-like *og-ag-ag* often in flight. **Habitat** Fields, wasteland, marshes; nests colonially in reedbeds, bushes or trees. **Note** Passage and winter hatched.

Eastern Cattle Egret *Bubulcus (ibis) coromandus* V

L: 50. W: 85. Very similar to Western Cattle Egret, but has ***longer bill and tarsi***. In summer plumage ***buff on head reaches cheeks and throat***. In winter plumage doubtless overlooked. **Note** Vagrant Oman, UAE.

Little Bittern
♂
♀
juv
ad
Cinnamon Bittern
ad summer
ad winter
imm
ad summer
Squacco Heron
imm
Indian Pond Heron
ad summer
ad summer
Western Cattle Egret
ad summer
ad winter
Eastern Cattle Egret

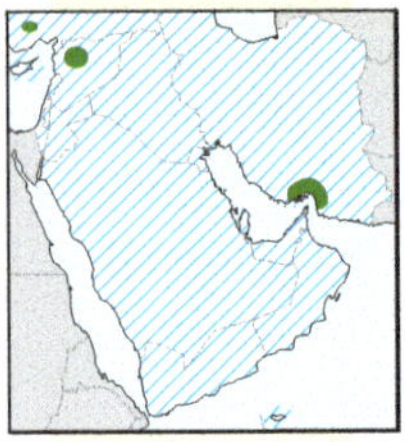

Western Great Egret *Egretta alba* — WV, PM, sv

L: 95. W: 155. ***Largest white heron with long, angular neck often stretched to full extent.*** In breeding plumage note scapular plumes, black bill with yellow base and black legs, yellowish above the joint. Winter adult and juvenile have yellow bill and blackish-green or brownish legs. ***Told in all plumages from Intermediate Egret by longer, thinner bill and black gape-line extending behind eye.*** Sedate in flight and on ground. **Habitat** Wetlands and rivers. Nests colonially in reeds or trees. **Note** Passage and winter hatched. Smaller eastern subspecies *modestus* rare breeder Iran, occasional visitor SE Arabia. [Alt: Great White Egret]

Intermediate Egret *Egretta intermedia* — V

L: 65. W: 110. All white; between Western Great Egret and Little Egret in size. Bill orange-yellow; in non-breeding season usually has black bill tip in Asian subspecies, *intermedia*, recorded UAE (though bill all-black in breeding season). Yellowish facial skin and blackish legs with brownish-grey joints and tibia. Most easily confused with Western Great Egret but note smaller size, shorter neck and ***shorter, stouter bill. Best told by very short gape-line, which does not extend behind eye (as it does in Western Great Egret).*** Can appear similar to white morph of Western Reef Heron but larger, with longer legs and blackish feet; bill also brighter yellow and straight. **Habitat** Wetlands, coastal and inland. **Note** Passage and winter hatched; vagrant UAE, Yemen.

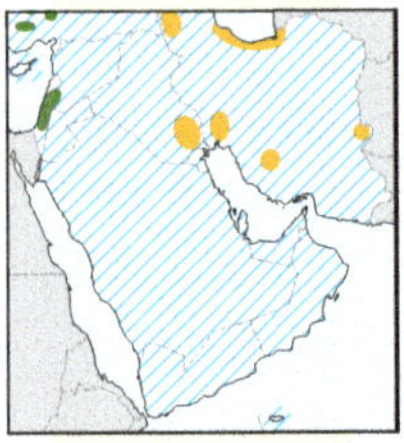

Little Egret *Egretta garzetta* — WV, PM

L: 60. W: 90. Graceful, clean white egret which, in adult plumage, has ***all-black bill and black legs with yellow feet.*** In breeding season shows long, delicate plumes on nape and mantle. Juveniles have brownish-green legs and pinkish base to lower mandible. May be confused with white morph of Western Reef Heron, which see. Told from cattle egrets by larger size, bill colour and lack of buffish wash to plumage. **Voice** Throaty grunted *raaak* in flight. **Habitat** Wetlands; nests colonially in trees and reedbeds. **Note** Passage and winter hatched.

Western Reef Heron *Egretta gularis* — RB

L: 60. W: 90. White, dark and intermediate morphs occur; subspecies *schistacea* (now called Indian Reef Heron) occurs. White morph similar to Little Egret but less elegant, ***thicker-billed with curved culmen giving slightly drooping look.*** Bill in all plumages pale brown to yellowish, usually with reddish flush in breeding season. Legs dark olive-brown, but the tarsus often greyer-greenish up rear, sometimes also on lower forepart or even entirely greenish, with the feet generally greenish or yellow (as Little Egret), very rarely rosy. Outside breeding season facial skin yellow or greenish-yellow (blue-grey in Little Egret). Dark morph ***slate-grey with white chin and throat*** and occasionally a few white flight feathers; juvenile dark morph is paler grey than adult with whitish on foreneck, breast and belly; like adult can show white feathers in wing. Juvenile white morph often has grey feathers in plumage. See also Intermediate Egret. Feeds from rocks or by slowly wading in shallow water, along tide edge or beach with occasional sudden dash after prey, often with wings flailing. **Voice** Guttural *grrurr.* **Habitat** Coastal, especially tidal flats; rarely inland; nests colonially in mangroves, low bushes on offshore islands.

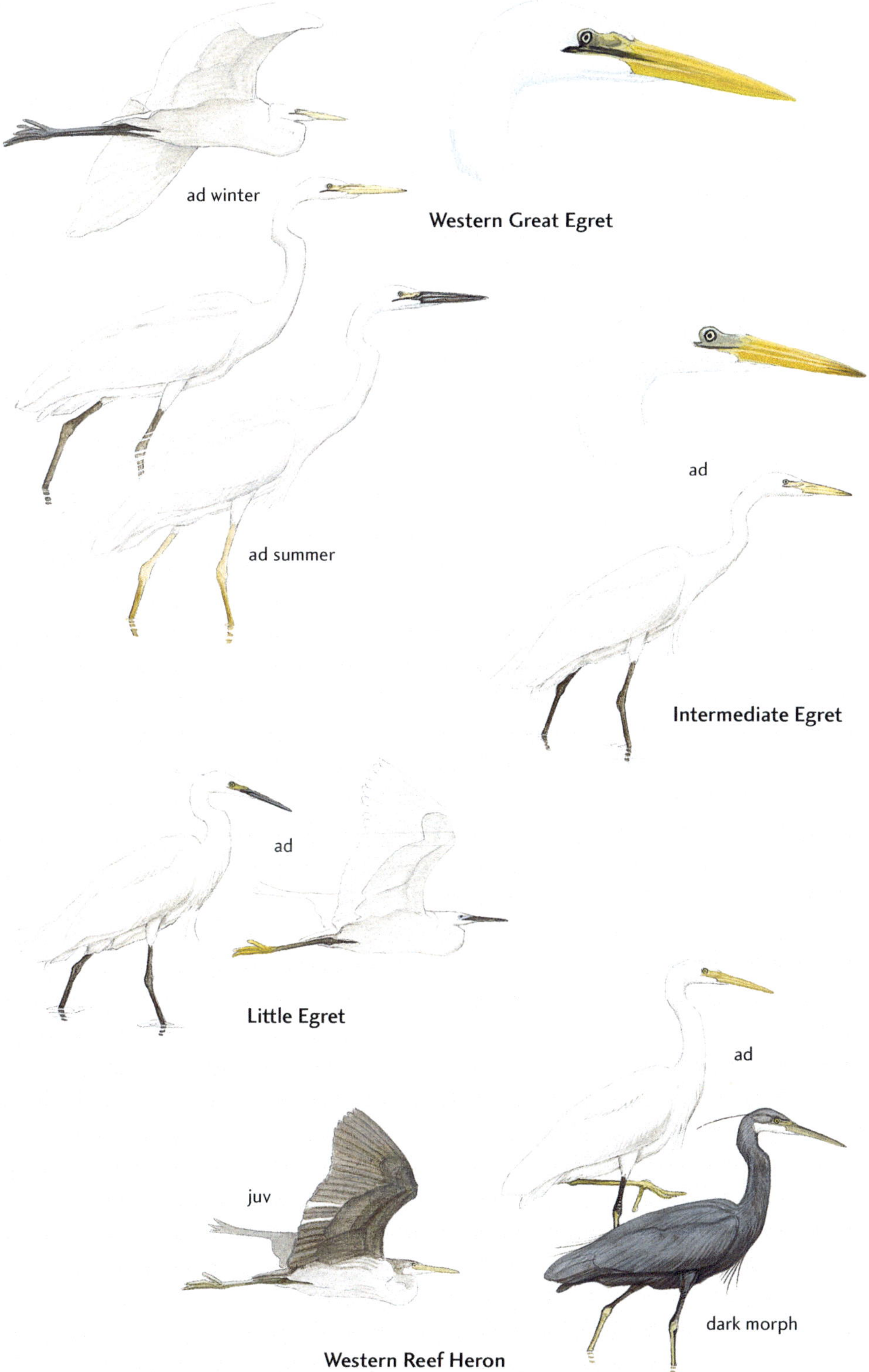
ad winter
Western Great Egret
ad summer
ad
Intermediate Egret
ad
Little Egret
ad
juv
dark morph
Western Reef Heron

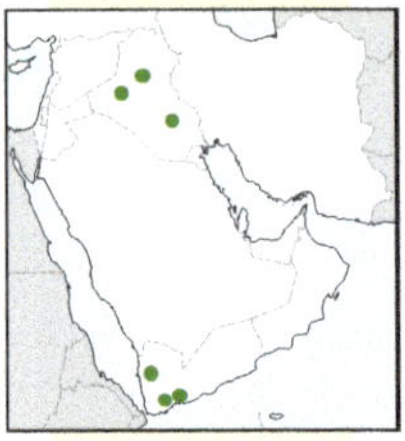

Black-winged Kite *Elanus caeruleus* V

L: 33. W: 76. Small, similar in size to Common Kestrel, but appears larger in flight; also hovers. Plumage essentially ***pale grey and white, with black primaries and large black patch on forewing of pale grey upperparts***. Underparts white; eyes red. Juvenile darker above than adult with white-tipped greater coverts forming narrow line. Note well-protruding, broad head, ***shortish tail*** with slight notch when closed, long and pointed but relatively broad-based wings. Rather owl-like flight with soft wingbeats; wings raised, rather like a harrier when soaring; persistently hovers. **Voice** In display, a thin whistling *wee-oo, wee-oo*. **Habitat** Open country with trees. **Note** May breed Iran; vagrant Kuwait, Oman, Qatar, Saudi Arabia, UAE.

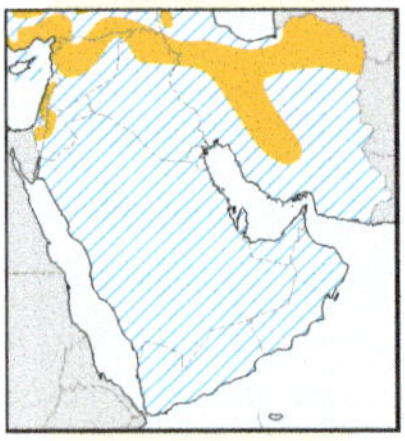

Lesser Kestrel *Falco naumanni* PM, wv

L: 28–33. W: 63–74. Very like Common Kestrel but slightly smaller and slimmer, with slightly narrower wings, more wedge-shaped tail and quicker wingbeats. ***Male is unmarked rufous above but greater coverts usually blue-grey. Head ash-grey without moustache or pale cheeks***. From below, ***white underwing contrasts with dark wing-tip and creamy-buff body***, both of which have small black spots; in some, underwing-coverts virtually unmarked. Female like Common Kestrel but on average has slightly whiter, less barred flight feathers below, sometimes fewer and finer spots on underwing-coverts and greyer uppertail-coverts. ***Female and juvenile can be identified by the wing-tip formula: primary 10 (outermost) longer than P8 and clearly longer than P7***, or, on close views, ***by pale claws*** (black in Common Kestrel). Hovers less persistently than Common Kestrel, mostly taking insects in flight. Gregarious at breeding sites and on passage. **Voice** Rasping, trisyllabic *chae-chae-chae*, very different from Common Kestrel. **Habitat** Hunts over open country; nests colonially in holes in buildings, cliffs, trees. **Note** Passage hatched.

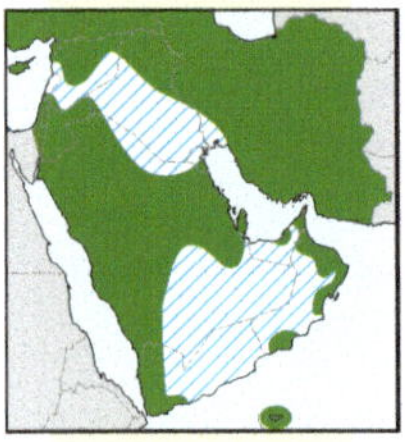

Common Kestrel *Falco tinnunculus* RB, PM, WV

L: 32–38. W: 70–78. Long, narrow, fairly pointed wings, long, slightly ***tapering tail, shallow loose wingbeats, persistent hovering and rufous upperparts, contrasting with darker flight feathers***. Compared to Lesser Kestrel, ***male has black spots on back and wing-coverts, lacks blue-grey greater upperwing-coverts***, different head pattern and more marked underwing. Female can approach male in greyness on head, tail-base and uppertail-coverts. Juvenile paler brown with thin white fringe to greater coverts. Active flight alternates with glides, some soaring and frequent hovering. ***At all ages told from Lesser Kestrel by black claws and wing-tip formula: primary 10 (outermost primary) shorter than P8 and equalling P7*** (useful when soaring at close range outside autumn period of primary moult). **Voice** Shrill *kee-kee-kee*, often repeated and heard mostly in breeding season. **Habitat** Open country with trees, mountains and semi-deserts; nests in hole or ledge on cliff or building; will use old nests of other species. **Note** Partial migrant.

Merlin *Falco columbarius* pm, wv

L: 25–30. W: 55–65. Female larger than male; smallest falcon in the region. ***Short, pointed wings, medium-length tail***, speedy flight with fast wingbeats, interspersed with short glides. ***Male told by blue-grey upperparts with blackish primaries, broad black tail-band and ill-defined head pattern***. Underparts buffy or whitish with dark streaks, or sometimes rich reddish spotting. Female and juveniles are brownish above, creamy below with dark streaks or dense dark spotting, with a ***diffuse moustache, barred primaries above and five pale/dark bands of equal width on uppertail***. In Siberian subspecies, *pallidus*, the male is distinctly paler blue-grey above with some rusty on neck, shoulders and mantle and underparts are whiter. Female and juvenile *pallidus* are rufous above with Common Kestrel-like dark bars (but kestrel's flight, proportions and denser tail-barring prevent confusion). Hunts usually low over ground with undulating flight, changing direction, followed by a straight attack. When perched, wings fall well short of tail tip. **Habitat** Open country; steppes and semi-deserts, marshes, farmland and plains. **Note** Passage and winter hatched; rare SW Arabia; vagrant Bahrain, Oman.

Black-winged Kite
juv
ad
Lesser Kestrel
♀
♂
♂
Common Kestrel
♀
♂
♂
Merlin
♀
♂
♂
♀ *pallidus*

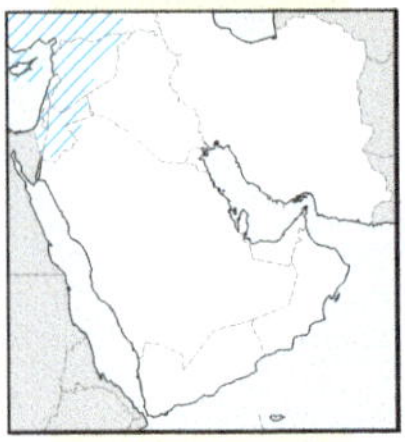

Red-footed Falcon *Falco vespertinus* V

L: 30. W: 73. ***Male uniformly slate-grey with silvery primaries above, red thighs and undertail-coverts and blackish underwing-coverts. Female has rusty-yellow underparts and head, dark eye-mask and barred tail; yellow-buff underwing has dark trailing edge*** (conspicuous in juvenile). First-spring male slate-brown with rufous and slate underparts, but underwing-coverts, flight and most tail feathers as juvenile. Some second-autumn adults' central primaries show variegated pattern as moult progresses. Less stocky than Eurasian Hobby in flight with loose, kestrel-like, wingbeats. Gregarious; hunts flying insects, alternating with spells of hovering. **Habitat** Plains with trees, grasslands. **Note** Passage hatched; vagrant Iran, Iraq, Kuwait, UAE.

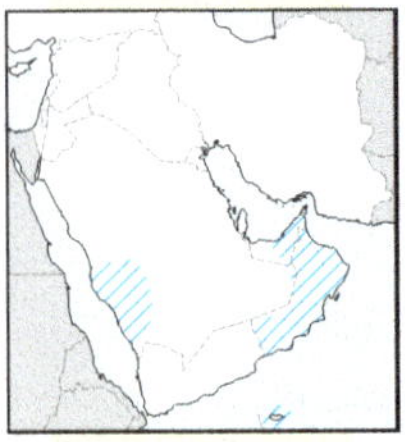

Amur Falcon *Falco amurensis* V (rare pm)

L: 26–32. W: 65–75. Resembles Red-footed Falcon in flight and silhouette, but ***male has white underwing-coverts and grey (not black) tail, paler cheeks and paler grey under- than upperparts. Female white below with warm buff thighs and streaked breast***, boldly barred flanks and ***lightly spotted underwing-coverts; dark crown and short moustache contrast with white cheeks*** (female Red-footed has rusty crown with dark eye-mask). ***Juvenile and immature resemble Red-footed but note white ground-colour below and darker crown.*** Second-year male like Red-footed until white underwing-coverts appear at one year. Female and juveniles told from Eurasian Hobby by barred uppertail, smaller moustache, paler under-surface, flight and hovering. **Habitat** Cultivations, lightly wooded areas. **Note** Known passage hatched, rare; vagrant Kuwait, Qatar, Yemen.

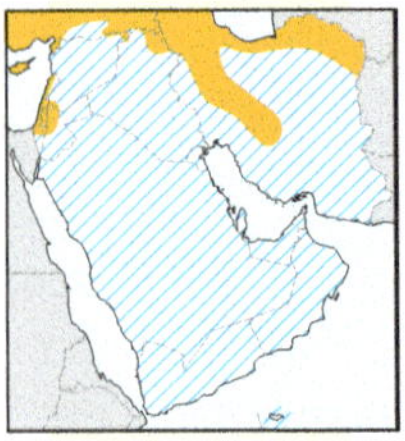

Eurasian Hobby *Falco subbuteo* PM

L: 32–36. W: 74–92. ***Scythe-like, pointed wings*** and relatively short tail. Adult has ***slate-grey upperparts, uniform tail, prominent moustache and conspicuous white cheeks***, densely streaked underparts with ***red thighs and undertail-coverts***. Juvenile browner above with pale feather fringes, lacks red thighs and undertail-coverts; told from juvenile Red-footed Falcon by more distinct breast streaking, darker head with more contrasting face-mask, unbarred uppertail and underwing pattern. Flight swift and agile; has strong steady wingbeats, short fast glides; accelerates when hunting birds, soars when catching insects; rarely hovers (briefly). **Habitat** Scattered woodland, cultivations, open country. **Note** Passage hatched.

Eleonora's Falcon *Falco eleonorae* V

L: 39. W: 97. ***Long-winged, long-tailed,*** recalling Eurasian Hobby but larger. Flight swift and agile or ***relaxed with slow wingbeats***. Pale morph recalls Eurasian Hobby but has ***darker underparts and dark, unmarked underwing-coverts contrasting with pale-based, unbarred flight feathers***; at distance looks dark below except for pale throat and cheeks. Dark morph (25% of population) ***uniform blackish-brown***; from male Red-footed Falcon by size, proportions, flight, underwing-pattern, dark primaries above and lack of red thighs. Juvenile (both morphs) paler below than adult pale morph. ***Told from Hobby by dark underwing-coverts contrasting with paler flight feathers, which have dark trailing edge***; also thinner moustache. Often hunts in flocks, especially at dusk; catches insects in flight, sometimes hovers or stoops. Breeds as Sooty Falcon. **Voice** Loud hoarse *kjie-kjie-kjie* when breeding. **Habitat** Rocky islands and sea-cliffs; often hunts over wetlands. **Note** Passage hatched; vagrant eastern Saudi Arabia, UAE.

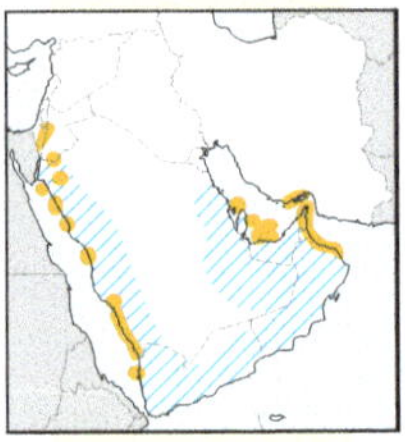

Sooty Falcon *Falco concolor* mb

L: 32–38. W: 85. Larger than Eurasian Hobby and long-winged like Eleonora's Falcon, but tail slightly shorter, with elongated central tail feathers. Adult from dark morph Eleonora's by ***slaty-grey upperparts with darker primaries and outer uppertail; blue-grey underparts*** (underwings paler) ***without Eleonora's contrasting underwing pattern***. Female darker than male. Juvenile from similar Eurasian Hobby and pale morph Eleonora's (adult and juvenile) by ***greyer upperparts with darker wing-tip, less clearly streaked underparts, the spot-streaks almost merging on upper breast; lightly marked dusky underwing*** has dark wing-tip and trailing edge. (Eleonora's has dark coverts contrasting with paler flight feathers; Hobby has uniform underwing); ***undertail finely barred except near tip*** (Hobby and young Eleonora's have undertail barred to tip). Flight recalls Eurasian Hobby but glides on level wings. When perched, wing-tip reaches tail-tip or slightly beyond. Breeds late summer, feeding young on autumn migrants. **Voice** Shrill ringing alarm *kee-kee-kee* at nest-site. **Habitat** Colonial on islands, inland desert cliffs; usually nests in hole. **Note** Passage hatched; vagrant Kuwait.

Red-footed Falcon
♀
juv
♂
♀
♂ 2nd year
Amur Falcon
juv
♀
♂
Eurasian Hobby
juv
ad
juv
Eleonora's Falcon
ad
pale morph
ad
dark
morph
Sooty Falcon
juv
ad

PLATE 16: LARGE FALCONS

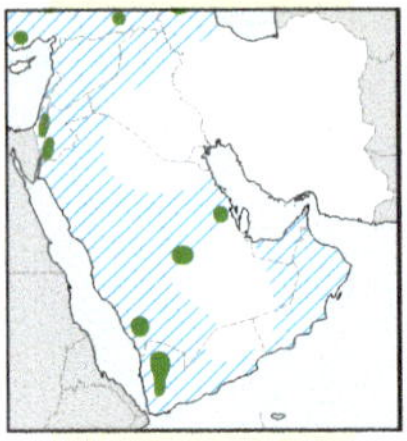

Lanner Falcon *Falco biarmicus* — V; E

L: 42–52. W: 95–115. Peregrine-sized, resembling Saker Falcon in plumage and shape; long wings, slightly blunt-ended when soaring; tail relatively long. ***Adult from Saker by barred, greyish upperparts, distinctly barred uppertail***, more contrasting head pattern: ***black forehead band, clear-cut narrow black eye-stripe, conspicuous moustache and spot-bars on flanks. Crown unstreaked creamy-buff*** (Near Eastern *tanypterus*), ***chestnut*** (SW Arabian *abyssinicus*) or ***pale rufous rear crown and nape*** (European *feldeggi*). Juvenile dark brown above with boldly streaked underparts and rear underwing-coverts; ***unbarred closed uppertail*** (unlike most Sakers). Contrasting underwing pattern and more densely streaked underparts, separates it from juvenile Peregrine and Barbary Falcons. Moderately slow, stiff wingbeats, faster when hunting; stoops or runs down prey; soars with wings level or slightly upcurved. **Voice** Slow, scolding *kraee-kraee-kraee* at breeding site. **Habitat** Mountains, plains and semi-deserts. **Note** Declining; dispersal hatched, but rare; escapes confuse true picture.

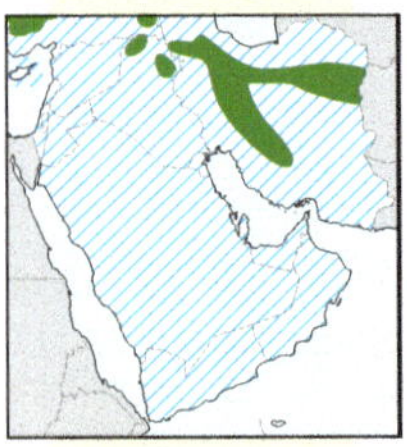

Saker Falcon *Falco cherrug* — pm, wv; E

L: 47–55. W: 105–125. Like Lanner Falcon but larger, heavier-chested with ***creamy crown*** (sometimes just nape), ***unbarred kestrel-like contrast above, less distinctly barred uppertail, poorly developed moustache and less contrasting head pattern*** (diffuse eye-stripe, no dark forehead-band). Whitish supercilium often more conspicuous and belly more spotted, but lacks Lanner's spot-bars on flanks. Saker subspecies (*saceroides*-type) are greyish above with dark bars, including uppertail; these ***best told by head pattern and size***. Juvenile similar to young Lanner, but outer tail feathers generally conspicuously spotted buff on outer webs (seen well in half-spread tail) and dark stripe behind eye less clear-cut; best told by size. When perched, ***wing-tip falls short of tail-tip*** (unlike most Lanners) and 'trousers' heavier, covering much of tarsus. Slow, flattish wingbeats; when soaring, wings flat or slightly upcurved. **Voice** Like Peregrine, but harsher; also thin, querulous note like cross between Curlew and Herring Gull of W. Europe. **Habitat** Wooded steppes, foothills, mountains, semi-deserts. **Note** Passage and winter hatched; declining and rare in all areas.

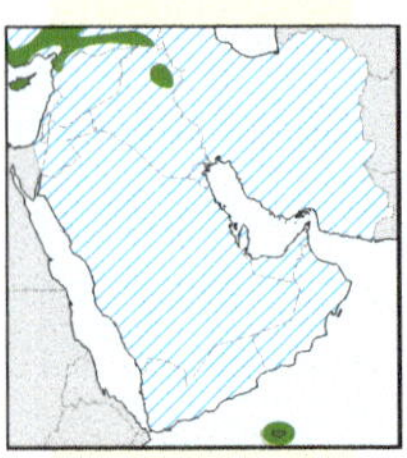

Peregrine Falcon *Falco peregrinus* — PM, WV; E

L: 40–52. W: 85–120. Large, stocky with relatively short tail and ***broad-based, sharply tapering wings, more pointed than Saker and Lanner*** when soaring. Adult told by ***black crown and bold moustache, contrasting with white throat and cheeks, barred underparts, white upper breast and uniform underwing***. Juvenile ***has smaller whitish cheek-patch*** (not reaching eye, unlike Saker and Lanner) and also uniform underwing (unlike Saker and Lanner). Young of migrant *calidus* are tricky, showing Saker-like head pattern and large size, but told by underwing pattern and wing shape. Subspecies breeding in region, *brookei*, is more compact, like Barbary Falcon, with salmon wash to breast and sometimes rufous wash on nape. Fairly quick, shallow, stiff wingbeats; impressive when hunting, with long, fast stoops. **Voice** Alarm loud, scolding *aack-ack-ack*. **Habitat** Mountains, forests, cliffs; outside breeding season also marshes, wastelands. **Note** Passage and winter hatched.

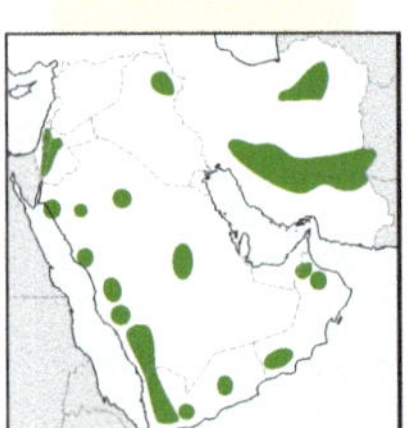

Barbary Falcon *Falco (peregrinus) pelegrinoides* — rb

L: 32–45. W: 80–100. Resembles Peregrine, especially ***brookei***, but slightly narrower-based wings give impression of longer tail. Adult told by ***rufous nape and rear eyebrow, narrower moustache and larger pale cheek-patch, almost reaching eye; more creamy, less barred underparts***, confined to flanks in E. Iranian *babylonicus*, which has redder crown; ***underwing whiter with more extensive dark wing-tips*** than Peregrine; ***often with dark 'comma' on greater primary coverts***; upperparts paler blue-grey, with darker end to tail. Juvenile like Peregrine, ***but narrower moustache, larger cheek-patch, tawny supercilium and rusty nape***; rustier underparts with thinner, more restricted streaks. Some juveniles have yellow cere and legs soon after fledging (in Peregrine blue-grey, usually becoming yellow in first-winter). From young Lanner by pattern of underwing and underparts. **Voice** Harsh *keck-keck-keck*, less hacking than Peregrine. **Habitat** Arid mountains, semi-deserts. **Note** Some autumn and winter dispersal; vagrant Kuwait, Qatar.

Lanner Falcon
tanypterus
juv
ad
ad
Saker Falcon
ad
milvipes
ad
ad
juv
cherrug
Peregrine Falcon
calidus
Barbary Falcon
ad
ad
brookei
ad
juv
ad
juv

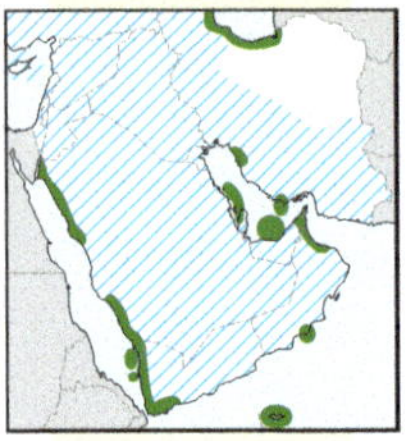

Osprey *Pandion haliaetus* RB, pm, wv

L: 56–61. W: 145–165. ***Large; long, narrow wings, distinctly angled when gliding, white undersurface with black carpals and band through centre of underwing, white crown and dark eye-mask.*** Variable dusky band across foreneck (usually boldest in female). Juvenile has whitish scales and white line on greater coverts above. Flies with steady, shallow wingbeats, glides on smoothly curved wings; may recall large soaring gull at distance. Hovers over water for fish and dives with splash, feet-first, almost disappearing. **Habitat** Always near water, inland or coastal; nests in trees, sea-cliffs, remote islands (often on ground), ruins, old wrecks, sometimes in scattered groups. **Note** Passage hatched, winters on coasts.

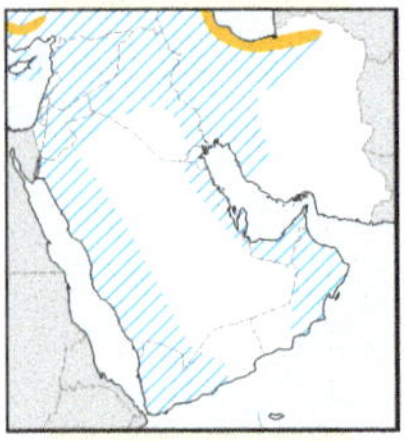

European Honey Buzzard *Pernis apivorus* pm

L: 55. W: 135–150. ***Recalls Steppe Buzzard in shape but slimmer with longer, narrower tail with rounded corners; head and neck narrower, protruding in cuckoo-like manner. Wingbeats more flexible and soars on flattish wings, and glides on slightly lowered wings.*** (Steppe Buzzard soars on raised wings.) Plumage variable; typical male has greyish head and upperparts; female browner. Below, some are dark, others largely white, ***but most are barred on body and coverts, and have black carpal patches; flight feathers show prominent black trailing edge and characteristic bars at base (more bars in female). In all morphs tail has dark band at tip and two bars at base.*** Cere grey, eyes yellow or orange-yellow (male). Juvenile dark brown, rufous-brown or creamy-white with streaked breast; ***usually, but not always, with dark carpal patches and narrow whitish crescent on uppertail-coverts; head often whitish with dark eye-mask***; may show pale band on underwing-coverts, separating secondaries from dark forewing (unlike Steppe Buzzard) and ***three evenly spaced bars on flight feathers*** (unlike adult). Juvenile with its more slender wings with curved rear-edge (bulging secondaries) and shorter tail has more of a Steppe Buzzard-like outline, but shape of tail and head, and soaring on flat wings important for identification. Migrates in flocks. ***See similar Crested Honey Buzzard for separation*** from that species. **Habitat** Woodland; widespread passage. **Note** Passage hatched, but rare E. Arabia.

Crested Honey Buzzard *Pernis ptilorhynchus* PM, WV

L: 65. W: 160. Resembles European Honey Buzzard but larger with noticeably broader body, slightly longer wingspan, and broader wings with bulging secondaries (all ages) and wing-tip showing ***six long 'fingers'*** (five in European Honey); also tail broader and shorter. Dark, pale and intermediate morphs occur (as European Honey). Adults have ***dark gorget across throat and lack contrasting black carpal patches; nape plumes sometimes visible when perched.*** Male has ***dark red eyes, undertail with, apart from broad black band at tail base, two black bands divided by broad pale band*** and ***black band on flight feathers reaches body*** (male European Honey has inner part of band hidden beneath greater coverts). Adult female's undertail pattern more like male European Honey, but innermost bar (at tip of tail-coverts) usually broader; secondaries crossed by three dark, evenly spaced bars (female European Honey usually has two bars with wider gap between dark trailing edge and first bar). Juveniles have underparts creamy to foxy or dark brown; dark eyes, yellow cere, and variable tail barring, typically four narrow bars of even width, all being characters similar to juvenile European Honey and thus best told by shape and structure, but note often has ***broad pale rump***, while none shows dark carpal patches (usually obvious in European Honey, except dark individuals). Adult Crested Honey Buzzard usually migrates in autumn with 4–5 new inner primaries (adult European Honey has from 0–3 renewed). Care is required with any wintering honey buzzard, as moulting Crested Honey may be missing outer primaries and then only show five fingers (as European Honey). **Habitat** Open woodland, wooded farmland. **Note** Rare passage (any habitat) and winter hatched; vagrant Oman, Saudi Arabia, Yemen.

Osprey
ad
ad
juv
dark morph ♀
typical ♂
♂
juv
pale morph ♂
typical ♂
juv
♂
European Honey Buzzard
ad ♂
Crested Honey Buzzard
juv

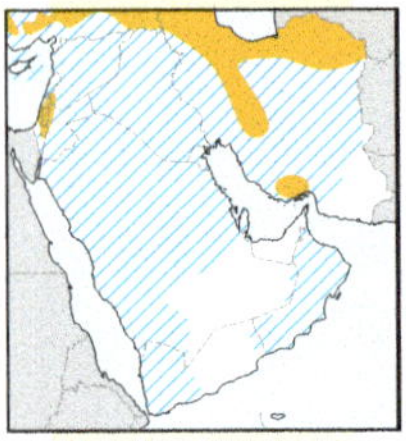

Black Kite *Milvus migrans* No records

L: 50–65. W: 125–150. Long winged with languid, elastic wingbeats. ***Tail long, and forked*** (often square-ended when fully spread). Adult has ***hardly any white on primaries below***. Juvenile shows dark eye-mask, pale feather tips on mantle and shoulders, boldly dark-spotted breast but paler belly and diffuse dark band to tail; also ***whitish tips to greater upperwing-coverts***. Soars and glides on slightly arched wings; manoeuvres tail when scanning for food. **Voice** Gull-like, whinnying *yiieerr*. **Habitat** Woodland, often near water; anywhere on migration, often gathering at rubbish dumps. **Note** Passage hatched; scarce in winter; birds in E. Arabia probably Black-eared Kite.

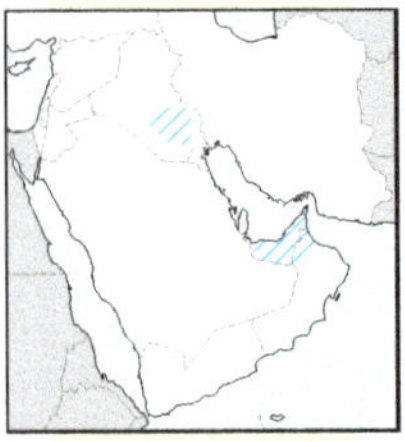

Black-eared Kite *Milvus (migrans) lineatus* pm. wv

L: 55–68. W: 135–162. Similar to Black Kite. Black-eared differs in having prominent white 'window' in ***bases of the primaries dark barred, broader hand with long sixth primary, all six 'free' primaries longer than in Black Kite and browner head (greyish in Black Kite) with more obvious dark eye-mask***. Underbody is streaked off-white to ochre (finer pale and black streaking in Black Kite). Juvenile similar to juvenile Black Kite but has broader hand with long sixth primary (like adult), whiter base to primaries, more noticeable streaking across breast and pale belly and vent. **Habitat** As Black Kite. **Note** Passage/winter hatched; uncommon UAE, vagrant Yemen.

White-eyed Buzzard *Butastur teesa* E/?V

L: 45. W: 100. Between a honey buzzard and a harrier when soaring, wings relatively narrow, held flattish, with tail half spread. When gliding wings angled and tail looks relatively long and narrow; in active flight has *Accipiter*-like wingbeats. Tail often twisted in kite-like manner. From above, adult has ***cinnamon-rufous tail with black band near tip***, and warm-brown primary patch and buffish panel across wing-coverts. ***Wing-tips blackish below***. At distance appears whitish below with darker breast. Young birds have paler head, whitish underparts (streaked at close range) and brown iris. Perches erect for long periods; ***then white throat with dark central streak and dark cheek-streak visible*** (latter narrow or absent in young birds). Often confiding. **Voice** Plaintive, mewing *pit-weer, pit-weer*. **Habitat** Dry open country with scrub and few trees; lightly wooded foothills. **Note** Vagrant Oman; UAE records considered escapes.

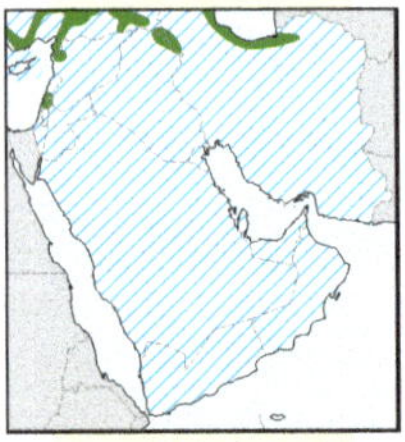

Steppe Buzzard *Buteo buteo vulpinus* pm

L: 48. W: 118. Fox-red, grey-brown and rare blackish morphs occur. Smaller than Long-legged Buzzard with ***shorter, narrower wings, shorter tails, stiffer wingbeats and glide on flattish wings*** (Long-legged has flexible wingbeats and kinked gliding profile). Fox-red morph has rusty-orange underwing-coverts framed by dark greater coverts, with narrow blackish comma-shaped carpal patch (patch usually large in Long-legged); rufous-brown uppertail usually more barred, head dark and pale primary patch above usually small. Juvenile has streaked breast, diffuse band on trailing edge of underwing, no broad dark band on tip of tail; often pale-headed, with pale upperwing-coverts and prominent primary patch above. Soars on raised wings; may hover when hunting. Migrates in flocks. **Voice** Mewing *peeeoo*. **Habitat** Woodlands, plains, mountain slopes with trees; anywhere on passage. **Note** Passage and winter hatched, but rarer or absent in winter.

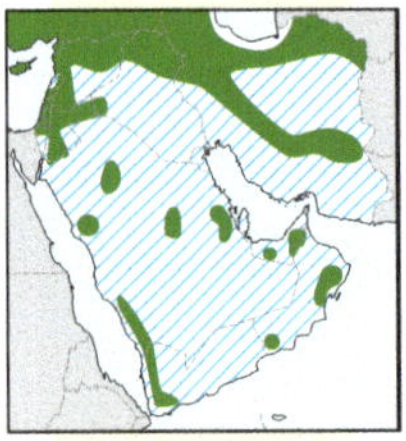

Long-legged Buzzard *Buteo rufinus* WV, pm, rb

L: 60–66. W: 130–155. Larger than Steppe Buzzard with ***longer wings and tail***, kinked wing position when gliding and flexible wingbeats; ***soars on raised wings***. Wide plumage variation: creamy-white, rufous-brown and blackish morphs on account of colour on body and underwing-coverts. The blackish morph can show coarse dark bars on flight and tail feathers. ***Typical Long-legged Buzzards have pale head and breast, becoming dark towards belly, pale sandy or rufous-brown upperwing-coverts contrasting with flight feathers, unbarred pale rusty-orange uppertail and large black carpal patches***. Juvenile has finely barred outer tail and diffuse dark trailing edge to underwings; some are almost white below with bold carpal patches and dark rusty-brown belly or belly sides. Sits prominently, soars or hovers when hunting. **Voice** Recalls Steppe Buzzard. **Habitat** Steppe, semi-deserts, mountains and woodland. **Note** Winter and passage hatched. Subspecies *cirtensis* breeds Arabia; larger nominate form visits.

Black Kite
ad
juv
ad
White-eyed Buzzard
ad
ad
juv
Black-eared Kite
ad
fox-red
morph juv
blackish
morph ad
fox-red
morph ad
grey-brown
morph imm
ad
pale
morph juv
Steppe Buzzard
ad
Long-legged Buzzard
ad
dark
morph juv

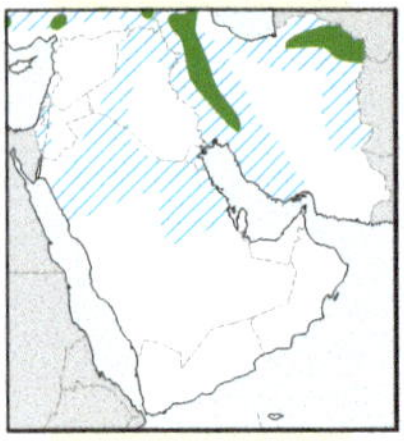

Cinereous Vulture *Aegypius monachus* E/?V

L: 105. W: 255–295. Very large. Readily told from Eurasian Griffon Vulture by all-blackish plumage without any contrast and parallel-edged wings held flat, or slightly downcurved, particularly when gliding; tail also slightly longer and less square-cut than in Griffon. Young birds blacker than adults, but in both pale legs stand out against black undertail-coverts. At close range adult has black and whitish head pattern; head blackish-brown in juveniles. Plumage blacker throughout than rather similar Lappet-faced Vulture. Told from dark eagles by larger size, longer and more deeply fingered wings, and less protruding head. The occasional wingbeat is slow and deep (like Lappet-faced). **Habitat** Desolate mountains (often extensively wooded), foothills, plains and semi-deserts; nests in tree, sometimes on cliff. **Note** Winter dispersal hatched; vagrant Oman. [Alt: Eurasian Black Vulture]

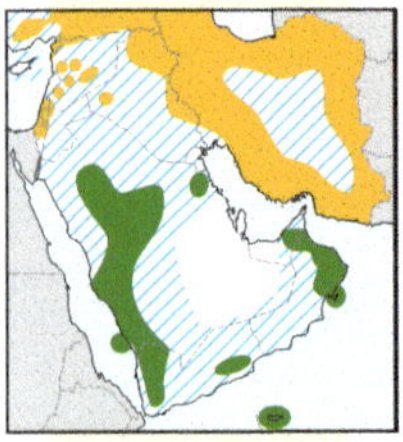

Egyptian Vulture *Neophron percnopterus* WV, sv, ?rb

L: 62. W: 155. A small vulture. Adult has ***white, wedge-shaped tail***, white underparts with black flight feathers (secondaries greyish-white above), ***small pointed head and thin bill***; colour pattern of plumage resembles pale morph Booted Eagle or Western White Stork below but shape quite different. Juvenile is mid-brown below with blackish ruff; dark brown above with creamy bars on wing-coverts, pale rump and whitish uppertail-coverts; ***wedge-shaped tail grey-brown, tipped paler***. Soars on flat to slightly arched wings; active flight has many deep wingbeats between glides. Often in flocks. **Habitat** Mountains, isolated jebels, wadis and open country; frequents village refuse dumps; sometimes foreshore; nests on cliff. **Note** Partial migrant south Iran; passage hatched; vagrant Qatar.

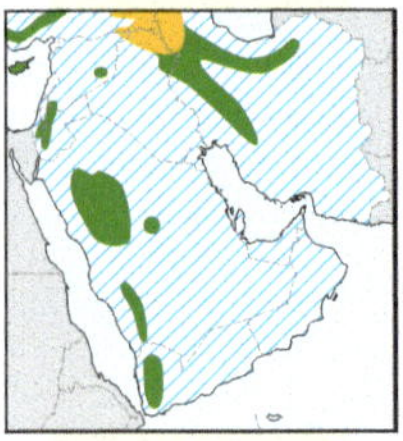

Eurasian Griffon Vulture *Gyps fulvus* V

L: 95–105. W: 245–270. Large; heavy with long, broad, deeply fingered ***wings with curved trailing edge***; short, broad, square-cut tail and slightly protruding narrow head. ***Soars effortlessly for long periods on raised wings***; active flight with very slow, deep wingbeats; glides on kinked wings. Adult ***gingery-buff above and below contrasting with dark flight feathers***. Juvenile even paler brownish-yellow on rear underwing-coverts, thus greater contrast with flight feathers. Gregarious. **Habitat** Mountains; occurs over all types of country in search for food; nests colonially in caves or on cliff-ledges. **Note** Passage, winter and dispersal hatched, but rare much of Arabia; vagrant UAE.

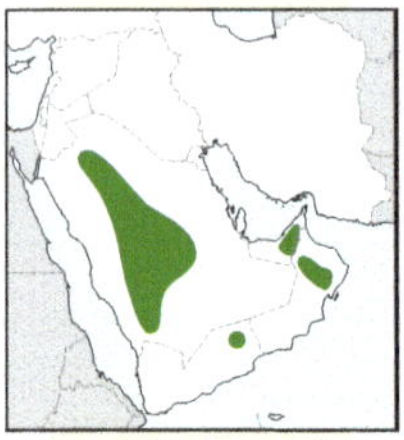

Lappet-faced Vulture *Torgos tracheliotus* sv, wv

L: 105. W: 255–290. Very large, heavy vulture, ***paler than Cinereous Vulture but darker than Eurasian Griffon***. Long, deeply fingered wings and short tail with distinctly pointed feather-tips; wings less parallel-edged than in Cinereous but less curved than in Griffon. ***From above, dull grey-brown wing-coverts contrast much less with flight feathers than in Eurasian Griffon*** (but bleach paler; in Cinereous virtually no contrast). From below, dark wing-coverts have variable whitish 'vulture streak' near leading edge and ***flight feathers and their coverts are clearly paler*** greyish, but wing-tip blackish (in Cinereous, black underwing-coverts contrast with paler flight feathers, and wing-tip not clearly darker). Dark brown underparts have whitish-brown mottling on breast, creamy upper flanks, browner upper thighs but paler lower thighs and ventral region, ***giving underparts a variegated appearance***. Immature birds are more uniform with less developed 'vulture streak' and less variegated underparts but usually have some pale on vent. When perched, adult identified by very heavy bill, feathered hindneck, ugly unfeathered pinkish head and foreneck and long, lanceolated breast feathers; ***lappets often inconspicuous***. Solitary or in pairs, but small parties sometimes at carcasses. **Habitat** Savanna, semi-desert steppe, desert with scattered trees, foothills, rocky wadis; huge nest built on top of acacia or *Maerua*. **Note** Formerly bred UAE; some winter dispersal; vagrant Kuwait

Cinereous Vulture

Egyptian Vulture

Eurasian Griffon Vulture

Lappet-faced Vulture

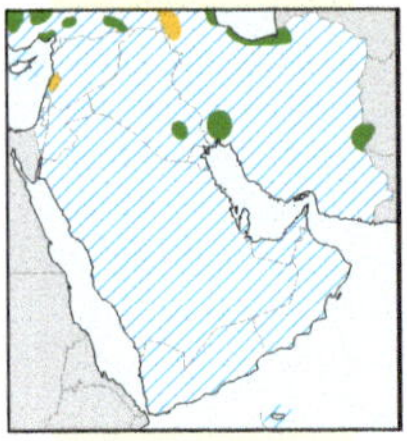

Western Marsh Harrier *Circus aeruginosus* PM, WV

L: 48–55. W: 115–130. Larger and broader-winged than other harriers; wavering low glides on raised wings when hunting. ***Male has tri-coloured wings***; underwing white but wing-tip black and ***rear-body red-brown***. Female dark brown with ***crown, throat and breast-spot yellow-white***. Juvenile blackish-brown, usually with rusty-yellow on head. Immature male has dirty-grey areas on upperwing and tail, rusty-brown body and underwing-coverts, and more extensive black wing-tips than adult. Rare dark morph is solidly blackish-brown, but adult male has distinct white base to flight feathers below. **Voice** High-pitched Lapwing-like *vay-ee* when displaying; also *ki-ki-ki* and feeble, high 'begging' whistles. **Habitat** Marshes, reedbeds, farmland. **Note** Passage and winter hatched.

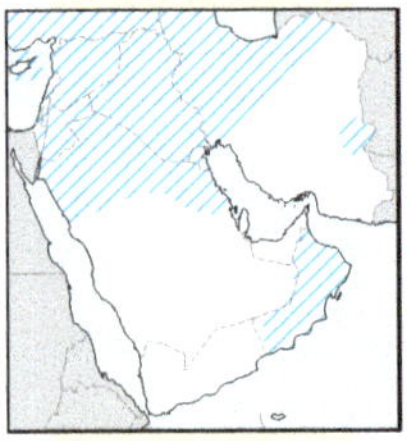

Hen Harrier *Circus cyaneus* V

L: 45–56. W: 100–120. Slimmer than Western Marsh Harrier with more buoyant flight. Male has ***clear-cut white uppertail-coverts, uniform pale grey upperparts, head and upper breast and extensive black wing-tips***. Second-autumn male can show black wedge on wing-tip like male Pallid Harrier (through primary moult). Female and juvenile brownish with white uppertail-coverts; streaked underparts whitish or rusty-yellow (warmest in juvenile); banding on secondaries below most distinct in female. Juvenile also has pale tips to greater upperwing-coverts and best separated from juvenile Montagu's and Pallid Harriers by proportionally ***shorter, broader wings with more ample rounded wing-tip*** (formed by four outermost primaries, but three in the other two species), ***less buoyant flight and streaked breast*** (unstreaked rusty yellow-brown in juveniles of the other two species). **Habitat** Marshes, meadows, farmland. **Note** Passage and winter hatched; rare south Arabia; vagrant Bahrain, UAE, Yemen.

Pallid Harrier *Circus macrourus* PM. WV

L: 40–48. W: 95–117. Proportions and flight similar to Montagu's Harrier. Male pale grey above ***without clear-cut white rump; whitish head and underparts with black wedge on wing-tip***. Female from Montagu's ***by pale, dark-streaked, collar*** (like female Hen Harrier), ***less spacing between dark bands on secondaries below, with pale bands becoming darker towards body;*** primaries below often pale, contrasting with darker secondaries and lacking distinct dark trailing edge; heaviest barring is on central primaries ***with bases often unbarred, creating pale 'boomerang' surrounding darkish coverts***; distal primaries with faint or no barring, except for narrow dark 'finger-tips' of longest primaries (unlike Montagu's). Except for pale leading arm, ***most underwing-coverts and axillaries rather dark-streaked*** and lacking distinct pattern (not bold rufous-barred as Montagu's). Streaks on underparts largely confined to upper breast, which contrasts more with paler rear-body than in Montagu's. Juvenile ***has broad, pale collar bordered by brown neck;*** primaries below rather evenly barred from base to tip ***though often with pale 'boomerang' at primary bases, 'fingers' never all dark as in most young Montagu's***. Male (9–12 months old) has paler head and breast than Montagu's; new central tail feathers show diffuse dark bands near tip (similar Montagu's has grey neck and breast, contrasting with paler belly and new central tail feathers plain grey). **Habitat** Steppes, grassland, agricultural fields, sand desert. **Note** May breed Iran; passage and winter hatched.

Montagu's Harrier *Circus pygargus* PM, wv, sv

L: 43–47. W: 97–115. Slender build, narrow wings and buoyant flight. Male has grey back and inner wing, silvery-grey outer wing with extensive black wing-tips; one black band on secondaries above and two below; red-brown streaks below dark grey upperbreast. Second-autumn male can show black wedge at wing-tip as result of primary moult (thus recalling male Pallid Harrier). Female has rufous-streaked underparts, well-spaced dark bands across pale secondaries and evenly barred primaries from base to tip with dark trailing edge to hand; close to pale underwing-coverts and axillaries show uniform bold rufous bars. Juvenile dark rufous to yellowish-ochre below, largely unstreaked; lacks distinct pale collar of young Pallid; 'fingers' and trailing edge of hand below dark, but hand otherwise pale with fine, regular barring from base to tip. Rare melanistic morph sooty-black with pale base to primaries below. **Habitat** Marshes, farmland; in winter/on passage any open country. **Note** Passage hatched, some winter south Arabia.

Western Marsh Harrier
♂
♀
♂
Hen Harrier
♂
♀
♂
juv
Pallid Harrier
♂
♀
♂
juv
Montagu's Harrier
♀
♂
♂
juv

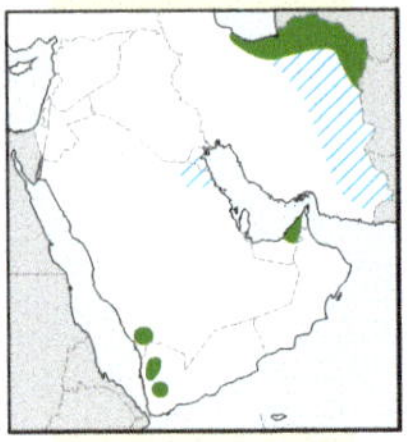

Shikra *Accipiter badius* rb (?E/I); V

L: 25–35. W: 50–65. Resembles rather small Eurasian Sparrowhawk in flight (blunt wing-tips) but ***slightly shorter tail has rounded corners*** (square-cut in Eurasian Sparrowhawk). Male of larger Iranian subspecies *cenchroides* ***pale dove-grey above, white below with faint orange barring (when close), narrow black wing-tips*** and obscure pale collar. Female pale brown above with blackish subterminal tail-band, darker barring below and wing-tips barely showing black. Note ***dark throat-stripe*** (absent in Eurasian Sparrowhawk), grey cheeks (rufous in Sparrowhawk) and absence of white supercilium. Smaller SW Arabian *sphenurus* male is darker blue-grey above with ***contrasting black wing-tips*** and lacks pale neck-collar; female shows less black on wing-tip. Juveniles (both subspecies) have ***dark longitudinal spots on underparts*** (largely barred in young Eurasian Sparrowhawk), banded uppertail and hardly any black on wing-tip. ***Differs from juvenile Levant Sparrowawk in smaller size, more rounded wings,*** and breast generally more spotted, less streaked. **Voice** Unhurried *ch-wick, ch-wick* recalls Tawny Owl; loud *kik-kooi* repeated at nest; in display a whistling *piu-piu-piu*. **Habitat** Light woodland, parks. **Note** Passage hatched, but rare; vagrant Oman.

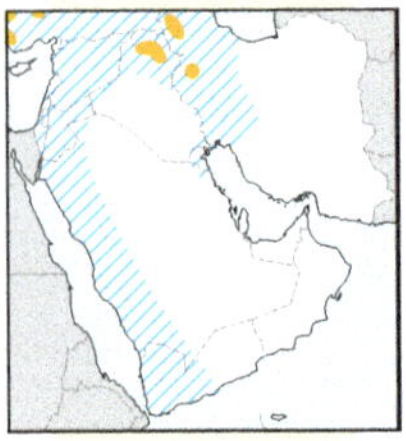

Levant Sparrowhawk *Accipiter brevipes* V

L: 32–39. W: 65–75. ***Larger than Shikra with longer, more pointed wings, giving falcon-like appearance when gliding***. Male blue-grey above (darker than Shikra) with ***blackish wing-tip*** and plain central tail feathers; female browner and closed uppertail has dark subterminal band. Below, ***male has white underwing with contrasting black tip, extending to inner primaries*** (more extensive than Shikra). Female, with more distinctly barred underparts, has less contrasting dark wing-tip. ***Black throat-streak, greyish cheeks and absence of supercilium separates from Eurasian Sparrowhawk***. Further told from Eurasian Sparrowawk by shape of wing-tip (four free outermost primaries, five in Eurasian Sparrowhawk), and often by more rounded tail-corners; adult female has dark outermost primaries (absent Eurasian Sparrowhawk) and lacks female Eurasian Sparrowhawk's fully banded central tail feathers above. Juvenile Levant is dark grey-brown above with closed tail banded, though often indistinctly; ***underparts have dark longitudinal spots, almost forming lines on breast,*** and dark throat-streak; ***longer, more pointed wings helps separate from young Shikra. Forms flocks on migration***. Secretive when breeding. **Habitat** Open country with deciduous woods, more widely on passage. **Note** Passage hatched but rare Saudi Arabia; vagrant Qatar, UAE.

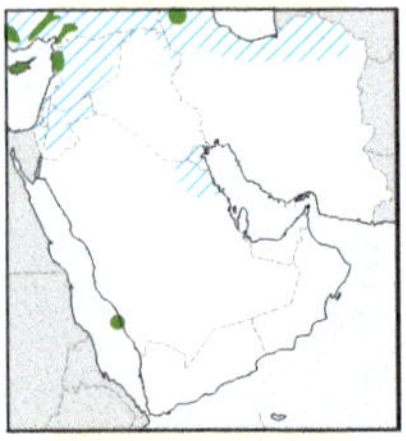

Northern Goshawk *Accipiter gentilis* V

L: 48–60. W: 90–125. ***Female much larger than male, with wingspan of Steppe Buzzard. Compared to female Eurasian Sparrowhawk male has deeper belly, slower, stronger, stiffer wingbeats and longer, broader-based, but more pointed wings***. Note shorter, broader-based tail, usually with rounded tip (thinner tail more square-cut in sparrowhawk) and ***more protruding head and neck***. Stronger, straighter glides than sparrowhawk and often soars on upturned wings. Adult dark grey above, darker head appears 'hooded' but supercilium white; underparts finely barred. Juvenile dark brown above with pale mottling on ear-coverts; rusty-yellow ***underparts boldly streaked darker***; lacks 'hooded' appearance of adult. Female told from large falcons by more rounded wings, bold tail-bands and flight. Hunts like Eurasian Sparrowhawk but also runs down prey on ground; display flight with soft harrier-like wingbeats in shallow waves. Treated warily by crows. **Habitat** Woods, particularly coniferous, often near open country. **Note** Winter hatched, but rare Kuwait, Saudi Arabia; vagrant Iraq, Oman, UAE.

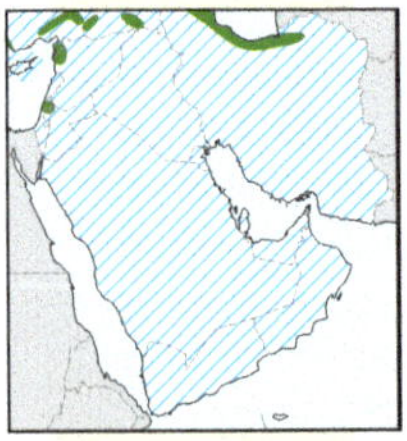

Eurasian Sparrowhawk *Accipiter nisus* PM, WV

L: 29–40. W: 60–80. Female much larger than male, approaching male Northern Goshawk in size, but wingbeats faster and lighter, body slimmer, less protruding head, wing-tips blunter and tail thinner, longer and more square-cut. Adult ash-grey above (female), bluer slate-grey (male), barred rufous or brown below; whitish supercilium in female (infrequent in male); ***pale underwing without dark tip***. Juvenile browner above with clear white supercilium; streaked or blotched throat and upper breast, otherwise barred below. Quick wingbeats interspersed with short descending glides (stronger, straighter glides in Northern Goshawk); display flight has slow harrier-like wingbeats, also occasionally when hunting. **Habitat** Woodland; open country with trees. **Note** Passage and winter hatched.

Shikra
ad
cenchroides
sphenurus
juv
Levant Sparrowhawk
♂
♀
juv
Northern Goshawk
ad
juv
Eurasian Sparrowhawk
♂
♀
juv

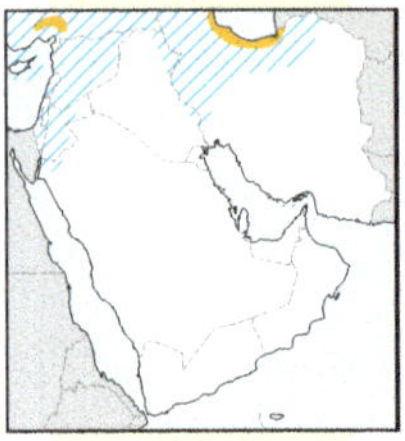

Lesser Spotted Eagle *Aquila pomarina* V

L: 62. W: 145–165. ***Medium-brown adult shows contrast between pale upperwing-coverts and darker brown mantle, underwing-coverts paler than flight feathers, a neat pale patch at base of primaries above*** and small creamy area on uppertail-coverts. Juvenile darker, warm brown below with ***flight feathers of same shade or slightly darker (coverts never darker than flight feathers***); unlike adult may show little contrast between mantle and wing-coverts; darker head has ***rusty-yellow spot on nape*** (absent in Greater Spotted, present in some adult Steppe Eagles); narrow white bar on greater upperwing-coverts and distinct whitish inner primary patch. ***Short or minute seventh primary, less deeply fingered wings and smaller hand separates from Steppe Eagle at all ages***; lacks dark band on trailing edge of underwing and tail of many adult Steppe. Wings relatively narrow, tail medium-long; active flight less heavy than larger aquilas. Soars and glides on arched wings with primaries lowered. On ground lacks heavy 'trousers'. ***Both spotted eagles have characteristic round nostrils*** (unlike Steppe). Migrates in flocks. **Habitat** Breeds in forests; open country on passage **Note** Passage hatched; vagrant Kuwait, Oman, Saudi Arabia, UAE, Yemen.

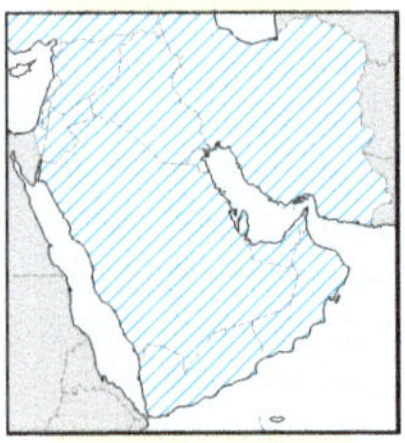

Greater Spotted Eagle *Aquila clanga* PM, WV

L: 65. W: 155–180. ***Typically darker than Lesser Spotted Eagle. Adult dark brown below, flight feathers similarly dark or a shade paler*** (reverse in Lesser Spotted); ***leading underwing-coverts sometimes blackish-brown*** (never so in Lesser); on upperwing, mid- to dark brown coverts sometimes contrast with darker mantle (like typical Lesser); no conspicuous pale primary patch above (unlike Lesser and Steppe Eagle). Adult Greater Spotted usually lacks band on trailing edge of underwing, seen on many adult Steppe; also lacks pale nape-spot of young Lesser and many adult Steppe. ***Juvenile is blackish-brown below with paler flight feathers*** (in Lesser Spotted coverts are brown but flight feathers never paler); ***blackish-brown upperwing has 1–3 white covert bars, often creating pale panel***; large, diffuse primary patch formed by whitish primary shafts and pale inner primaries (patch smaller, more conspicuous in Lesser). Infrequently, young Greater Spotted is abnormally coloured on body and wing-coverts: i.) '*fulvescens*' type – illustrated; ii.) yellow-brown above and below; iii.) yellow-brown above, normal below; iv.) yellow-brown below, normal above; v.) underwing-coverts greyish, or dark, mottled paler, underbody darker; or normal upper- and underwing, but contrasting paler underbody. Irrespective of age, ***secondaries below may have thin dense bars*** (broader in Lesser Spotted, more well-spaced in Steppe). Adult has relatively broad and parallel wings with slightly broader hand and deeper fingers than Lesser. Juvenile has narrower hand than adult with trailing edge fairly strongly incurved at body (visible when tail closed). Hand slightly shorter, less ample than Steppe (Greater Spotted has shortish seventh primary) and bill generally smaller. **Habitat** Usually near wetlands, coastal or inland, also rubbish tips. **Note** Passage and winter hatched, but often rare.

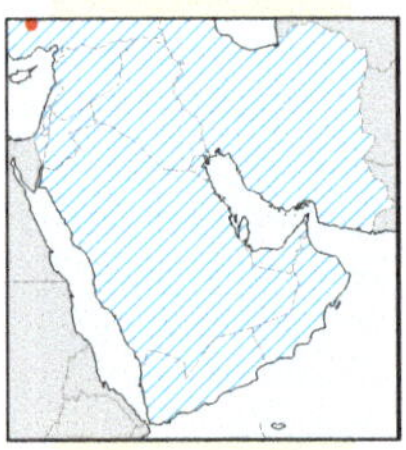

Steppe Eagle *Aquila nipalensis* pm, wv

L: 75. W: 175–210. Adult dark brown with ***uniform underwing and paler or darker flight feathers with well-spaced dark bars and clear-cut band on trailing edge*** (pattern sometimes diffuse); ***large dark carpal patch typical, except in darkest birds***. Above, coverts often palest part of wing; usually large, dark-barred, pale primary patch (patch virtually absent in adult Eastern Imperial and Greater Spotted Eagles); grey-brown tail often boldly barred and with broader band at tip (absent in spotted eagles). ***Juvenile pale brown with broad white band through underwing;*** above, note large primary patch and dark rump, which separates from most young Imperial Eagles. Subadult usually has darker body than underwing-coverts, very like some immature Lesser Spotted, but told by remains of white underwing-band or well-spaced flight feather barring, long deeply fingered wings, ample hand (long fourth primary) ***and longer, heavier bill (with nostril peanut-shaped)***. Flight heavy; often soars on flexed, flattish wings but can soar and particularly glide on arched wings with lowered hand. When perched, large heavy 'trousers' unlike the spotted eagles; ***long yellow gape flange to rear of eye*** separates it from other *Aquila* eagles. **Habitat** Steppe, semi-desert, hills, marshes; also rubbish dumps. **Note** Passage and winter hatched, but often rare.

Lesser Spotted Eagle

Greater Spotted Eagle

Steppe Eagle

Pallas's Fish Eagle *Haliaeetus leucoryphus* V

L: 80. W: 190–220. A large, often vocal eagle, usually found near wetlands. Adult has ***white tail with broad black terminal band***; bill dark grey (yellow in White-tailed). Juvenile told from White-tailed Eagle by ***paler head with dark patch behind eye, uniform pale brown underparts, broad pale band through underwing-coverts, contrasting with dark brown leading coverts and distinct white patch or white streaks on primaries below***; up to three years old the paler head and underparts emphasise dark eye-patch and, often, band round foreneck; underwing like juvenile but ***tail-centre distinctly white-mottled, forming pale band*** in some; differs from young Golden Eagle by pale axillary patch and centre of underwing, paler underparts, more ***parallel-edged wings held flattish when soaring*** and by longer neck. When perched, bare tarsus and pale loral-patch separates Pallas's from *Aquila* eagles. **Habitat** Wetlands and rivers; also coasts. **Note** Vagrant Iran, Iraq, Oman, Saudi Arabia, UAE.

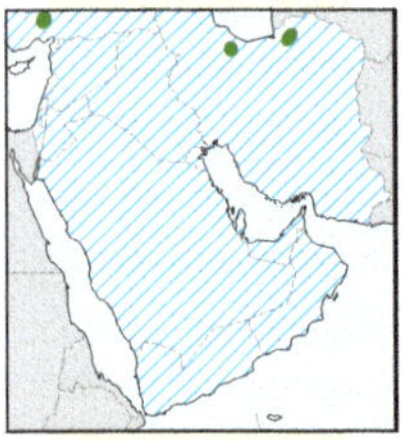

Eastern Imperial Eagle *Aquila heliaca* pm, wv

L: 72–83. W: 190–210. Adult told from Golden Eagle by ***blackish-brown plumage, contrasting yellow-white hindneck, pale uppertail with broad black band and white 'braces'*** (can be hard to see); also, in flight ***parallel-edged wings held flattish*** and, often, ***closed narrow tail when soaring***. Juvenile has ***dark-streaked breast forming pectoral band which contrasts with unstreaked yellow-buff rear-body, and distinct pale wedge on inner primaries below;*** yellow-brown upperparts show 1–2 complete whitish bars on coverts and ***creamy lower back and rump***; lacks white band through underwing of young Steppe. Immature below, ***still streaked*** (much as juvenile) ***or mottled blackish-brown and yellowish*** with rear-body clearly paler, ***possibly also retaining pale inner primaries***; adult head and tail pattern start to show early. Rather long-winged with ample hand, deep-fingered wing-tip (long seventh primary), well-protruding head and relatively long tail. Juvenile has broader, more 'S'-curved rear edge to wings. Wings sometimes slightly lifted when soaring, but arched during fast glides. Perched juveniles/immatures show ***pale lower underparts*** and, like adult, rather long protruding head (compared with other *Aquila* eagles). Tawny Eagle lacks streaks below of young Imperial (but seen in some African birds, which could occur in Arabia). **Habitat** Open plains and foothills with woods (nests in large or small tree); in winter also steppes, marshes, wooded desert or semi-desert, dumps. **Note** Passage and winter hatched, but rare.

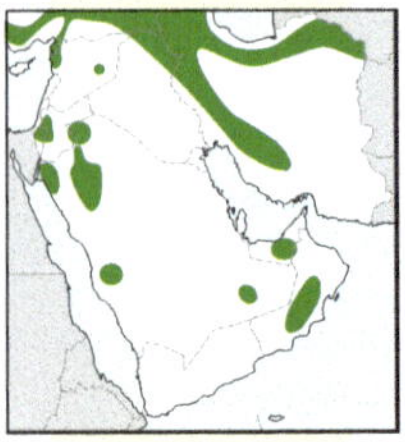

Golden Eagle *Aquila chrysaetos* rb, wv

L: 78. W: 190–230. Powerful flight with ***flexible wingbeats; soars and glides on markedly raised wings***, when note ***fairly long tail and slightly 'S'-curved rear-edge of wings*** (more pronounced in juvenile). Dark brown adult has ***rusty-yellow hindneck, pale panel across upperwing-coverts***, dark-barred, black-tipped greyish flight feathers which show as ***greyish area on outer wing above*** and greyish tail with blackish band at tip. (Adult Imperial Eagle is blacker, including outer wing above, has flatter, more parallel-edged wings when soaring and narrower tail.) ***Juvenile and immature have white patches in primaries and inner tail, the latter with broad black band at tip***, unique in the *Aquila* eagles. Birds older than one year show pale panel across upperwing-coverts. Often hunts in tandem. **Habitat** Barren or wooded mountains, plains and semi-deserts with trees; nests on rocky ledge, sometimes in tree. **Note** Some winter dispersal; vagrant Kuwait.

Pallas's Fish Eagle
ad
juv
Eastern Imperial Eagle
juv
juv
ad
sub ad
sub ad
ad
ad
ad
Golden Eagle
ad
juv
sub ad

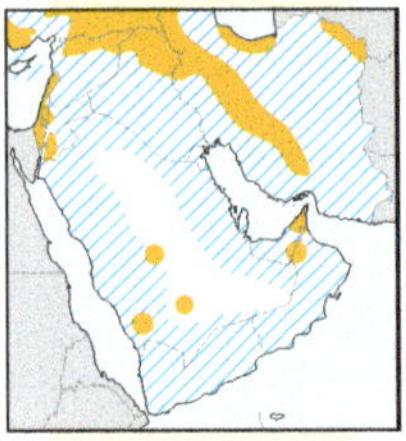

Short-toed Snake Eagle *Circaetus gallicus* pm, wv, rb

L: 64–73. W: 165–180. Large, long-winged eagle with broad head, very pale underparts and square-cut tail with evenly spaced dark bands. Whitish underparts variably spotted and barred; some are nearly all whitish, others with contrasting dark head and upper breast; lacks dark carpal patches. Flies with slow, flexible wingbeats, soars on flat or slightly lifted wings and hovers regularly. Separated from Osprey by broader wings, lack of dark carpal patch and different flight action. Pale morph Steppe Buzzards and European Honey Buzzards usually have dark carpals, blacker wing-tips, different spacing of tail-bands and are much smaller with quicker wingbeats. **Voice** Whistling, disyllabic *kee-yo* with long ascending start and short descending finish. **Habitat** Open wooded plains, stony foothills, semi-deserts; nests in tree or on cliff. **Note** Arabian breeders may be resident; passage hatched, few winter Arabia. [Alt: Short-toed Eagle]

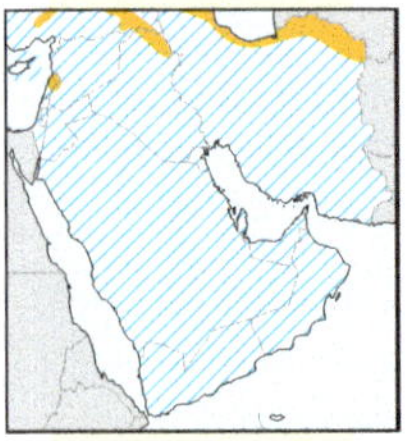

Booted Eagle *Aquila pennata* pm, wv

L: 43–53. W: 110–130. Two distinct colour morphs. Size of Steppe Buzzard, but outline and wing position close to Black Kite; ***tail square-cut***. More ample, deeply fingered wings than Steppe Buzzard. Pale morph has ***creamy-white underparts with contrasting blackish flight feathers***, kite-like panel on upperwing, ***pale scapulars (seen head on as 'landing lights') and uppertail-coverts and diagnostic white spots at base of neck***. ***Lacks dark carpal patch of most pale Steppe Buzzards and European Honey Buzzards***; also has darker base to flight feathers and paler inner primaries. Dark morph similar above to pale morph but underparts dark brown, or rufous with black band through centre of underwing. When perched, ***feathered tarsi also separates Booted Eagle from these and Long-legged Buzzards***. Has deeper, more powerful wingbeats and steadier glides than Steppe Buzzard; soars on flat wings; does not hover. **Habitat** Deciduous and pine forest; more open country outside breeding season. **Note** Passage hatched; few winter in Near East and Arabia.

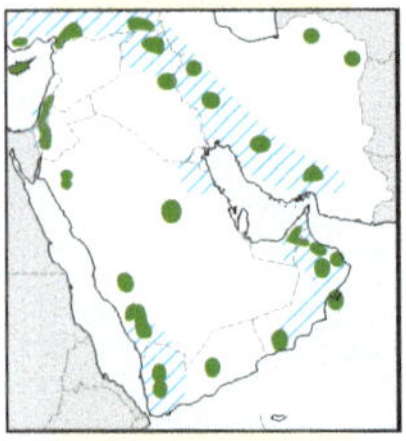

Bonelli's Eagle *Aquila fasciatus* rb, wv

L: 60–70. W: 150–165. In flight recalls large, thick-set Honey Buzzard. ***Adult identified by dark underwings contrasting with whitish underbody, pale tail with black band at tip and white patch on mantle***; at close range ***white leading edge of wing***. Pale rusty-buff juvenile lacks black tail-band, and flight feathers are pale with fine dark barring; ***paler translucent primaries contrast with blackish wing-tip***; when present, ***narrow dark bar on rear underwing-coverts diagnostic***, but in others confined to dark 'comma' on primary coverts; upperwing cinnamon-brown with large, pale primary patch. Soars on flat or slightly arched wings, often with long, almost square-cut tail held closed (may be twisted independently); glides with carpals pressed forward, trailing edge of wings straight (recalling European Honey Buzzard). Often hunts in pairs; stoops at great speed. **Habitat** Rocky mountains, forested foothills; in winter plains and semi-deserts. **Note** Winter dispersal hatched.

Brahminy Kite *Haliastur indus* E/?V

L: 48. W: 135. Size of Black Kite, with similar wing position though will soar with wings in shallow 'V'. Adult has ***white head, neck and breast***; otherwise red-brown above and on underbody with paler rufous underwing-coverts; flight feathers and undertail creamy-buff; blackish wing-tips most conspicuous from below. Juvenile lacks white head and neck, is darker brown above and below, with greyish secondaries below but ***conspicuous whitish primary patch***. Sometimes recalls Marsh Harrier when foraging low over ground or water, including sea. **Habitat** Open country, coasts. **Note** UAE record presumed escape.

ad
ad typical
ad pale morph
ad dark morph
Short-toed Snake Eagle
ad dark morph
ad pale morph
Booted Eagle
ad
juv
Bonelli's Eagle
juv
ad
Brahminy Kite

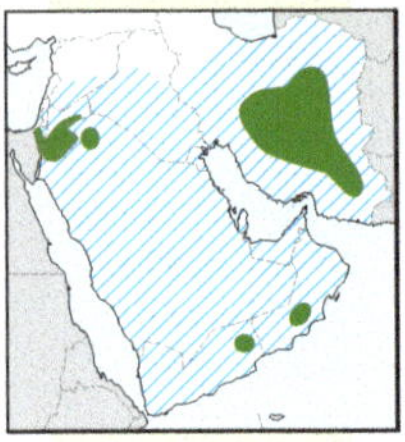

Macqueen's Bustard *Chlamydotis macqueenii* pm, wv

L: 60. W: 150. A large bustard, with ***black frill down side of neck in all plumages***. Flight rather slow, wingbeats shallow, with ***long-tailed and narrow-winged appearance*** and ***white patch confined to outer primaries only***. Shy, prefers sneaking away without flying. **Habitat** Stony or sandy steppes, semi-desert; also marginal cereals and other crops. **Note** Formerly more widespread breeding range; has bred Kuwait, Syria; may breed Iraq; reintroduced in some sites in Saudi Arabia and UAE; vagrant Cyprus, Lebanon, Qatar, Socotra. Now separated from Houbara *Chlamydotis undulata*, which occurs across North Africa.

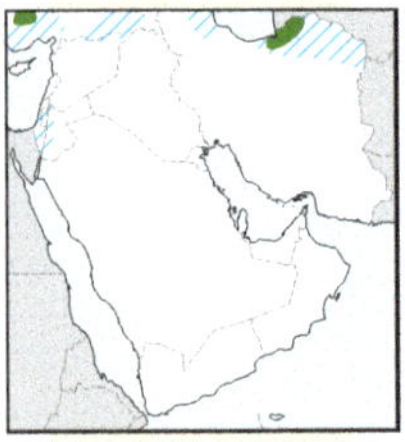

Little Bustard *Tetrax tetrax* E/I; ?V

L: 43. W: 110. Small, with ***small head and long neck***. Male has ***grey, black-and-white neck pattern***; non-breeding male, female and juvenile lack these striking neck marks. Flight fast with rapid, stiff, winnowing wingbeats showing ***almost completely white wings with black mainly confined to outermost primaries***. Male displays in spring with inflated neck and brief leaps in the air. Often in flocks, especially in winter; rather shy, often in cover of grass or low vegetation. **Voice** Male displays with a short *prrrt* call. In flight, male's wings make whistling noise. **Habitat** Grassy plains, large cereal fields or fodder crops. **Note** Formerly bred Syria; winter hatched, but rare Turkey; vagrant Cyprus, Iraq, Lebanon, Oman. Recent releases UAE.

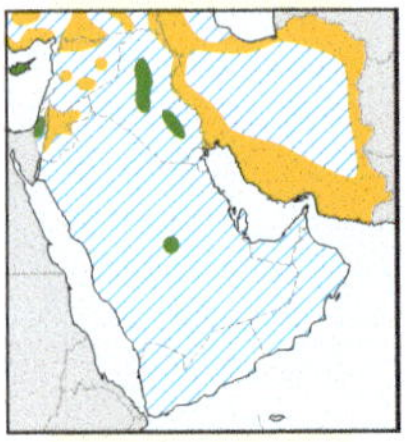

Eurasian Stone-curlew *Burhinus oedicnemus* WV, pm

L: 42. W: 81. Large, streaked, curlew-coloured wader with ***short bicoloured bill and large, staring yellow eye***. Found in dry habitats, often 'frozen' motionless or walking slowly in hunched posture. Flies with ***stiff wingbeats and shows two small white 'windows' in primaries and a paler wing-panel, bordered in front with a dark and a white line***. Adult Spotted Thick-knee is darker and boldly spotted (juvenile more streaked). Often encountered in flocks. More active by night than day. **Voice** Vocal mainly at night, reminiscent of Eurasian Curlew's *cur-lee* with emphasis on higher-pitched second syllable; also loud Oystercatcher-like *ku-beek, ku-beek*. **Habitat** Open plains, steppe and semi-desert, also extensive arable land; among scattered trees and light scrub in hotter climates. **Note** Passage and winter hatched, but absent winter Iran, Turkey.

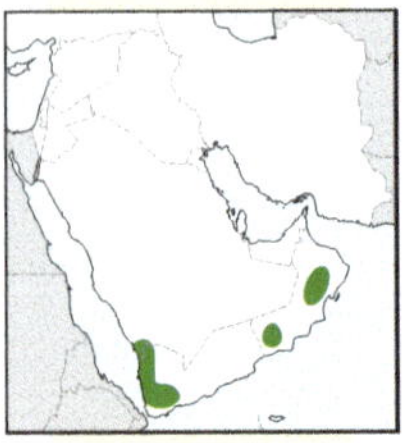

Spotted Thick-knee *Burhinus capensis* E/?V

L: 43. Resembles Eurasian Stone-curlew but upperparts buffier, ***spotted black in adult*** (but streaked in juvenile like Eurasian Stone-curlew), the spotting being particularly obvious on the paler coverts; ***lacks black-and-white bars on coverts. In all ages tertials and tail are diagnostically barred***. Flight pattern shows two prominent white patches on black primaries, and underwing usually shows a strong dark bar along central wing. Prefers to stay near cover of bushes. More active by night than day. **Voice** Usually at night: a whistled *ti-ti-ti-tee-tee-tee ti ti ti*, growing to a crescendo, then dying away. **Habitat** Savannas and scrub, rocky river beds, broken ground, more bushy than Eurasian Stone-curlew frequents; nests near cover of bushes. **Note** Reports from UAE deemed escapes. [Alt: Spotted Dikkop]

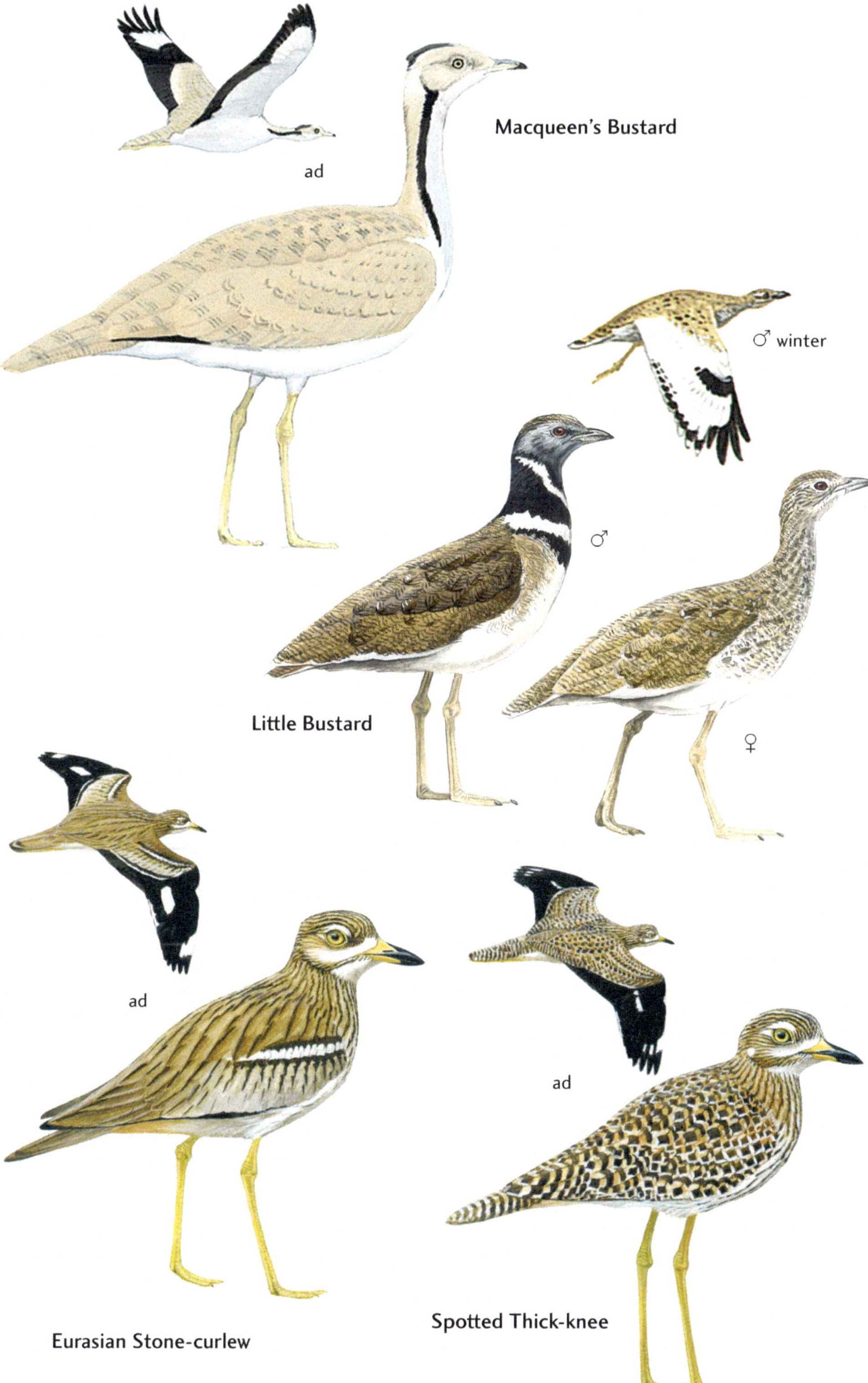
Macqueen's Bustard
ad
♂ winter
♂
Little Bustard
♀
ad
ad
Spotted Thick-knee
Eurasian Stone-curlew

PLATE 26: WATER RAIL & CRAKES

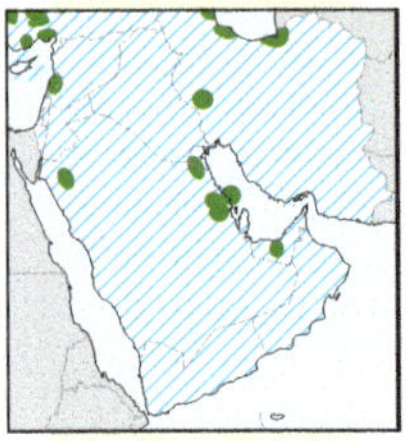

Water Rail *Rallus aquaticus* — pm, wv

L: 26. W: 41. Secretive; noticeably smaller than Common Moorhen, but ***larger than the*** *Porzana* ***crakes*** and easily told by ***slender, long red bill***. Adult has dark, ***mottled brown upperparts***, uniform slate-blue sides of head and underparts with ***heavy black-and-white bars on flanks and conspicuous white undertail***. Juvenile has browner face, mottled grey-blue underparts and blackish bill. Tail often cocked and jerked when walking. When flushed, flies only a short distance on fluttering, rounded wings with long legs dangling. **Voice** Grunting, groaning, whining and ***stomach-churning sounds*** from vegetation; sometimes ***like a squealing pig***. In spring, male (and female) utters for hours a rhythmic *trüt-trüt, trüt,* sometimes ending with a trill. **Habitat** Dense aquatic vegetation, ponds, ditches. **Note** Has bred UAE. Passage and winter hatched.

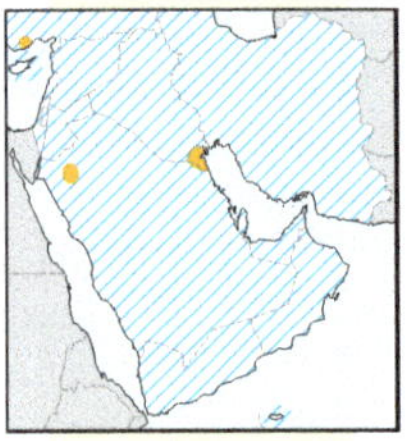

Little Crake *Porzana parva* — pm

L: 19. W: 37. Smaller than Spotted Crake and separated by ***uniform blue-grey underparts (adult male) and heavily barred undertail-coverts***. Less compact than Baillon's Crake with longer legs and neck, and ***much longer wing-projection***. Tertial pattern differs: pale buff fringes to inner webs form ***broad creamy line along inner aspect of folded wing;*** (in Baillon's paler edges to tertials never form broad continuous line). Male also told from Baillon's by ***less barring on flanks***, less spotted upperparts, red base to bill, and green legs. ***Female has brown-buff underparts***, white chin and throat and some grey on cheeks and supercilium. Juvenile lacks grey head pattern and has stronger flank-barring than adult; told from juvenile Baillon's by ***structure, tertial pattern*** and less barred underparts. (Any small crake with buff underparts seen in the region in mid-winter and spring will be female Little Crake). **Voice** Male's song loud, accelerating croaking *kwak… kwak… kwak, kwak, kwak-kwak-kva-kva-kva-kva-kva*. Female's call short, accelerating, with vibrant terminal trill *kwek, kwek-kverrrrr*. **Habitat** Swamps and wetland fringes; fondness for high reeds in deeper water and lagoons with floating vegetation. **Note** Passage hatched but rare throughout much of region; some winter Iraq and south Arabia.

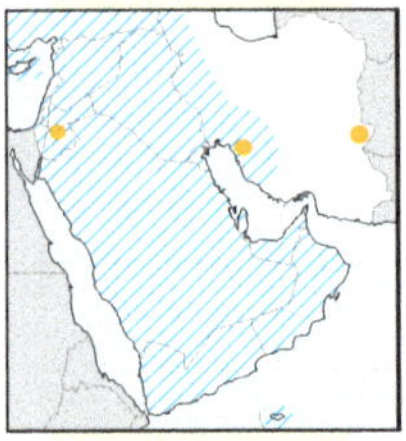

Baillon's Crake *Porzana pusilla* — V

L: 18. W: 35. Resembles Little Crake but ***more compact and with very short primary projection***. Upperparts warmer rufous with distinct, but small, irregularly scattered white spots; ***tertials do not form continuous pale line along inner aspect of folded wing*** as in Little Crake; underparts bluish-grey in both sexes with ***heavier black-and-white barring on flanks. Uniform green bill without red base***, and dirty-olive legs and feet. Juvenile more strongly barred below than Little Crake and best told by short primary projection and absence of broad, pale tertial-line. Most skulking of all the crakes. **Voice** Song a series of dry, rattling frog-like sounds lasting 1–2 seconds and repeated at intervals of 1–2 seconds, *trrrrr, trrrrr, trrrrr*. **Habitat** Dense vegetation (sedges, rushes), small pools, wetland edges. **Note** Passage hatched but rare or sometimes vagrant; some winter south Arabia.

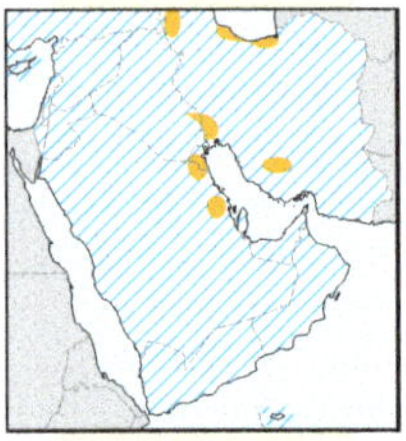

Spotted Crake *Porzana porzana* — pm, wv

L: 23. W: 40. Small, round-bodied crake, slightly larger than Little Crake. ***Note heavily white-spotted plumage and*** short, red-based, yellow bill; flanks and vent strongly barred black and white, but ***buff undertail-coverts*** (barred in Little and Baillon's Crakes), visible when walking with tail cocked. Juvenile lacks grey head pattern, has whitish throat and bright brownish underparts with whitish spots. Secretive, moves with slow, stalking steps and sudden crouching run. In short flight often dangles legs. **Voice** Song (both sexes) ***a rhythmical, far-carrying whistle* whitt, whitt,** repeated each second, mainly from late dusk onward through night. **Habitat** Swamps, overgrown ditches, margins of ponds. **Note** Has bred UAE. Passage hatched, some winter.

Corncrake *Crex crex* — pm

L: 29. W: 50. Skulking, rarely seen out of cover unless flushed; larger than Common Quail. In flight can suggest young gamebird, but note ***chestnut wing-coverts*** and often dangling legs. Otherwise located by ***characteristic song*** in breeding season. **Voice** Male's breeding call a far-carrying, disyllabic rasping, *arrp-arrp, arrp-arrp*. **Habitat** Meadows, lush vegetation, and crops; avoids standing water. **Note** May breed Iran; passage hatched, but rare; some winter south Arabia.

Water Rail
ad
juv
ad
Little Crake
♂
♀
juv
ad
Baillon's Crake
ad
juv
ad
Spotted Crake
ad
juv
ad
ad
Corncrake

PLATE 27: GALLINULES & COOTS

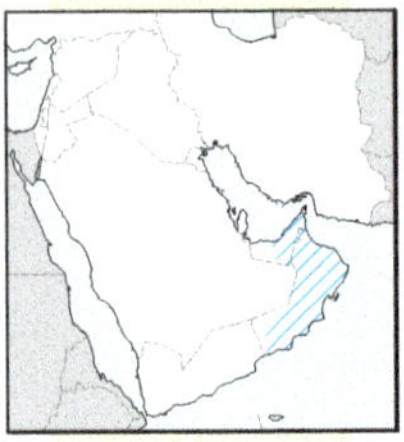

White-breasted Waterhen *Amaurornis phoenicurus* — wv

L: 32. W: 49. Slimmer than Common Moorhen; ***slaty-brown and white, with rufous-chestnut vent and undertail-coverts***; crimson eye set in white face is very conspicuous; in breeding season note reddish base to upper mandible. Juvenile has white face obscured by slate-brown. Skulking, but often easily seen in open, walking with jerking tail, displaying undertail-coverts. Climbs in tangles and creepers, sometimes into canopy. Flight with dangling legs; occasionally swims. **Voice** Very vocal in breeding season; call a prolonged ululating *kaargh-kaargh* or breaking into *kurrwah-kurrwagh-kurrwagh, krrr-kwok-kwok-krr-oowark-oowark*. **Habitat** Ponds and tangled cover. **Note** Winter hatched, but rare; vagrant Qatar, Saudi Arabia, Yemen.

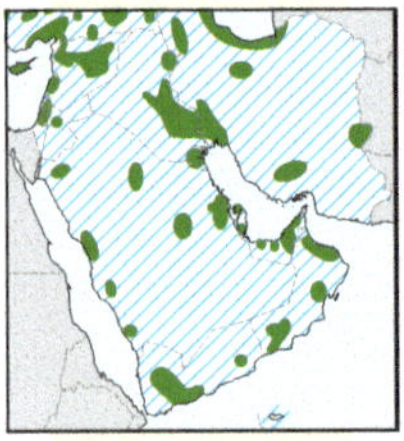

Common Moorhen *Gallinula chloropus* — RB, WV

L: 33. W: 52. Dark, hen-like waterbird, with prominent ***red bill and shield, and white flank line. Constantly jerks tail to show white undertail-coverts***. Swims with vigorous nodding movements and body tilted forward. Juvenile paler, grey-brown with dark bill; distinguished from young Eurasian Coot by ***white flank line and undertail pattern***. Seeks cover readily and patters over water when disturbed; often walks openly in wet grassland, marshes. **Voice** Sings at night with persistent clucking *kreck-kreck-kreck*. Many calls can be confused with those of Eurasian Coot; sometimes a short variable *kek* or *kr-r-eck*; also characteristic sudden, loud, gurgling, *grrll* and a 2- or 3-note *kwett, kwette-wett*. **Habitat** Freshwater wetlands and poolsides with cover. **Note** Passage and winter hatched.

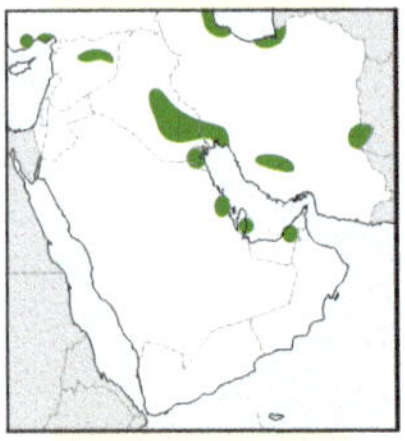

Purple Swamphen *Porphyrio porphyrio* — E/I (rb), V

L: 47. W: 95. Large and heavy; twice the size of a Common Moorhen. ***Uniformly bluish-purple with grey head and neck, huge red bill and frontal shield; red legs and long toes***. Swims with body tilted forward, ***white undertail-coverts often striking***. In flight like huge Common Moorhen but with blue wings and long red legs. Juvenile drabber, with greyish-blue underparts, dull red legs and greyish bill. Shy and often in cover. **Voice** Loud clucks, clanks, bugling and deep mooing notes; a low *chock-chock*; also *tschak-tschak* and nasal bugled song *quin quin krrkrr, quin quin krrkrr*. **Habitat** Swamps with extensive reedbeds, borders of lakes fringed with tall, dense cover. **Note** Vagrant Bahrain, Oman. [Alt: Purple Gallinule]

African Swamphen *Porphyrio madagascariensis* — E/I

L: 47. Similar to Purple Swamphen, but body ***and head bluish-purple, with mantle and much of wings (scapulars and tertials) clearly dirty moss-green***. Immatures duller with lighter greyish-blue underparts, though still greenish (often muted) on upperparts. **Voice & habitat** As Purple Swamphen. **Note** Escapes apparently bred recently UAE.

Red-knobbed Coot *Fulica cristata* — V

L: 39–44. Adult very similar to Eurasian Coot, but differs in less rounded head shape, variably dusky grey-blue bill but white frontal shield and ***two prominent red knobs on forehead*** in breeding season; these are shrunken and inconspicuous at other times. Immatures and winter adults best identified by shape of ***feathering at bill base, being rounded in Red-knobbed*** but a sharply pointed wedge in Eurasian Coot. In flight wings lack white trailing edge seen in Eurasian Coot. *Warning*: Hybridisation with Eurasian Coot can occur; the offspring show characters of both species. **Voice** Shrill nasal *kerre, krrk* or rolled *krre-krre-krre*, quite unlike calls of Eurasian Coot. **Habitat** Marshes, lakes and lagoons with cover close by. **Note** Vagrant Oman, UAE.

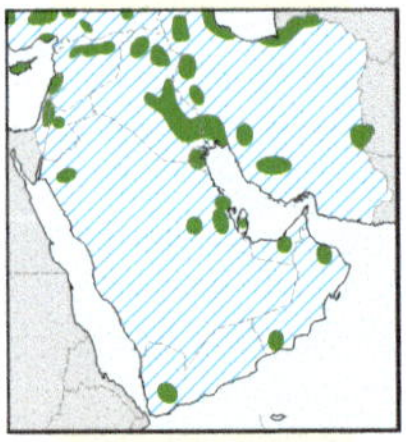

Eurasian Coot *Fulica atra* — WV, pm

L: 36–42. W: 75. ***Adult sooty-black with white bill and frontal shield***; hunch-backed on water. Upright stance out of water when note ***long greenish legs and lobed feet***. Flight stronger and heavier than other rails, more duck-like, on rounded wings and with long pattering run across water before take-off, with ***narrow white trailing edge on inner wing***. Long toes trail behind tail-tip. Juvenile duller and paler with nearly white underparts and smaller frontal shield. Dives well, but only for a short time. Markedly gregarious especially in winter. **Voice** Commonest call is a short staccato *kewk*, also an explosive high *pitts*. **Habitat** Lakes, reservoirs, ponds with grassy margins; sometimes salt-water in winter. **Note** Has bred UAE. Passage and winter hatched.

White-breasted Waterhen
ad
Common Moorhen
ad
juv
ad
juv
Purple Swamphen
Red-knobbed Coot
ad
ad winter
ad summer
African Swamphen
ad
Eurasian Coot
juv

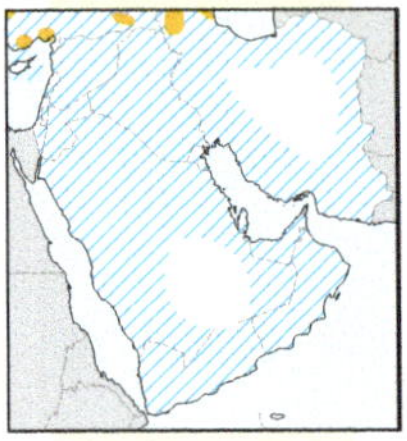

Eurasian Oystercatcher *Haematopus ostralegus* — PM, WV

L: 43. W: 83. Large ***black-and-white wader with long, red bill and rather short, red-pink legs (adult)***. Flight strong and direct; shows ***conspicuous white wing-bar***, white rump and ***terminal black tail-band***. Non-breeding adult has white neck collar and duller bill-tip. Juvenile and immature have duller black upperparts, dark tip to bill, greyish-pink legs and white neck collar. Gregarious. **Voice** Noisy; common call far-carrying *kleep-kleep* and a disyllabic *pick-pick*. **Habitat** Lakes and rivers; in winter mainly coastal. **Note** Passage hatched, rare inland; winters all coasts, some remaining in summer.

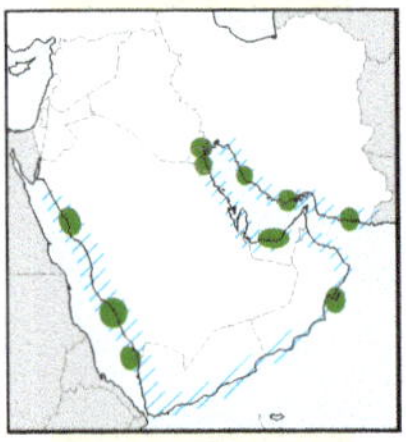

Crab-plover *Dromas ardeola* — RB, PM, WV

L: 39. W: 77. Distinctive with ***large head and straight, massive bill***. Adult winter and immature have dark streaks on crown and nape. Juvenile shows streaking on rear crown, greyish mantle, scapulars and tail, and at distance could be mistaken for a gull. ***In flight, shows white wings with blackish flight feathers and long trailing legs***; flight slow with stiff wingbeats and head sunk into shoulders. Often gregarious. **Voice** Noisy; shrill *tchuk-tchuk-tchuk* near nest; alarm a sharp *kjep*; in flight *chee-rruk*, often by night. **Habitat** Coasts, mudflats, coral reefs; never inland. Nests in tunnel excavated in sandy ground; colonial. **Note** Disperses to hatched areas.

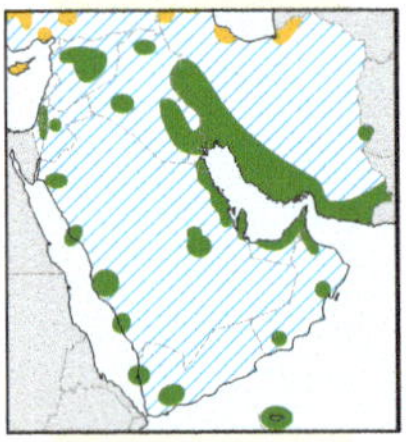

Black-winged Stilt *Himantopus himantopus* — RB, PM, WV

L: 37. W: 75. ***Exceptionally long, deep-pink legs and slim black-and-white body***; walks with high, graceful carriage. ***In flight, white with uniform black wings and obvious long trailing legs***. Female has slightly browner mantle and scapulars. Non-breeding adults have dusky head and neck marks. Juvenile and immature brownish above with greyish crown and hindneck and white trailing edge to wings in flight. **Voice** Noisy in breeding area; a variable sharp *kek*, high-pitched continuous *kikikikik* or *kee-ack*. **Habitat** Shallow fresh or brackish water, estuaries. **Note** Range expanding; passage and winter hatched.

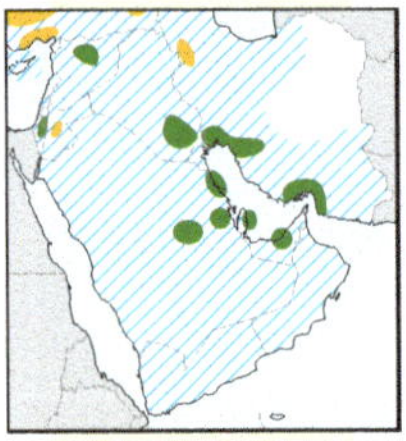

Pied Avocet *Recurvirostra avosetta* — wv, pm, rbR

L: 44. W: 79. Slender, pied wader with mainly white appearance. ***Black crown and hindneck and long, thin, upcurved bill*** distinctive. Flight stiff-winged and flickering, showing ***white wings with prominent black markings*** and long trailing grey-blue legs. Juvenile tinged brown on black parts, and white is mottled buff in upperparts. Walks steadily and delicately with head down when feeding in shallow water. Gregarious. **Voice** Noisy on breeding ground; most common call a repeated *bluit-bluit*. **Habitat** Saline, mudflats, estuaries and sandbanks; nests colonially near shallow water. **Note** Passage and winter hatched.

Greater Painted Snipe *Rostratula benghalensis* — V

L: 25. W: 52. Skulking; when flushed note dangling legs, short tail, rounded wings and rather slow flight. ***Bill pale, shorter than Common Snipe's and slightly decurved at tip***. Distinctive head markings, buff 'V' on mantle and ***white line in front of wing***; dark foreparts contrast sharply with white belly. ***Male has prominent buff eye-ring and streak behind eye***. In flight, shows golden-buff spotted barring on forewing and flight feathers and black wing-bar; also ***Ruff-like white rump pattern***. ***Larger and brighter female mainly greeny-bronze*** with ***chestnut head and neck with distinctive white marks*** (similar to male). **Habitat** Marshes with muddy patches, reedbeds; will feed on open fields. **Note** Vagrant Iran, Oman, SW Saudi Arabia, UAE.

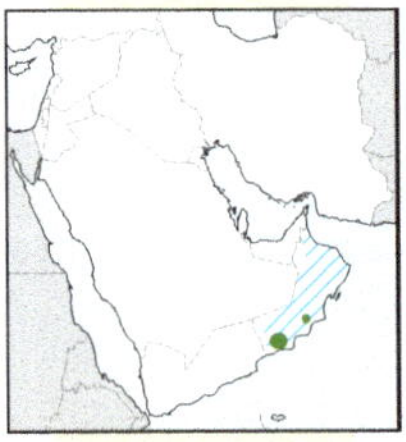

Pheasant-tailed Jacana *Hydrophasianus chirurgus* — No records

L: 31 (48cm with full tail). Rail-like and often seen walking ***on floating vegetation on very long toes***. Breeding adult has long, ***black, downcurved tail***, chocolate-brown body, strikingly patterned head with ***white face and foreneck, and golden-yellow hindneck, edged black***. ***White wings with black tips conspicuous in flight***. In non-breeding plumage tail is short, ***underparts turn white, but a dark breast-band remains, running up the neck and joining the eye-stripe***. Juvenile resembles non-breeding adult but breast-band flecked white, and head and neck pattern are duller. Low, rapid flight with dangling legs; landing with raised wings. **Habitat** Ponds, creeks and marshes with patches of open water and floating vegetation. **Note** Rare in hatched area; vagrant Qatar, Saudi Arabia, Yemen including Socotra.

ad winter
Eurasian Oystercatcher
ad summer
Crab-plover
ad
juv
juv
ad
Black-winged Stilt
ad
Pied Avocet
♀
ad winter
juv
♂
♀
ad summer
Pheasant-tailed Jacana
Greater Painted Snipe

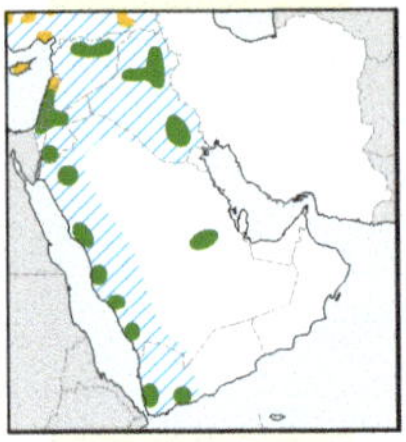

Spur-winged Lapwing *Vanellus spinosus* V

L: 26. W: 75. Elegant, long-legged plover with ***black head and underparts and contrasting white cheeks and sides of neck***. In flight, tri-coloured wing pattern: conspicuous white band between black flight feathers and sandy wing-coverts; also broad black tail-band; from below black belly and flight feathers contrast with white underwing-coverts. Juvenile similar to adult. Legs black. **Voice** Noisy in breeding area; alarm call a shrill Oystercatcher-like *dwitt-dwitt* or *kwitt-kwitt*. **Habitat** Fresh and saline marshes, irrigated land, with short vegetation. **Note** Dispersal/passage hatched; occasional NW Iran; vagrant Gulf States. [Alt: Spur-winged Plover]

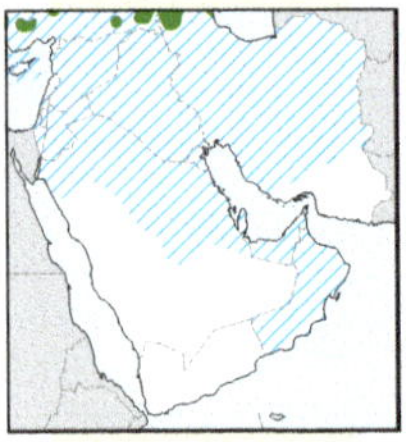

Northern Lapwing *Vanellus vanellus* wv, pm

L: 30. Unmistakable with ***long, thin, upturned crest and greenish upperparts***. Underparts clean white with broad black breast-band and rich burnt-orange undertail. ***In rather flappy flight, broad, rounded wings show no wing-bar***. In winter upperparts have narrow pale scaling to feathers. Juveniles lack tall crest, instead having scruffy, spiky tuft on nape. Often in large flocks outside breeding season. **Voice** Loud, shrill *peeo-vit* uttered in tumbling display flight over breeding grounds. **Habitat** Open fields, marshes, shallow pools and coastal flats. **Note** Passage and winter hatched, scarce in south.

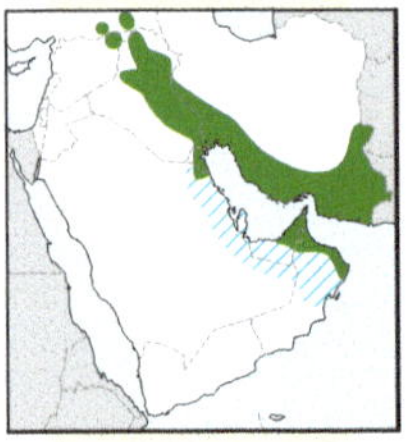

Red-wattled Lapwing *Vanellus indicus* RB

L: 33. W: 80. Rather large, colourful plover; easily identified by ***red bill, eye-ring and wattles, bright yellow legs, black head and centre of breast***. Flight light with slow wingbeats, showing similar wing pattern to Spur-winged and White-tailed Lapwings, but ***tail has black subterminal band with broad white terminal band***; yellow feet project distinctly beyond tail. Juvenile much duller with chin and throat almost white, black areas grey-brown, and wattle is tiny or absent. Fairly confiding. **Voice** Noisy; loud and shrill alarm notes rendered as *did-he-do-it, pity-to-do-it*. **Habitat** Open country, usually near fresh water, and showing preference for grassy fields and agricultural land. **Note** Dispersal as hatched. [Alt: Red-wattled Plover]

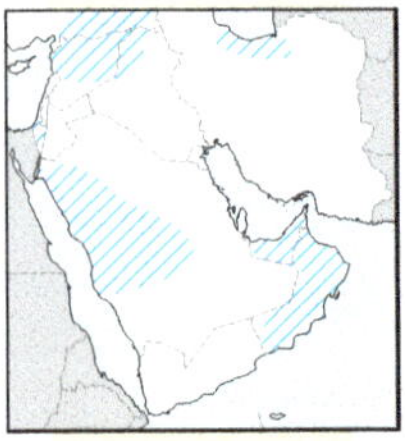

Sociable Lapwing *Vanellus gregarious* pm, wv

L: 29. Lapwing-sized, with more upright stance, especially when alert. In breeding plumage easily told by ***long white supercilium, joining on nape, black crown and chestnut-black belly***. In winter, loses belly-patch, becomes mottled on breast and supercilium is less distinct (but white forehead usually quite prominent). Juvenile, which is browner than adult, has buffish wash to forehead and supercilium and pale feather-edgings on upperparts. Legs dark. In flight, shows fairly rounded wings, with conspicuous black, white and brownish upperwing pattern and black band on tip of tail. Typical plover actions on ground, making short runs with head tucked into body, stopping to peck at ground or stand with head erect. Often in flocks, on passage and in winter sometimes with Northern Lapwings, golden plovers and coursers. **Voice** Harsh *chark-chark-chark* flight call. **Habitat** Steppes and bare or cultivated fields (e.g. with winter cereals); rare on coast. **Note** Passage hatched; occasional in winter; rare Arabia; vagrant Bahrain, Kuwait, Yemen. [Alt: Sociable Plover]

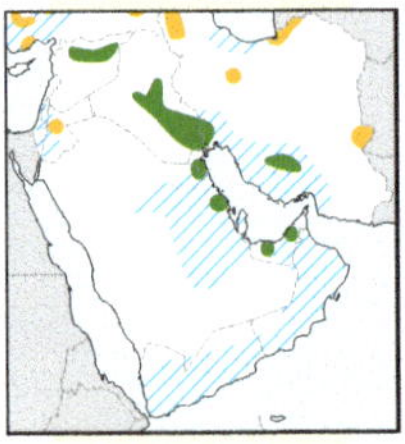

White-tailed Lapwing *Vanellus leucurus* RB, WV

L: 28. Slender and graceful, most closely resembling Sociable Lapwing in winter plumage but smaller and readily told by ***plain head*** (accentuating dark eye), ***longer, deep waxy-yellow legs*** (which protrude in flight) and, ***in flight, all-white tail***. Juvenile paler on neck and breast, dark-mottled on upperparts with dark cap and faint brown tip to tail. When feeding, tips down so steeply that it almost stands on its head! **Voice** High-pitched *kee-vee-ik*, persistently repeated at breeding site. **Habitat** Fresh or saline pools, marshes and wet plains; may nest colonially. **Note** Range expanding; passage/winter hatched. [Alt: White-tailed Plover]

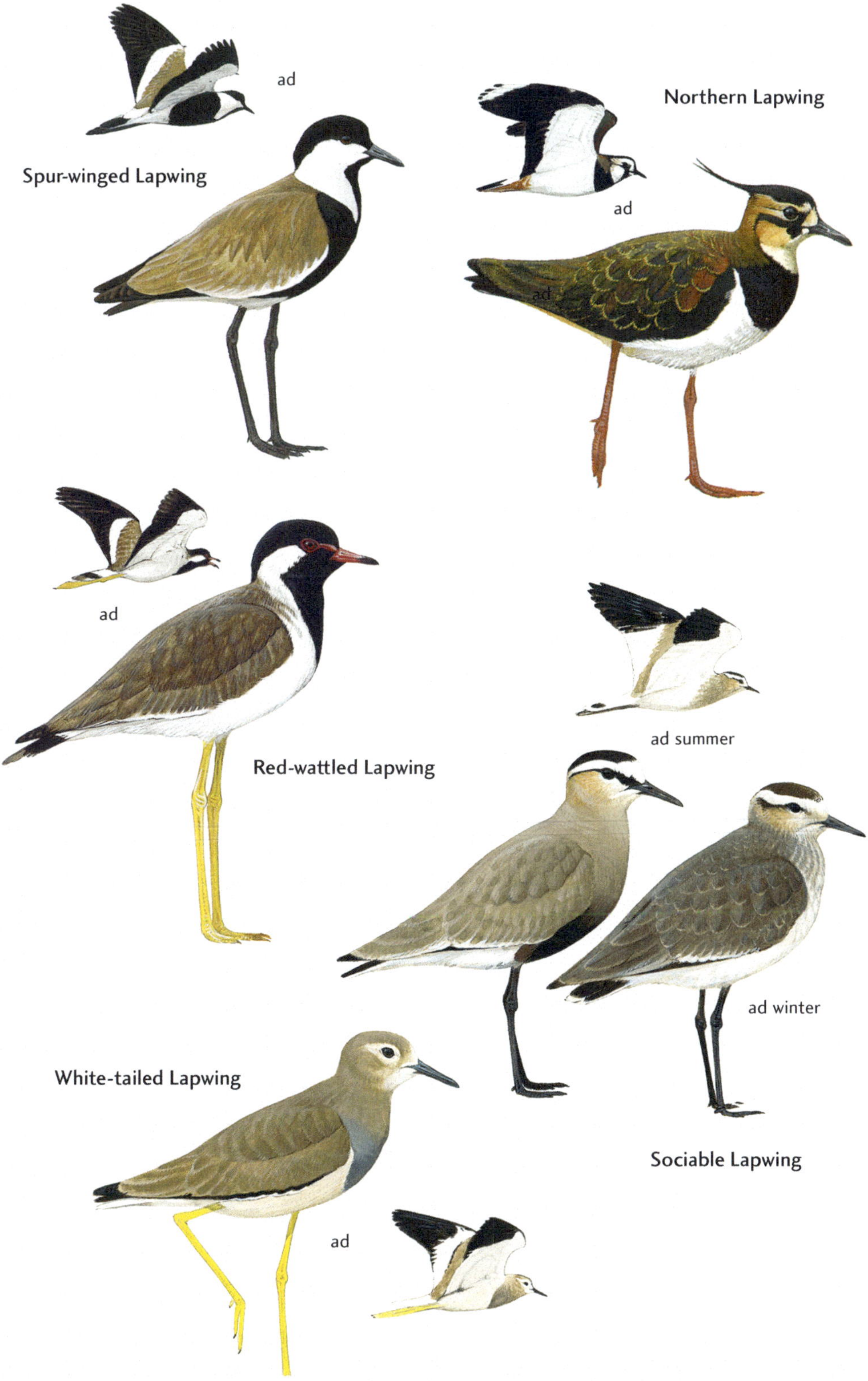
ad
Northern Lapwing
Spur-winged Lapwing
ad
ad
ad
ad summer
Red-wattled Lapwing
ad winter
White-tailed Lapwing
Sociable Lapwing
ad

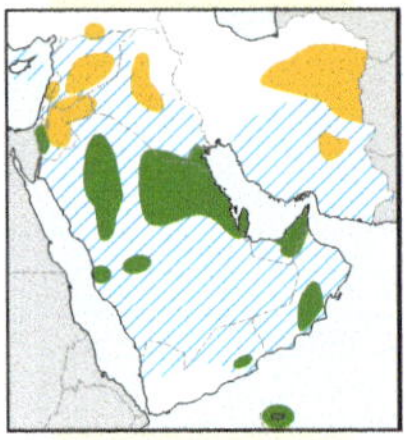

Cream-coloured Courser *Cursorius cursor* RB, WV

L: 24. W: 54. ***Sandy-buff with distinctive black-and-white head markings joining in 'V' on nape, bill short and downcurved.*** Erect posture; runs quickly with plover-like behaviour and sudden stops, prefers running away rather than flying. Flight rapid with long, slightly rounded wings and legs protruding well beyond tail. Wing pattern distinctive, ***black outer wing above and black underwing contrast sharply with rest of plumage.*** Juvenile lacks grey and black on crown; upperparts, head and breast have faint brown spots or irregular dark subterminal lines, and primaries are fringed buff. Sometimes in flocks post-breeding. **Voice** Flight call a short *kwit-kwit*. **Habitat** Sandy or stony semi-desert with scanty vegetation, marginal cultivation, arid flat country; post-breeding also short grassland, fields and turf farms. **Note** Post-breeding dispersal and passage hatched.

Eurasian Golden Plover *Pluvialis apricaria* pm, wv

L: 28. W: 71. Resembles Pacific Golden Plover but ***separated by heavier, short-necked and pot-bellied appearance, comparatively shorter legs and bill, white underwing and voice; in flight feet do not extend beyond tip of tail.*** Flight pattern rather uniform with faint wing-bar mainly on primaries, no white in tail. Non-breeding and immature similar to Pacific Golden Plover, but yellow-buff supercilium less distinct, with ***tertials falling well short of wing-tip.*** Larger Grey Plover has black axillaries, stronger bill, white rump and grey appearance. **Voice** Barely disyllabic ***melancholy whistle*** püyh ***or repeated*** pyü-pü (often hard to place). **Habitat** Grassland, ploughed land, stubble, coast. **Note** Passage and winter hatched; scarce UAE; vagrant Bahrain, Kuwait, Oman, Saudi Arabia.

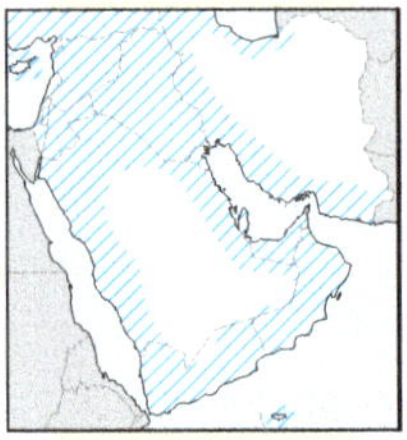

Grey Plover *Pluvialis squatarola* PM, WV

L: 29. W: 77. Larger than the golden plovers with ***heavier head and longer, stouter bill. Greyish appearance*** lacking obvious yellow or greenish tones in upperparts, except for faint yellow-buff tinge in juvenile. ***Black axillaries diagnostic in flight (all ages).*** Breeding plumage recalls Eurasian Golden, but more white on head and nape and ***coarsely speckled black-and-white upperparts;*** larger white breast-side patch ***without white flank-line. White wing-bar and rump*** obvious in flight. **Voice** Flight call a mournful trisyllabic whistle, *dee-oo-wee* (second note lower-pitched), often repeated. **Habitat** Tidal flats and saltings; occasionally inland wetlands. **Note** Passage hatched, rare inland, winters all coasts, with some remaining in summer.

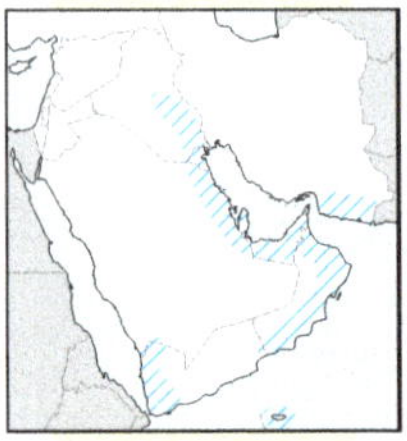

Pacific Golden Plover *Pluvialis fulva* PM, WV

L: 24. W: 66. Similar to slightly larger Eurasian Golden in all plumages. ***Best separated by slimmer build, dark underwing and voice.*** Compared to Eurasian Golden, wings often protrude 1–2 centimetres beyond tail-tip, ***bill is finer and longer, legs are clearly longer,*** particularly thighs, making it appear more elegant. ***In flight, toes extend beyond tip of tail and wings appear longer and narrower; underwing and axillaries are greyish-brown in all plumages*** (white in Eurasian Golden). Adult in breeding plumage brighter than Eurasian Golden, more golden on mantle and scapulars and often bright white spangling in the wing-coverts; non-breeding and immature have more distinct yellow-buff supercilium than Eurasian Golden. Upperwing as Eurasian Golden. **Voice** Soft disyllabic call *gru-it* (resembling Spotted Redshank). **Habitat** Mudflats, grassland, cultivated fields. **Note** Passage and winter hatched; vagrant west to Cyprus and Turkey.

Eurasian Dotterel *Charadrius morinellus* V

L: 21. W: 60. Brownish, dry-country plover; in non-breeding plumage recalling winter Eurasian Golden Plover, but smaller and with ***whitish supercilia, meeting in 'V' on nape, and narrow, whitish upper breast-band*** in all plumages. Adult breeding female (male duller) has striking ***white supercilium and throat contrasting with blackish cap*** and greyish neck, ***upper breast bordered white, with chestnut and blackish below.*** Adult non-breeding and immature with ***brown, scaly upperparts and buff underparts. No wing-bar in flight,*** but shows white tip to tail, and buffish-grey underwing; white shaft on outer primary sometimes rather distinctive. **Voice** Trilling, rather dry Dunlin-like *dryrrr.* **Habitat** Steppes and poor arable land. **Note** Passage and winter hatched; vagrant Bahrain, Oman, UAE.

Cream-coloured Courser
ad
Eurasian Golden Plover
1st-winter
Grey Plover
winter
ad summer
winter
ad summer
winter
Pacific Golden Plover
juv
ad ♀ summer
winter
Eurasian Dotterel

Common Ringed Plover *Charadrius hiaticula* PM, WV

L: 19. W: 52. Small plover with ***orange legs and black-tipped bill, black breast-band and white hindneck collar***. In flight, shows ***conspicuous white wing-bar***. In non-breeding adult black is replaced by dark grey-brown, supercilium and forehead are tinged brown, and bill becomes all dark. Juvenile similar but is paler and duller, often with broken breast-band; upperparts with buff feather-fringes. Distinguished from Little Ringed Plover by white wing-bar, lack of pale orbital ring, and ***pale supercilium in juvenile. Separated from other plovers by obvious white hindneck-collar and call***. Mainly nominate subspecies occurs around Mediterranean basin, but further east most often only the subspecies *tundrae*, with darker upperparts, encountered. **Voice** A soft, rising, disyllabic whistle *tooip*. **Habitat** Sandy, muddy and stony shores, both coastal and inland. **Note** Passage and winter hatched; some oversummer.

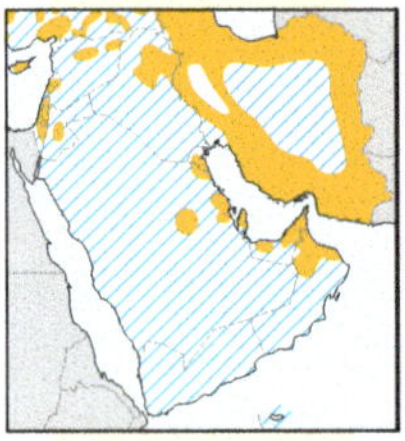

Little Ringed Plover *Charadrius dubius* MB, PM, wv

L: 15. W: 45. Small, ***slim, long-winged plover*** with horizontal stance. In breeding plumage similar to Common Ringed Plover, but breast-band narrower and ***lacks white wing-bar in flight***; note ***yellow orbital ring, dark bill, white line behind black forecrown***, and muddy-coloured or pinkish legs. Adult non-breeding and juvenile have almost plain brown forehead with ill-defined pale patch behind or above eye and duller orbital ring; breast-band is broken or absent. Juvenile also has yellow-buff tinge to face and throat. ***Separated from all other plovers by call and lack of wing-bar***. Mostly solitary or in pairs, sometimes small groups. **Voice** Flight and alarm call a loud, plaintive almost monosyllabic *diu*; in wavering display flight, with slow wingbeats, gives *pree-pree-pree* and tern-like *krre-u krre-u*. **Habitat** Mainly freshwater, in particular gravelly river islands and sandy borders of lakes; also coastal in winter. **Note** Passage hatched; some winter, mainly in south Arabia.

Kittlitz's Plover *Charadrius pecuarius* V

L: 13. W: 42. Structure and actions resemble Kentish Plover, but smaller with brownish, ***buff-fringed and dark-centred feathers on upperparts, long legs, striking head markings and creamy-buff wash on breast and belly***. In adult breeding, white supercilium and black eye-stripe both meet at base of hindneck. In non-breeding and juvenile, head markings are much duller, lacking black and with white tinged buff. In flight, rather dark with blackish forewing and prominent, but short, white wing-bar, especially visible on inner primaries; tail pattern similar to Kentish; toes project well beyond tail-tip. **Habitat** Muddy wetland margins. **Note** Vagrant Bahrain, Saudi Arabia, UAE.

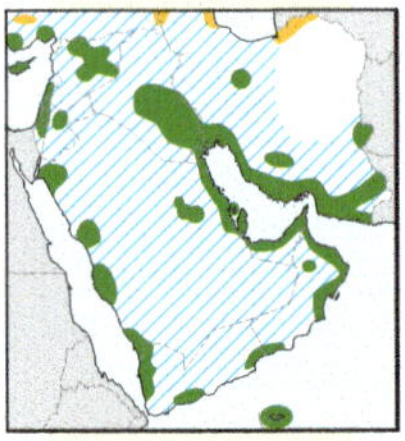

Kentish Plover *Charadrius alexandrinus* RB, PM, WV

L: 16. W: 44. Small, sandy plover with white underparts, rather long, blackish legs, ***conspicuous white hindneck-collar and lacking complete breast-band***; in flight, a clear, white wing-bar and ***broad, white sides to tail***. Adult breeding male has variable, ***rufous cap, black frontal (forehead) bar, and lateral black breast-patches***. Breeding female, adult non-breeding and juvenile are duller, lacking black in plumage, and resemble non-breeding Common Ringed and Little Ringed Plovers, but separated by rather long, blackish legs, white breast, more white in tail, and call. By late-summer plumage of some individuals may bleach to entirely pale sandy-whitish above, with breast-patches also often entirely missing. **Voice** Flight call a soft *kip* or *twit*, recalling Little Stint; song a rattled repetition of *tjekke-tjekke* or *jid-id-jid... eer*. **Habitat** Shingle, sandy and muddy beaches, mudflats, mainly coastal but also inland, often on saline lagoons. **Note** Passage and winter hatched.

Common Ringed Plover
ad summer
winter
Little Ringed Plover
ad summer
winter
Kittlitz's Plover
ad summer
imm
ad summer
Kentish Plover
♂ summer
winter

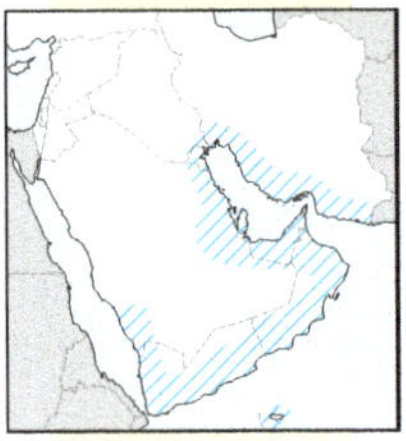

Lesser Sand Plover *Charadrius atrifrons* PM, WV

L: 20. W: 51. This species is split into two groups: the '*atrifrons* group' and the '*mongolus* group'; only *atrifrons* has been recorded in the region. Very similar to Greater Sand Plover in all plumages, behaviour and shape; lone individuals can be difficult to identify. Best separated by size (***body slightly larger than Common Ringed Plover; Greater Sand distinctly larger***); ***shorter bill***, less pointed and with more swollen tip (but can overlap with Greater Sand); smaller, more rounded, less angular head; ***shorter, darker, more greyish legs*** (yellowish-green in Greater Sand). At rest stance often more upright than Greater Sand. Breeding male has ***all-black forehead and face-mask, and broad reddish-chestnut breast-band*** (Greater Sand never shows completely black forehead, and breast-band is usually narrower); female has black of head reduced (like Greater Sand). Birds of '*mongolus* group' very difficult to identify except in breeding plumage, when both male and female have prominent white forehead, and male often shows narrow, black margin to chestnut breast-band. Non-breeding and immature plumage (both groups) very like Greater Sand Plover, then structural features are important. ***Separated from 'ringed' plovers by size and lack of white hindneck collar***; from Caspian Plover by shorter legs and wings, bolder white wing-bar, white underwing, and less bold supercilium. In flight, legs reach to or slightly beyond tip of tail (in Greater Sand legs show fairly prominently beyond); both Sand Plovers show clear, but variable, white wing-bar: in Lesser Sand Plover this being of more even width (in Greater Sand often most prominent on inner primaries). **Voice** Quieter than Greater Sand Plover, having a short, sharp and less trilling *chitik, chi-chi-chi, chik-tik*; also *kruu-kruit* or *drriiiit*. **Habitat** Tidal mudflats and sandy coasts. **Note** Passage and winter hatched, many over-summer; rare inland.

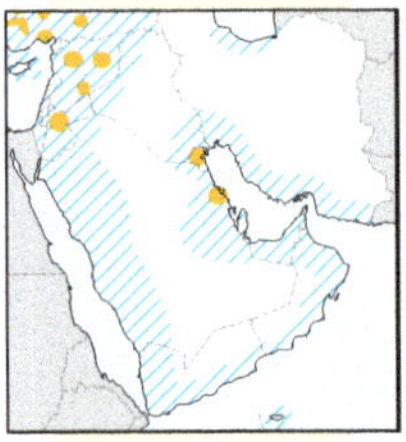

Greater Sand Plover *Charadrius leschenaultii* PM, WV

L: 24. W: 56. Resembles larger version of Lesser Sand Plover, and lone individuals can be difficult to identify. Note especially the larger size (***obviously larger than Common Ringed Plover); longer, more pointed bill, larger, more angular head and large eye; yellowish-green longer legs, especially thighs (tibiae), with toes projecting well beyond tail-tip in flight***. Caspian Plover is slimmer with proportionally longer legs, and wings protrude well beyond tail when perched; head is more rounded with broader supercilium, underwing is dusky, and it has only a faint white wing-bar. Display flight recalls that of the ringed plovers, with dry chortling song incorporating Ruddy Turnstone-like call-note. **Voice** When flushed a trilling *kyrrr, kirr* or *trrr*; in songflight *huit-huit-huit* or ascending *dui-dui-tui-dit*. **Habitat** Breeds on inland sand- and mudflats, mainly near water; otherwise mainly coastal. **Note** May breed Iran; passage and winter hatched. Some oversummer Arabian coasts.

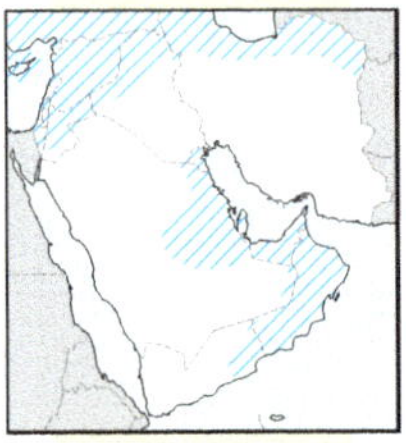

Caspian Plover *Charadrius asiaticus* pm

L: 19. W: 58. ***Slim, delicate plover*** mostly recalling sand plovers in size and general appearance, but separated by ***long, attenuated body with wings projecting well beyond tail-tip; proportionally longer legs***, long neck and smaller head with rather fine tapering bill; ***broader white supercilium gives a capped appearance***. Male breeding shows distinct ***blackish lower border to rufous breast-band***. In flight appears long-winged with faint wing-bar, only visible on inner primaries, and toes clearly projecting beyond tail; ***underwings are dusky (not white)***, and tail is dark with less white at sides and tip. **Voice** Flight call a short, sharp, *tyup*, sometimes repeated, and occasionally combined into a rapid series of rattling notes *tptptptptp*. **Habitat** Fields, grassy plains, semi-desert; also coastal areas. **Note** Generally a rare migrant, but regular Arabia; vagrant Yemen.

Lesser Sand Plover
winter
ad summer
Greater Sand Plover
winter
ad summer
Caspian Plover
winter
♂ summer

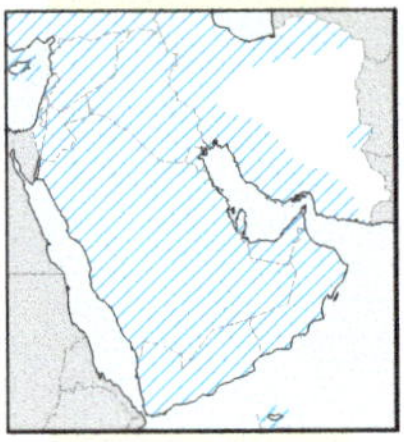

Jack Snipe *Lymnocryptes minimus* pm, wv

L: 18 (incl. bill 4). W: 40. ***Very small with short bill.*** Well-camouflaged and difficult to flush, usually ***rising silently at about one metre or less, in low, slightly jerky flight and settling almost immediately; two bright yellow lines on upperparts and short, pointed tail*** distinctive. On ground note green-glossed back, absence of central crown-stripe and dark-striped flanks (not barred). Separated from other snipes by size and short bill. Usually solitary. **Voice** Usually silent when flushed. **Habitat** Wetland margins. **Note** Passage and winter hatched, but mostly scarce.

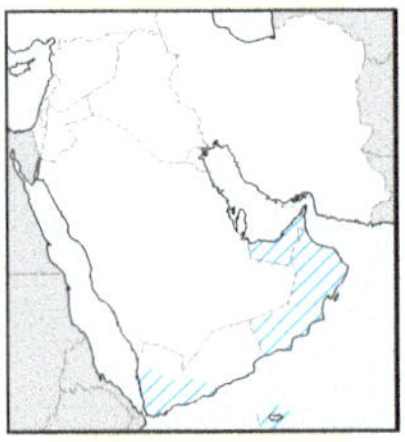

Pin-tailed Snipe *Gallinago stenura* PM, WV

L: 26 (incl. bill 6.5). W: 45. Very similar to Common Snipe; best distinguished in flight, when note paler grey-buff panel on wing-coverts, ***absence of white trailing edge to secondaries***, slightly more rounded wing-tips, and ***completely barred underwing-coverts making underwing look dark greyish*** (Common Snipe shows broad white bands, creating much paler underwing). On ground appears generally colder and darker above, with slightly shorter, broader-based bill, ***bulging supercilium in front of eye*** (more parallel-sided in Common Snipe), ***more scalloped scapular pattern*** due to similar width and colouration of pale edges on inner and outer webs (in Common Snipe scapulars have broad white edges on outer webs only, contrasting with brown inner webs, creating white, often diagonal, stripes) and ***barred median wing-coverts*** (more spotted in Common Snipe). When flushed, has slower take-off with slightly heavier, less erratic, flight; flies lower and drops after short distance. **Voice** A ***short, sneezed dry* etch**, often given just once; close to Common Snipe's but shorter, without inflection. **Habitat** Wet fields, marshes, more often on drier ground than Common Snipe. **Note** Passage and winter hatched, but uncommon; vagrant Iran, Saudi Arabia.

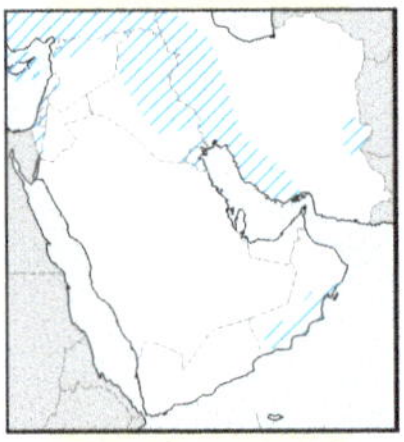

Great Snipe *Gallinago media* V

L: 28 (incl. bill 6). W: 49. ***Larger, heavier*** and darker than Common and Pin-tailed Snipes with ***shorter bill and closely barred underparts and underwings***. When flushed, ***rises at short distance, flies low and straight without zigzagging; flight heavy with rather slow wingbeats***; ***pot-bellied appearance*** reminiscent of a small Woodcock. ***In flight note absence of white belly-patch*** (shown by Common and Pin-tailed Snipes), ***dark underwing*** (similar to Pin-tailed, but unlike Common Snipe); ***white wing-bars on greater and median coverts, including primary coverts, more white on tail-corners*** and absence of white trailing edge to secondaries (conspicuous in Common Snipe). **Voice** When flushed low-pitched, rather weak, muffled *orrk*. **Habitat** Marshes, stubble fields, wet or dry rough grassland. **Note** Passage hatched but rare; rare/vagrant Saudi Arabia; vagrant UAE, Yemen.

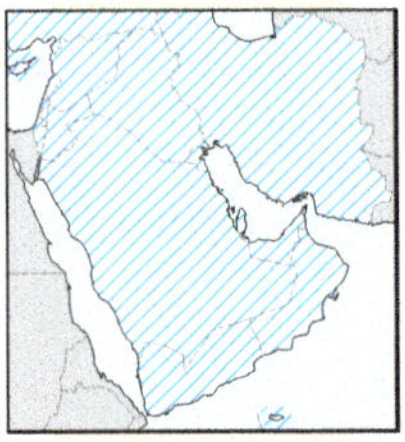

Common Snipe *Gallinago gallinago* PM, WV

L: 26 (incl. bill 7). W: 45. Medium-sized with distinctly yellow-striped head and upperparts, dark-striped breast, barred flanks and ***very long bill***; probes mud with vibrating movements. Often squats low; usually not seen until flushed at 10–15m distance; ***rises explosively, immediately uttering several harsh calls, while zigzagging to a good height***. In flight shows narrow ***white trailing edge to wings*** (lacking in Pin-tailed Snipe) and ***white belly***. See Pin-tailed Snipe for further differences. **Voice** When flushed utters a few harsh *ärrtch* notes with slightly rising inflection. In aerial display produces distinctive reverberating sound ('drumming'). **Habitat** Wet grasslands, marshy water-margins. **Note** Passage and winter hatched.

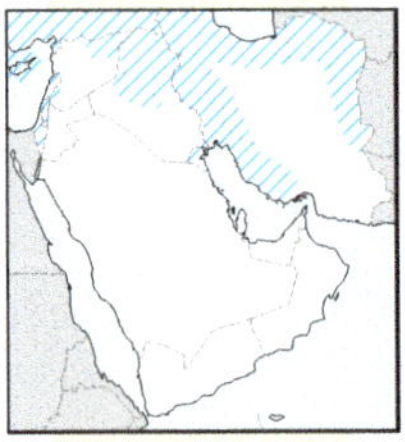

Eurasian Woodcock *Scolopax rusticola* V

L: 34. W: 58. Plumper and larger than the snipes; mainly in woodland. Usually seen when flushed; rises silently with zigzagging flight between trees, showing long bill, round body, broad wings, red-brown rump and tail with black subterminal band. On ground note large eyes set far back, barred crown (striped in snipes), and completely barred buff underparts. Often only flushes almost underfoot. **Voice** Displays at dusk in level, direct flight low over treetops with distinctive call: one or more deep croaks followed by sharp *twzzip*. **Habitat** Woodland floor, scrub, thick cover. **Note** Winter hatched; vagrant Bahrain, Oman, Saudi Arabia, UAE.

Jack Snipe
Pin-tailed Snipe
Great Snipe
Common Snipe
Eurasian Woodcock

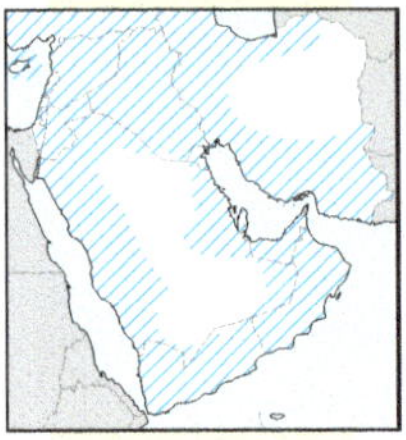

Black-tailed Godwit *Limosa limosa* PM, WV

L: 42. W: 77. Large with long bill and legs. Similar to Bar-tailed Godwit, but slightly larger and more erect due to ***longer legs and neck; bill slightly longer and straighter and totally different upperwing pattern*** (also visible from below). In summer plumage, both sexes have varying amount of ***rusty-orange on head, neck and fore-body*** with ***diffuse, dark bars on lower breast, fore-belly and flanks***. Female duller, sometimes predominantly greyish. Non-breeding plumage uniform grey, washed ochre on neck and breast in juvenile and with pink-based bill. In all plumages easily ***distinguished in flight by broad, white wing-bar, white tail with black terminal band and trailing legs and feet***. Often in flocks, including when feeding. **Voice** All calls rather nasal and scolding; alarm call *titi-teev* or, from more excited birds, *wicka-wicka-wicka*. **Habitat** Muddy freshwater margins, grassland, marshes, estuaries, tidal creeks. **Note** Passage and winter hatched, some oversummer Arabia.

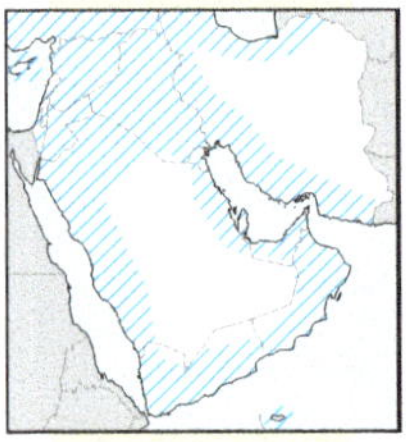

Bar-tailed Godwit *Limosa lapponica* PM, WV

L: 38. W: 75. Resembles Black-tailed Godwit especially on ground in non-breeding plumage, but note more compact, less erect appearance with ***shorter legs and neck***, and ***slightly upturned bill***. Easily separated from Black-tailed in flight by lack of obvious wing-bar (and resembling Whimbrel). In breeding plumage has ***deep rusty-red head and entire underparts***; larger female is buffish or faintly rusty, ***lacking breast- or belly-bars of Black-tailed Godwit***. Non-breeding/winter plumage buffish-grey with dark shaft-streaks above and on breast (almost uniform smooth grey in Black-tailed). Juvenile similar but more buffish with darker upperparts lacking shaft-streaks and breast markings. Roosts but does not feed in flocks. **Voice** In flight, or when flushed, soft, low-pitched, nasal *beb-beb*, or sharper *ke-kek*. Often silent. **Habitat** Mudflats, sandy beaches, estuaries. **Note** Passage and winter hatched; many oversummer coastally.

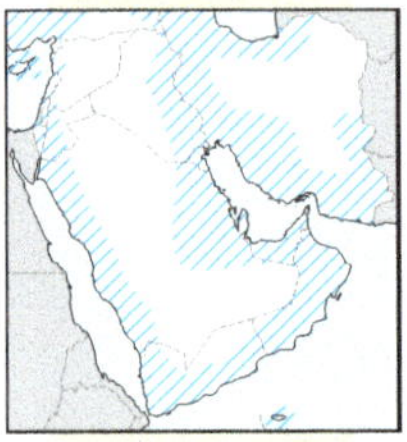

Whimbrel *Numenius phaeopus* PM, wv

L: 41. W: 83. ***Smaller and slightly darker than Eurasian Curlew*** with faster wingbeats. ***Bill usually shorter and more decurved near the tip*** (though overlaps with young Eurasian Curlew in length) and head shows ***dark crown with pale central stripe and dark eye-stripe***. Similar in flight to Eurasian Curlew; body and upperwing usually appearing darker than in Curlew. **Voice** Flight call ***characteristic, fast series*** of whistled notes, a tooted *bi-bi-bi-bi-bi-bi*, quite unlike the soft rising whistle of Eurasian Curlew. **Habitat** Estuaries, sandy beaches, rocky shores, coral reefs, grasslands. **Note** Passage hatched, rare inland; winters Iranian and Arabian coasts, where some also summer.

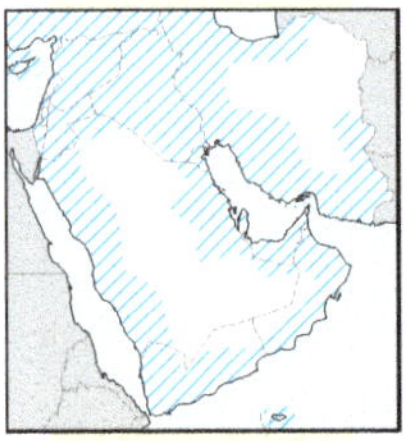

Eurasian Curlew *Numenius arquata* PM, WV

L: 55. W: 90. Large, streaked, brownish wader with ***long, decurved bill***. Male smaller than female, with shorter bill, which in young male overlaps with Whimbrel's, but is more evenly decurved. Flight with rather slow, gull-like wingbeats, showing barred tail and white wedge on rump and lower back (similar to Whimbrel), but ***feet project***, unlike in Whimbrel. Otherwise told from Whimbrel by size, ***usually longer bill, uniform head pattern and flight call***. The subspecies *orientalis*, the most frequent in Arabia, has white underwing and much longer bill – not to be confused with extralimital Eastern Curlew *N. madagascariensis*, which has brownish underparts and rump and much darker underwing. **Voice** Flight call a drawn-out, melodic, slowly rising whistle, easily imitated, *cour-leee*. **Habitat** Tidal mudflats and sands, rocky shores, coral reefs; also inland on muddy or grassy wetland margins. **Note** Passage and winter hatched, some summer Arabia.

Black-tailed Godwit
ad summer
winter
winter
ad summer
winter
Bar-tailed Godwit
ad
Whimbrel
ad
orientalis
Eurasian Curlew

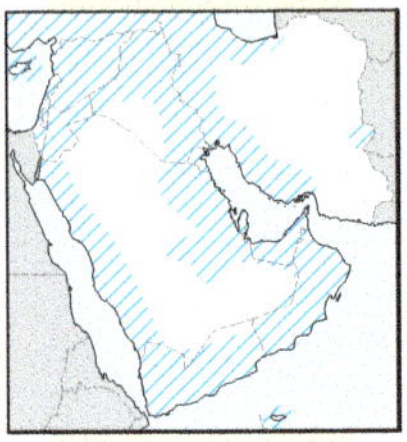

Spotted Redshank *Tringa erythropus* pm, wv

L: 30. W: 64. Medium-sized, ***rather slim wader with red legs***. In winter plumage recalls Common Redshank, but larger with paler grey upperparts, white underparts, and ***longer, finer bill***, slightly drooping at tip. ***Upper mandible wholly black***, lower red with black tip. In breeding plumage ***black, finely spotted white***. In flight, shows all-dark wings, ***distinctive white rump extending in wedge up back***, and feet projecting well beyond tail. **Voice** Flight call distinct piercing ***disyllabic* tju-it**. **Habitat** As Common Redshank, but freshwater and lagoonal more than open tidal settings. **Note** Passage and winter hatched.

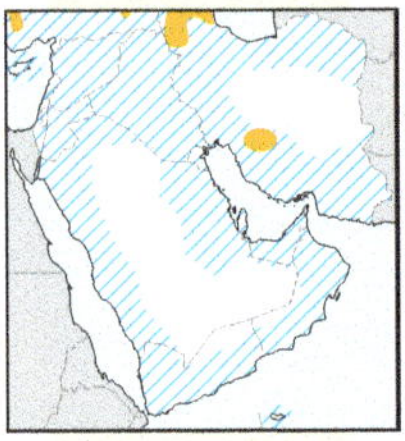

Common Redshank *Tringa totanus* PM, WV, sv

L: 28. W: 62. Greyish with ***bright red legs***, in flight showing ***broad white trailing edge to wings and white on rump extending up back***. Breeding adult dark-spotted, with red base to bill. Winter plumage uniform grey above, paler grey below with ill-defined spotted breast. Juvenile has buff-spotted upperparts and all-dark bill. Larger and slimmer juvenile Spotted Redshank has greyer upperparts, heavily barred or vermiculated underparts, finer, longer bill and different wing pattern. **Voice** Calls distinctive; ***disyllabic* djü-dü**, with stress on first syllable; alarm a persistent *tjü-tjü-tjü....* **Habitat** Breeds in damp grassland; on passage and winter mainly coastal shores, mudflats; also inland wetlands. **Note** Passage and winter hatched.

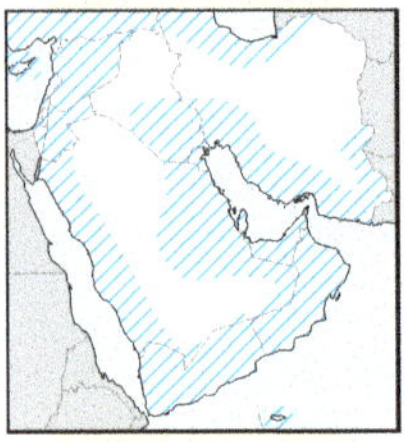

Ruff *Philomachus pugnax* PM, WV, sv

L: 28 (male); 22 (female). W: 56 (male). Male about size of Common Redshank, but more upright with ***longer neck, smaller head*** and proportionally ***shorter, slightly drooping bill***; often looks hump-backed and pot-bellied. Flight lazy, sometimes interrupted with glides; shows narrow, wing-bar and characteristic ***oval white patches at tail-base***. Adult in summer plumage variable, upperparts, breast and flanks show mix of black, brown, chestnut, ochre and white, heavily barred or blotched (male may show large ruffs and ear-tufts in spring). Plumage greyer in winter, when ***lores always pale***, face often whitish and rear belly and undertail always white; a few males show white on head and breast. Bill mostly blackish-brown, but can be yellow or pinkish, tipped dark; ***legs vary from orange-red to greenish-grey. In juvenile, upperparts scaly, blackish-brown distinctly fringed buff, and head, neck and breast yellowish-brown, tinged orange***, legs are yellowish-brown or greenish. Often in loose flocks. **Habitat** Inland wetland fringes, wet grassland, coastal lagoons. **Note** Passage and winter hatched.

Buff-breasted Sandpiper *Tryngites subruficollis* V

L: 19. W: 45. Slender, buffish Ruff-like wader with small, rounded head with buff face and prominent black eye. ***Short, straight bill and bright yellow-ochre legs; upperparts distinctly scaly, and underparts from head to undertail buff with fine spotting on breast-sides***. In flight, appears rather plain with ***no white in wings or tail***. Juvenile similar to adult. From young female Ruff by smaller size, shorter straight bill, rounded head, clean buff face, spotted breast-sides, brighter yellowish legs and different upperpart pattern in flight. **Habitat** Fields, short grass, in preference to shores. **Note** Vagrant Oman, Saudi Arabia, UAE.

Pectoral Sandpiper *Calidris melanotos* V

L: 21. W: 42. Long-winged brown sandpiper, ***larger than Dunlin*** with short, slightly decurved bill and, when not feeding, often with rather long-necked appearance. Told in all plumages by ***clearly demarcated streaked breast contrasting with white belly, dull yellowish legs and faint wing-bar***. **Voice** Loud *kreet* in flight. **Habitat** Marshes, pools with grassy edges. **Note** Vagrant UAE, Cyprus, Turkey.

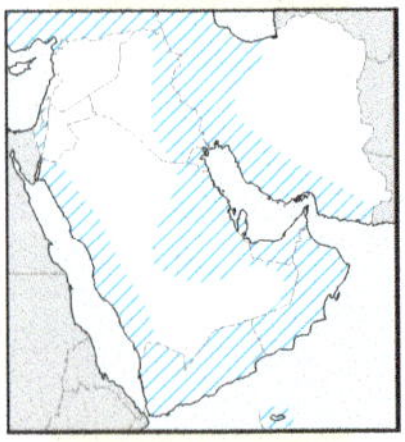

Terek Sandpiper *Xenus cinereus* PM, WV

L: 23. W: 58. Rather squat, front-heavy wader with short ***orange-yellow legs and long upcurved bill***. Grey head and upperparts ***bordered by dark shoulder bar, carpal patch and primary-line***. ***In flight, wings show white trailing edge***, much narrower than that of Common Redshank. In juvenile and winter adult, black shoulder bar is faint or absent. ***Frequently bobs rear end*** (reminiscent of Common Sandpiper) and when feeding often runs fast with head lowered. Feeds singly, but may roost in group. **Voice** Flight call clear, fluty *tjiy-tjiy* or *dwitt-dwitt*, softer than Common Redshank, sometimes recalling Ruddy Turnstone in character. **Habitat** Tidal mudflats, saltmarsh and mangrove creeks, coral reefs; scarce inland on passage. **Note** Passage hatched; winters Arabian coasts (where some summer).

Spotted Redshank
ad summer
winter
winter
ad summer
Common Redshank
winter
ad
Ruff
Buff-breasted Sandpiper
juv
Pectoral Sandpiper
winter
winter
ad
♂ summer
ad summer
winter
Terek Sandpiper

Marsh Sandpiper *Tringa stagnatilis* PM, WV

L: 23. W: 57. ***Resembles small, slim Common Greenshank with proportionally longer legs and straight, thin, fine-pointed bill. Face almost white, usually with distinctive supercilium. Flight pattern similar to Greenshank's with dark wings, contrasting white tail and wedge up back; protruding legs distinctive beyond tip of tail***. Flight action rapid and more similar to Wood Sandpiper. Winter plumage rather uniform, paler grey than Greenshank with almost white face and distinctive supercilium. In breeding plumage, becomes markedly black-spotted on head and upperparts; legs usually more yellowish. **Voice** Flight call a clear *djeeu-djeeu* often repeated, weaker and less shrill than Greenshank. **Habitat** Freshwater wetlands, tidal creeks and flats. **Note** Passage and winter hatched.

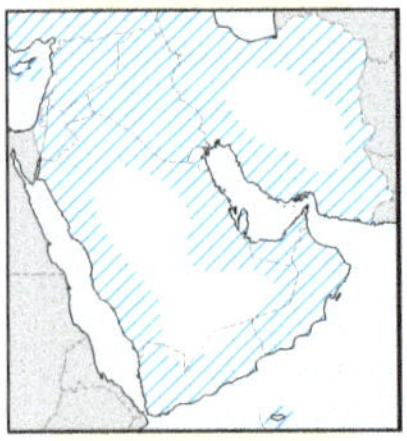

Common Greenshank *Tringa nebularia* PM, WV

L: 32. W: 69. ***Larger than Common Redshank with long greenish legs and fairly long, slightly upturned greyish-green bill***. Flight action slow and jerky with long, ***dark wings contrasting with paler head and neck, and conspicuous white tail and wedge up back***. In winter plumage resembles Marsh Sandpiper but darker grey, and ***darker face lacks pale supercilium***. Summer plumage has black feather-spotting on upperparts and distinctly spotted breast and flanks. Juvenile uniformly patterned with buff-fringed grey-brown upperparts, white underparts and streaked grey head, neck and breast. Rather active, often running when feeding in shallow water. **Voice** Flight call characteristic ***shrill trisyllabic* djiu-djiu-djiu, *with equal stress on all syllables***. Marsh Sandpiper's call is similar but thinner and less shrill. **Habitat** Coastal shores, mudflats, inland wetlands. **Note** Passage and winter hatched; some oversummer UAE.

Lesser Yellowlegs *Tringa flavipes* V

L: 23–25. ***Marginally smaller than Common Redshank and clearly more slender with long primary projection. Legs long, bright yellow***. Plumage finely spotted above, with breast diffusely streaked. In flight wing and tail pattern, especially square white rump, recalls Wood Sandpiper. Bill fine, straight and all dark. Legs may be paler yellow, even orangey, in winter or in young birds. Supercilium short, reaching only to eye (unlike Wood Sandpiper). **Voice** Clear *tew* close to that of Greenshank and Marsh Sandpiper. **Habitat** Mainly freshwater edges. **Note** Vagrant Oman, UAE.

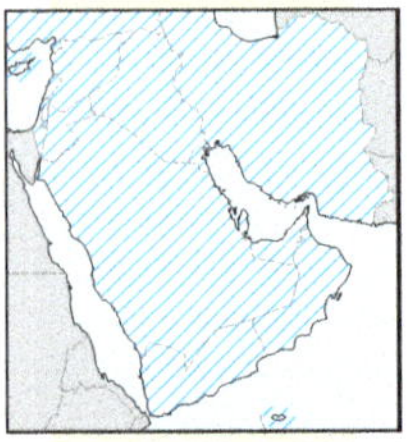

Green Sandpiper *Tringa ochropus* PM, WV

L: 23. W: 59. Distinguished by call and contrasting black-and-white plumage. Shy, often first seen when flushed, when note ***black wings above and below contrasting sharply with white belly, and white rump***. At rest blackish upperparts and breast contrast with clear white belly and flanks. Juvenile darker, more uniform and buff-spotted, head rather dark with obvious pale eye-ring and short supercilium in front of eye. Often in small flocks. **Voice** Flight call ***sharp, melodic* dlo-eed-witt-witt**, loud and far-carrying. **Habitat** Muddy streams, small pools, wadis, edges of freshwater wetlands. **Note** Passage and winter hatched.

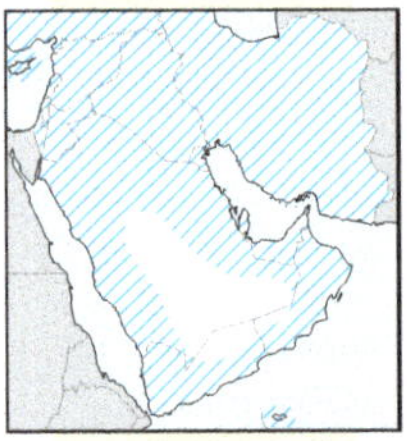

Wood Sandpiper *Tringa glareola* PM, wv

L: 20. W: 56. Resembles Green Sandpiper but ***upperparts paler, less contrasting and boldly speckled whitish and with conspicuous white supercilium***; also rather long, yellowish-green legs and more elegant appearance. In flight brownish upperparts contrast with white rump; ***differs from Green Sandpiper in whitish (not dark) underwing, longer feet-projection*** and call. Juvenile has buff-spotted upperparts, somewhat recalling young Common Redshank. Occurs singly or in small flocks. **Voice** When flushed, ***characteristic, far-carrying series* jiff-iff-iff-iff**. **Habitat** Freshwater marshes with muddy margins. **Note** Passage hatched, rare or absent in winter except south Arabia.

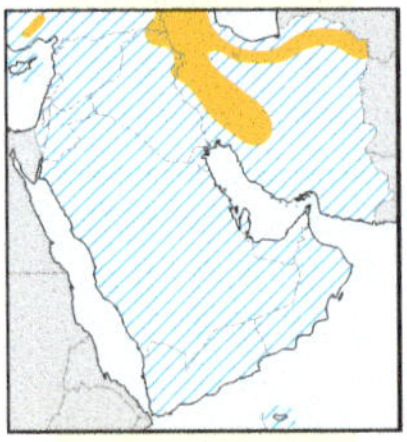

Common Sandpiper *Actitis hypoleucos* PM, WV

L: 20. W: 40. Short-legged wader with almost uniform brown upperparts, clean white underparts running up in a wedge between wing and brownish breast-sides; rather long tail projecting well beyond wing-tips at rest; note constantly bobbing rear-body. Unique flight action low over water with vibrating, shallow and stiff wingbeats, alternating with short glides. In flight shows distinct, but rather thin, white wing-bar (also obvious from below) and brown rump and tail with pale outer edges. Juvenile has wing-coverts barred buff and dark. Feeds singly, sometimes roosts in small groups. **Voice** Flight call characteristic series of piping notes, descending *hee-dee-dee-dee*. **Habitat** Winter and passage on edge of any wetland. **Note** Passage and winter hatched.

Marsh Sandpiper
ad summer
winter
ad summer
1st-winter
winter
Common Greenshank
Lesser Yellowlegs
ad summer
winter
Green Sandpiper
ad summer
ad summer
juv
Wood Sandpiper
winter
Common Sandpiper

Red-necked Stint *Calidris ruficollis* V

L: 14. W: 30. Similar to Little Stint but with slightly shorter bill and longer wings and tail, giving a more attenuated appearance, although these features are subtle. Readily identified in summer plumage (but confusable then with larger Sanderling, especially if no size comparison), Red-necked having ***neatly defined rusty-red neck and upper breast (without black streaks shown by Sanderling)***; ***white lower breast flecked with arrow-shaped spots***. From summer-plumaged Little Stint by breast pattern (lacks dark streaking though the rufous-red shown by Little), pale grey wing-coverts (rusty fringes in Little Stint) and call. Juvenile difficult to distinguish from juvenile Little Stint; most reliable features are ***call, diffuse streaking on breast-sides (which lack rufous wash shown by juvenile Little Stint) and pale greyish tertials (black in Little Stint)***. Adults in winter similar to Little Stint; best told by call. **Voice** Call short, dry *dreeet* or *dreeej*. **Habitat** Coastal flats and wetlands. **Note** Vagrant Iran, UAE.

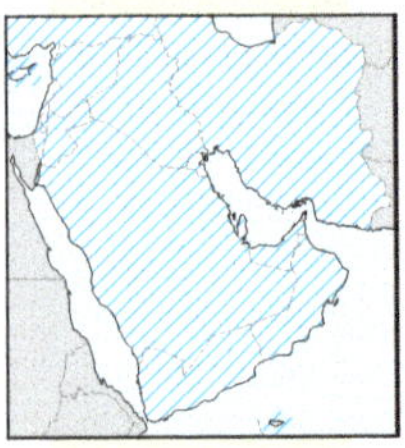

Little Stint *Calidris minuta* PM, WV

L: 13. W: 30. Abundant in region. ***Almost always has dark bill and legs***, the latter separating from Temminck's Stint. In summer plumage, colour on face, neck, breast and scapulars varies from dull orange to warm buff; but ***always shows pale 'V' on mantle and dark centre to crown***. In winter, upperparts become grey, usually with dark shaft-streaks to feathers. Juvenile, which looks very white below, has warm rufous tone to upperparts, a distinct white 'V' on mantle, diffuse greyish neck-collar and rufous, streaked breast-sides. Feeding action rapid. **Voice** Flight call a short *tip*, *chit* or trill. **Habitat** Coastal flats and inland wetlands. **Note** Passage and winter hatched; some oversummer Arabia.

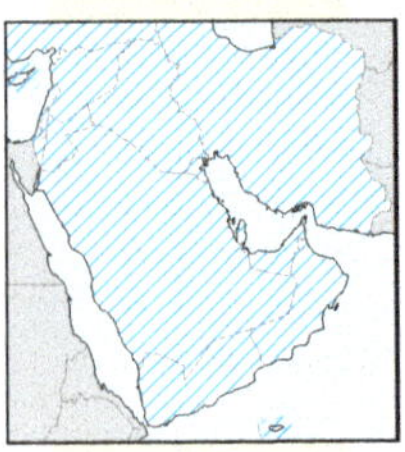

Temminck's Stint *Calidris temminckii* PM, WV

L: 13. W: 30. Similar in size to Little Stint but with more elongated body and ***shorter, yellowish-green legs***; ***white sides to tail prominent in flight***. In summer, mainly grey-buff with ***dark centres to many scapulars***, and lacking rufous-orange tones. In winter the dark scapulars are lost and then looks plain buff-grey, but note defined grey breast. Juveniles have narrow buffish fringes to scapulars and coverts with some dark markings on upper scapulars. ***Lacks white 'V' on mantle in all plumages***. Often keeps tipped forward, legs flexed. Sometimes in small groups; often towers high in erratic flight when flushed. **Voice** Flight call distinctive ringing trill, *tirrrrr*, quite unlike Little Stint. **Habitat** Inland pools, marshes, muddy coasts. **Note** Passage hatched; winters mainly S. Iran/Arabia.

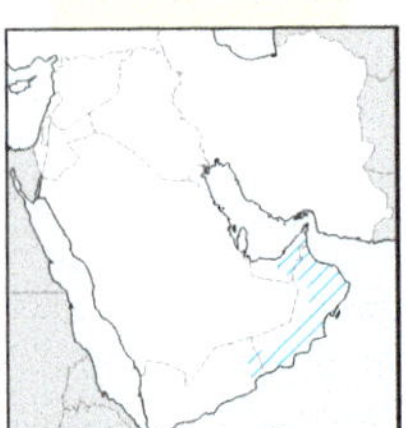

Long-toed Stint *Calidris subminuta* pm

L: 14. W: 30. Similar to Little Stint in size but with ***longer neck*** (noticeable when standing upright) and ***longer legs, which are dull yellowish or yellowish-brown***; the long toes can be difficult to see. Leg colour similar to Temminck's Stint, from which told by longer legs, more upright posture and plumage. In summer, rufous and well-streaked, with noticeable supercilium creating capped appearance. In winter, note dark feather centres to upperparts and fine streaking on head and breast, unlike the plain grey of Little and Temminck's Stints. When flushed, shows faint wing-bar and often towers high like Temminck's Stint. **Voice** Short, soft *prrt* or *tit-tit-tit*, in flight. **Habitat** Freshwater margins, coastal pools, mudflats. **Note** Passage and winter hatched, but scarce; vagrant Bahrain, Iran, Saudi Arabia, Yemen.

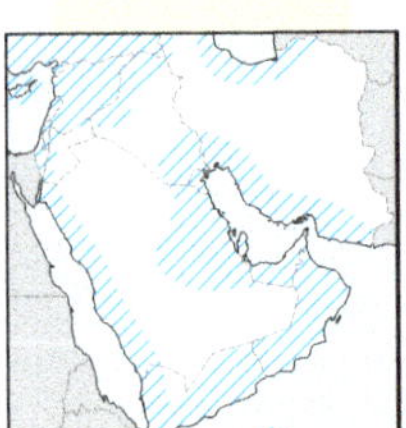

Sanderling *Calidris alba* PM, wv

L: 20. W: 40. Slightly larger than Dunlin with ***shorter, straighter bill***. In winter has ***very pale plumage with dark mark at bend of wing***. In summer and juvenile plumage more easily confusable with other small waders, particularly stints. ***Note lack of hind toe***. In summer plumage can be quite ***rusty on head and prominent breast-band, but always shows dark scalloping in the red of the breast***, a feature that helps separate from smaller, vagrant Red-necked Stint. Juveniles ***spangled black and white on upperparts***, sometimes with buff wash on breast-sides. ***White wing-bar noticeable in flight***. ***Often runs fast***, particularly on open shoreline ahead of wave-wash. **Voice** Usual flight call a loud *plit*. **Habitat** Sandy beaches, mudflats. **Note** Passage hatched, rare inland; coastal in winter.

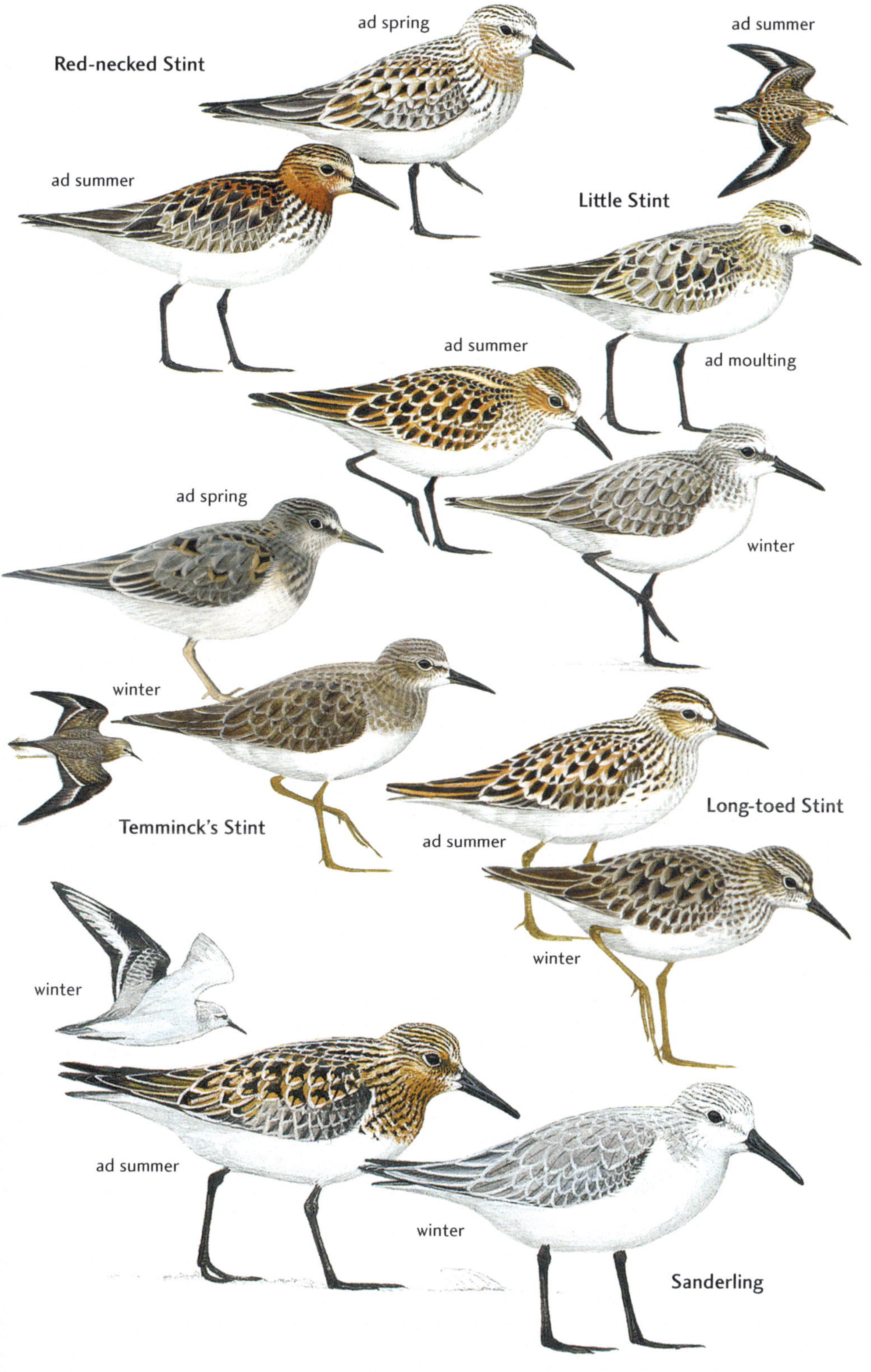
Red-necked Stint
ad spring
ad summer
ad summer
Little Stint
ad summer
ad moulting
winter
ad spring
winter
Temminck's Stint
Long-toed Stint
ad summer
winter
winter
ad summer
winter
Sanderling

Great Knot *Calidris tenuirostris* pm, wv

L: 27. W: 55. Like large Red Knot, similarly stocky build but more tapering at rear. In winter plumage told from Red Knot by ***longer, slightly more decurved bill with heavier base, larger greyish spots on underparts, less defined supercilium (more extensive greyish lores) and more obvious tail pattern (white rump and uppertail-coverts contrasting with dark upperparts and tail)***; upperparts darker than Red Knot with dark streaking to centres of grey feathers. In breeding plumage easily told by ***dense blackish spotting on breast*** and flanks, dark streaks on mantle and hindneck, and ***chestnut centres to scapulars***. Juvenile also told from similar Red Knot by ***heavily marked breast, contrasting with pale belly, and pale-fringed wing-coverts with dark shaft-streaks***. In first-winter some juvenile coverts often retained, helping identification. **Voice** Usually silent, occasionally a soft *prrt*. **Habitat** Coastal mudflats. **Note** Passage and winter hatched, but often rare; vagrant Bahrain, Qatar, Saudi Arabia, Yemen.

Red Knot *Calidris canutus* V

L: 24. W: 50. In winter told by combination of stocky build (noticeably larger and plumper than Dunlin), pale grey upperparts and straight, rather stout bill (about length of head); in flight, rather long winged, rump greyish-white and tail grey. Legs grey-green. In summer plumage, brick red below with black, white, grey and buff mottling above; told from summer Curlew Sandpiper by shorter straight bill, larger size and tail pattern. Juvenile has poorly marked breast (unlike Great Knot); wing-coverts have dark subterminal markings and pale tips, but no dark shaft-streaks as in Great Knot. **Voice** Short nasal *wut* or *wut wut*, rather quiet. **Habitat** Coastal mudflats. **Note** Winter hatched; widespread vagrant Iran, Iraq, Near East and Arabia.

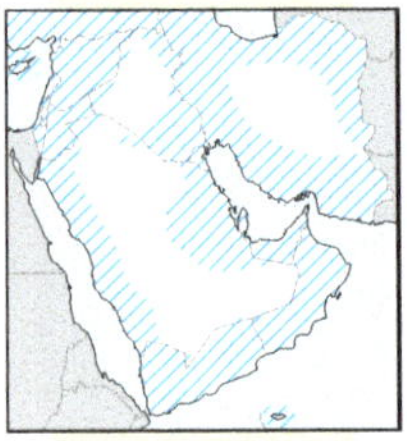

Curlew Sandpiper *Calidris ferruginea* PM, WV

L: 19. W: 40. Told from Dunlin in all plumages by ***white rump, longer, more decurved bill and longer legs***, giving a more elegant appearance. Easily told in breeding plumage by ***chestnut-red face and underparts*** (often with white feather-fringes). In winter plain grey above, white below with a light suffusion to breast-sides and ***noticeable white supercilium***; then also told from Dunlin by cleaner appearance with whiter underparts. In first-autumn has rather scaly grey-brown upperparts, noticeable white supercilium and yellowish-buff wash to breast. Usually in small flocks. **Voice** Trilling, *trururip* or almost disyllabic *churrip* in flight. **Habitat** As Dunlin. **Note** Passage hatched, winters mainly coastal Arabia (some oversummering).

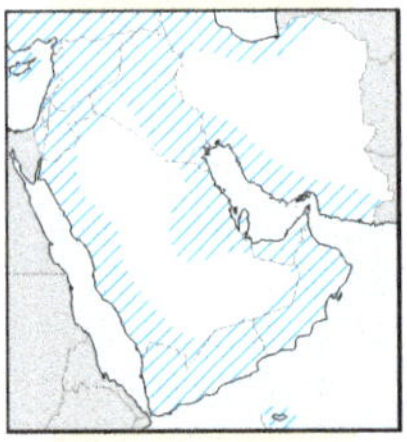

Dunlin *Calidris alpine* PM, WV

L: 18. W. 38. Larger than the stints with ***longer, slightly downcurved bill***. In summer plumage easily told by ***black belly-patch***. In winter, this patch is lost as are rufous tones to upperparts, then has grey upperparts with narrow pale fringes to coverts and scapulars, grey breast and white belly. From similar Curlew Sandpiper by different rump pattern (***white with dark centre***), lack of white supercilium, darker upperparts and breast, and shorter bill and legs. In first-autumn note chestnut on coverts, white 'V' on mantle, and lines of dark spotting on flanks below finely streaked breast. **Voice** Reedy *kreep* in flight distinctive. **Habitat** Coastal mudflats; also inland wetlands. **Note** Passage and winter hatched; some summer south to Arabia.

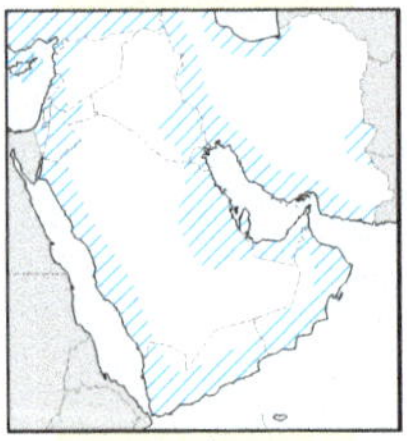

Broad-billed Sandpiper *Limicola falcinellus* pm, wv

L: 17. W. 35. Slightly smaller than Dunlin, from which told by ***longer, broad-based bill with downward droop near tip*** (head-on, tip also seen to be swollen), ***shorter yellowish-grey legs*** and ***double supercilium***. In summer upperparts rather dark with white 'V' on mantle, white line on scapulars and ***white underparts with dark-spotted and streaked breast and flanks***. In winter, greyer and supercilium less obvious; then shape and length of bill, leg colour and, if present, dark area on carpals important for identification. In first-autumn resembles adult but streaking below finer and confined to breast. Often has slow-moving and crouching feeding action. In rather erratic flight appears small, heavy-fronted and, in breeding plumage, dark with thin wing-bar. **Voice** Flight call rather weak with a dry, slightly buzzing character, ascending a little at end, *brlliid*. **Habitat** As Dunlin. **Note** Passage hatched; winters mainly coastal Arabia.

winter
ad summer
Great Knot
winter
ad summer
Red Knot
♂ summer
winter
Curlew Sandpiper
winter
♀ summer
ad summer
winter
Dunlin
ad summer
winter
Broad-billed Sandpiper

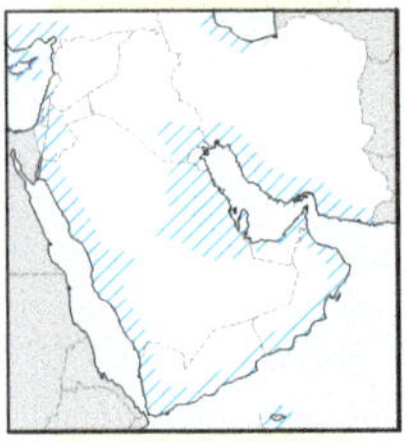

Ruddy Turnstone *Arenaria interpres* PM, WV

L: 23. W: 53. Stocky, ***short-billed and short-legged wader***; distinguished in breeding plumage by ***striking head and breast markings, chestnut shoulders and wing-coverts, and bright orange legs***. Flight is strong and direct and shows distinctive pied and chestnut appearance, ***black breast is a striking feature***. In winter and juvenile plumage, head and neck are much darker, and usually lack chestnut; also juvenile shows bold buffish fringes on most coverts. Walks with a rolling gait, and turns or flips over stones and seaweed for food items. **Voice** Calls short and staccato; when feeding low *chuk*; in flight *krytt-te- krytt-te-krytt* or *kritt-it-it*. **Habitat** Rocky or sandy coasts; rarely inland lakes. **Note** Passage and winter hatched; occasional inland. Some remain in summer.

Wilson's Phalarope *Steganopus tricolour* V

L: 23. W: 41. Resembles Marsh Sandpiper more than the other two phalaropes, but has shorter, brighter legs, ***long needle-like black bill, and white rump*** (not white wedge extending up back). Feeds as much on land as in water and very active. In flight shows ***white rump, plain greyish upperwing without wing-bar*** and feet projecting beyond tip of tail. In winter plumage mainly ***grey with white underparts and yellow legs***. In breeding plumage, albeit unlikely to be seen in region, has a colourful and striking head and neck pattern, especially in female. **Habitat** Fresh and brackish waters. **Note** Vagrant Oman, UAE.

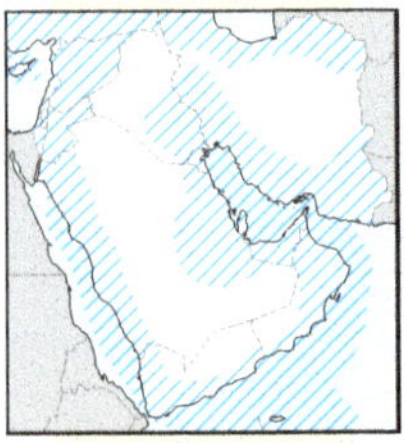

Red-necked Phalarope *Phalaropus lobatus* PM, WV

L: 18. W: 36. Small, elegant wader ***which swims*** high on water, often spinning to whirl up food items, which will be taken with ***needle-thin, black bill***. In breeding plumage has striking head and neck pattern. In winter plumage upperparts are pale grey with ***black mask through eye and black on hindcrown*** conspicuous. Juvenile has similar head pattern, but ***upperparts are dark brown with prominent ochre bands***. Flight fast and jerky, wings show distinct white wing-bar; ***often suddenly settling on water***. Separated from Grey Phalarope in all plumages by smaller size and very thin black bill; in winter Grey Phalarope is also paler and more uniform above. Gregarious; often in large flocks at sea. **Voice** Flight call short, sharp *kritt* or *kitt* recalling Sanderling, but quieter and finer. **Habitat** Maritime; inland lakes on passage. **Note** Passage hatched, winters Arabian Sea.

Grey Phalarope *Phalaropus fulicarius* V

L: 21. W: 42. Slightly larger and more robust than Red-necked Phalarope, with shorter, thicker neck; ***bill distinctly thicker, less pointed and sometimes pale (yellowish) at base***. Red breeding plumage with black-and-white head pattern is not likely to be seen in the region. Winter plumage recalls a very small gull when feeding on water with ***unmarked grey upperparts*** (paler, more uniform than Red-necked). First-winter similar to adult, but black on hindneck sometimes remains in winter. Behaviour and flight pattern similar to Red-necked Phalarope, but wingbeats less jerky. **Voice** Flight call a sharp *pik*. **Habitat** Essentially maritime, but vagrants may visit lakes, pools or puddles, large or small. **Note** Widespread vagrant in Middle East. [Alt: Red Phalarope]

Ruddy Turnstone
winter
ad summer
winter
Wilson's Phalarope
1st-winter
juv/1st-winter
Red-necked Phalarope
winter
juv
♀ summer
winter
winter
Grey Phalarope
winter
♀ summer

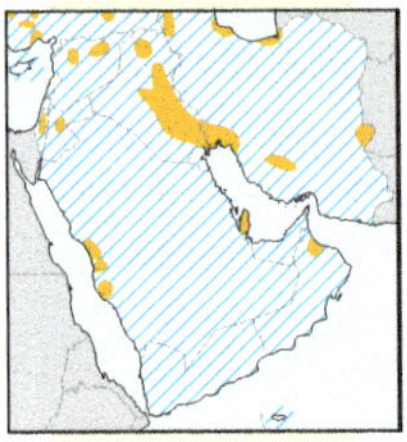

Collared Pratincole *Glareola pratincola* PM

L: 25. W: 63. Distinctive, aerial wader, resembling marsh terns, having ***graceful fast flight, long pointed wings, deeply forked tail and short bill***. Usually seen in loose flocks chasing winged insects. On ground, plover-like with quick tripping actions on short legs, often with upright stance, head held high. ***Adult has creamy-buff throat bordered black***; tail and wing-tips are equal. In flight, uniform dark olive-brown above (slightly darker flight feathers) with ***narrow, but distinctly white trailing edge to secondaries***, contrastingly, white rump and belly; ***underwing-coverts reddish-brown***, but often look shadowy black. ***Note: Some, juveniles/worn adults may lack or have only thin white trailing edge to wings***. (Black-winged Pratincole has generally darker upperparts, lacks white trailing edge to wings, and has black underwing-coverts.) Adult non-breeding and juvenile lack distinct black throat line, and juvenile's brown feathers have pale tips and fringes; outer tail feathers are shorter than in adult. **Voice** Most characteristic call a tern-like, sharp, chattering *kikki-kirrik* and a short *check* or *che-keck*. **Habitat** Sun-baked mudflats and flat, firm plains with low vegetation, grasslands; often near water. Nests colonially. **Note** Has bred UAE; passage hatched, few winter south Arabia.

Oriental Pratincole *Glareola maldivarum* V

L: 23. W: 63. Difficult to identify. Has the distinctive features of, and resembles, Collared and Black-winged Pratincoles closely. Colour of upperparts and ***lack of white trailing edge to secondaries (a thin pale fringe may show in fresh plumaged adult or juveniles) are similar to Black-winged, but shares reddish-brown underwing-coverts with Collared Pratincole***. Told from both in adult plumage by ***obviously shorter tail, the tip of which falls short of wing-tip***, being best judged when settled, but shortness of tail certainly noticeable in flight. Adult has underside of outermost tail feather with restricted black tip (distal half black in Collared). Combination of features as given here may sometimes be required for conclusive identification; beware active or suspended moult, or damaged tail, in Collared Pratincole may produce apparently 'short-tailed' individuals. **Habitat** As Collared Pratincole. **Note** Vagrant Iran, UAE.

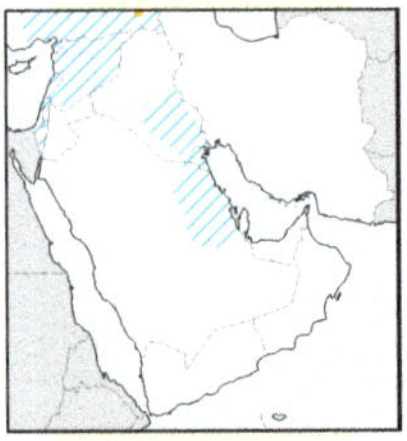

Black-winged Pratincole *Glareola nordmanni* V

L: 24. W: 64. Structure and behaviour as Collared Pratincole and often difficult to distinguish between the two. Adult slightly darker above than Collared, with no contrast between flight feathers and coverts, ***lacks white trailing edge to secondaries, and underwing-coverts are jet-black*** (although these coverts are reddish-brown in Collared Pratincole this can be hard to assess – and they often appear dark); tail fork of Black-winged shallower. When perched, wings project beyond tail tip (in Collared approximately equal), and shows less red at base of bill. Non-breeding adult and juvenile resemble corresponding Collared Pratincole but identifiable in flight by wing pattern. **Voice** Call resembles Collared's, though a little sharper, and in breeding area alarm call a shorter *pwik* or *pwik-kik-kik*. **Habitat** Much as Collared, but favours steppe. **Note** Passage hatched, but rare; vagrant Bahrain, Iran, Oman, Qatar, UAE, Yemen.

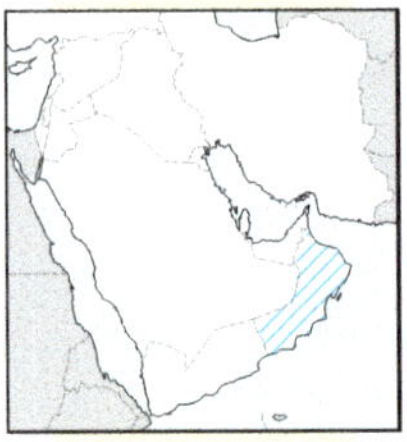

Small Pratincole *Glareola lactea* V

L: 17. W: 45. Small, ***swallow-like pratincole***. In flight, appears sandy-grey, black and white with diagnostic wing pattern: striking, ***broad white secondaries with narrow black trailing edge and black primaries; underwing black with striking, broad white panel on secondaries and inner primaries***; slightly forked black-and-white tail. Flight is swift, bouncy and rather swallow-like, on long and pointed wings. Juvenile resembles adult. **Voice** Harsh notes recall Black Tern, also a Little Tern-like *tuck-tuck-tuck* or *ke-terrick-ke-terrick*; birds in feeding flocks have high-pitched rolled *prrip*. **Habitat** Dry sandy or muddy areas adjacent to wetlands, over which it hawks for insects. **Note** Passage hatched, but rare; vagrant Bahrain, Iran, UAE, Yemen.

Collared Pratincole
ad
juv
ad
Oriental Pratincole
ad
Black-winged Pratincole
ad
imm
ad
Small Pratincole
ad

PLATE 41 : DARK-HOODED GULLS

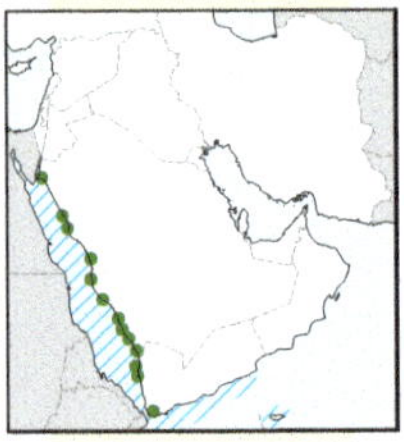

White-eyed Gull *Larus leucophthalmus* V

L: 40. W: 108. From heavier Sooty Gull by ***long and slender all-dark bill*** (dark red with black tip in adult; blackish in immature), which droops at the tip (bill stouter and bicoloured in Sooty); ***hood and bib black*** (dark brown in Sooty), ***upperparts greyer, with conspicuous white eye-ring at all ages*** (faint and narrow in Sooty). Immature birds similar to Sooty Gull though upperpart feathering less conspicuously pale-fringed. At distance can be mistaken for a skua. **Voice** As Sooty Gull but less harsh and deep. **Habitat** Coastal; nests colonially on low-lying islands. **Note** Dispersal throughout hatched area; vagrant Iran, Oman, UAE.

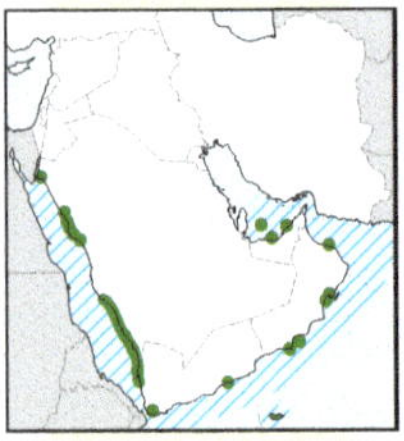

Sooty Gull *Larus hemprichii* MB, PM, wv

L: 44. W: 112. Told from similar White-eyed Gull by ***dark brown hood and bib*** and ***thick, straight bill, which is bicoloured (at all ages)*** – in adult being yellow-green with black band behind red tip; immature pale blue-grey with black tip. ***Narrow white crescent above eye*** at all ages (seldom also below). Immature brownish-grey on head, breast and flanks; upperparts brownish with buff feather-edges. Rather long-winged, relatively short-tailed and front-heavy but with steady flight; underwing dark. At distance over sea can be mistaken for skua; often piratical on other gulls and terns. **Voice** Loud mewing *kaarr* or *keee-aaar*; also high-pitched *kee-kee-kee*. **Habitat** Coastal; often near ports and fishing villages; nests on islands or cliffs, mostly colonial. **Note** Dispersal throughout hatched area; vagrant Bahrain.

Mediterranean Gull *Larus melanocephalus* V

L: 40. W: 106. Size between Common Black-headed and Common Gulls. Breeding plumage adult from former by ***all-white primaries, longer thicker bill, paler grey upperparts*** and ***more extensive black hood*** (not dark brown); in winter head mainly white with ***pronounced dark patch behind eye***. Juvenile and first-winter told from similar-aged Common Gull by ***darker, longer, heavier, slightly drooping bill, blacker outer and paler inner primaries*** and ***pale mid-wing panel***. ***Fully spread dark outer primaries show white subterminal spots*** (absent in Common Gull, which lacks dark patch behind eye and white eye-crescents). In first-winter ***back and scapulars much paler than Common Gull***. Subadults have black subterminal spots on wing-tip. **Voice** Deep, nasal, far-carrying *ga-u-a* with middle note highest. **Habitat** Coastal; nests on islands in fresh and saltwater lakes. **Note** Passage and winter hatched; vagrant Kuwait, Saudi Arabia (Gulf), UAE.

Franklin's Gull *Larus pipixcan* V

L: 32–36. W: 81–93. Small and compact, size of Common Black-headed Gull, but with no real confusion species in region, though could perhaps be mistaken at rest for Sabine's Gull. Adult has dark grey upperparts, ***white tips to primaries, black subterminal tips to outer primaries and white band separating the black subterminal tips from the grey primary bases, red legs and bill, black hood (in summer) and broken white spectacles***. Immature with dark rear crown and ear-coverts (as winter adult), dark bill, dark grey mantle, and white spectacle visible behind eye. White-eyed Gull and Sooty Gull are both larger with some grey on underparts, at least on neck-sides, in all plumages (Franklin's all white below). **Voice** Unlikely to be heard in region; soft *krruk* or shrill *guk*. **Habitat** Coastal. **Note** Vagrant UAE.

Little Gull *Hydrocoloeus minutus* V

L: 26. W: 63. ***Much smaller than Common Black-headed Gull, with tern-like flight***; wing-tips fairly rounded though more pointed in juvenile. Adult told by ***blackish underwing with conspicuous broad white trailing edge***, unmarked pale grey upperwings and, ***in summer, by extensive black hood*** (less extensive, dark brown in Common Black-headed). In first-winter, underwing pale and ***upperwing has blackish diagonal band***, recalling similar, much larger Kittiwake, but ***top of head sooty-grey*** and ***secondaries have narrow dark band*** (Kittiwake has white crown, black band on hindneck and white secondaries). Attains partly black hood in first-summer but has remains of diagonal band; subadult has black subterminal spots on wing-tip. Feeds largely by dipping to water surface. **Voice** Hard, rather sneezed, marsh tern-like *kjeck-kjeck*. **Habitat** Coastal and inland waters. **Note** Passage and winter hatched; vagrant Bahrain, Kuwait, Saudi Arabia, UAE.

White-eyed Gull
ad
1st-winter
ad
juv
Sooty Gull
ad
1st-winter
ad
juv
Mediterranean Gull
ad summer
juv
ad winter
1st-winter
1st-winter
Franklin's Gull
ad winter
juv
ad winter
ad summer
ad summer
Little Gull

Brown-headed Gull *Chroicocephalus brunnicephalus* V

L: 43. W: 100. Slightly larger and heavier than Common Black-headed Gull, with stronger bill and pale eyes. Told by ***broad black wing-tip breaking white leading edge above, dark grey leading primaries below*** and ***small white mirror near wing-tip of both surfaces***. In summer, hood paler brown than Common Black-headed with clearer black rim at rear; in winter, head pattern much like Common Black-headed; but ***iris yellow***. First-winter birds have similar wing pattern, but ***broader black wing-tip without white mirror; also bold blackish trailing edge to wings***. Iris dark in juvenile but becomes pale during first winter or spring, when most of inner wing and tail moulted to adult pattern. **Voice** Deep *grarhh*. **Habitat** Coastal, lakes. **Note** Vagrant Iran, Oman, UAE.

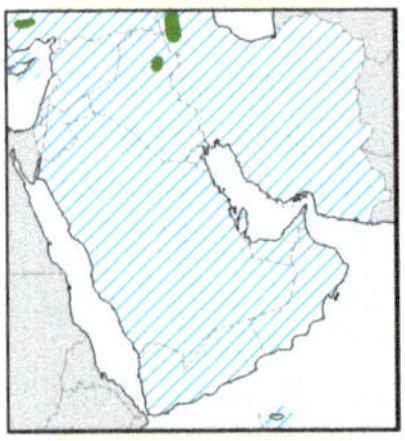

Common Black-headed Gull *Chroicocephalus ridibundus* WV, PM

L: 38. W: 93. Medium-small gull, told (except from Slender-billed Gull) by ***broad white leading edge of primaries, contrasting below with dark grey remaining primaries***. In summer, has ***dark brown hood*** (looking blackish except close to); in winter, head white with black spot on ear-coverts and, often, vague bar from eye over crown and from ear-coverts over nape. Juvenile and first-winter have adult-type primary pattern; ***told from similar Slender-billed Gull by slightly smaller size, shorter bill with darker tip, shorter neck, less bulky breast, more distinct markings on head and always dark eye***. Flight light and buoyant. **Voice** Harsh *krreeea*. **Habitat** Coastal and inland waters; nests colonially. **Note** Passage and winter hatched; a few summer S. Arabia.

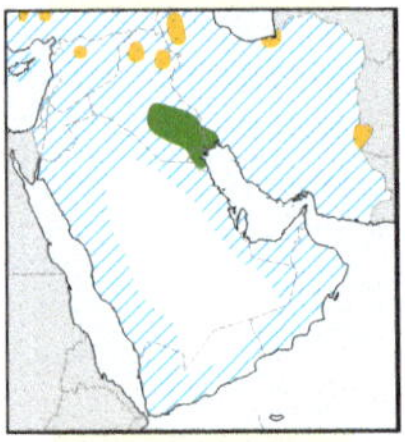

Slender-billed Gull *Chroicocephalus genei* PM, WV, sv

L: 43. W: 100. Larger than Common Black-headed Gull, ***bill distinctly longer, forehead rather sloping*** with long feathering at base of upper mandible accentuating length of bill. Wing pattern similar to Common Black-headed, but ***head completely white in summer, when breast often has rosy tinge***. Legs longer; when alert ***appears curiously long-necked***. In winter, often shows ***small greyish spot on ear-coverts*** (larger, blacker in Black-headed), ***but no vague bar over crown or nape*** (as shown by many Black-headed Gulls); ***iris pale in winter*** but dark in many breeding birds (always dark in Black-headed). ***Dark red bill often appears blackish at distance***. Juvenile and first-winter also told from Common Black-headed by pale bill with poorly marked, or unmarked tip (tip always dark in Black-headed) and different head pattern (same as winter adult); pale iris starts to show during first winter. ***In flight rather hunch-backed, with more protruding neck bulging downwards,*** useful characters with experience. **Voice** Hoarse nasal *yaarr*, deeper and harsher than Common Black-headed, can recall squeal of distant Water Rail; also abrupt cheery *yap* or *yirp*. **Habitat** Coastal and inland waters; nests colonially. **Note** Mainly summer visitor to breeding sites; passage and winter hatched.

Sabine's Gull *Xema sabini* V

L: 34. W: 89. Smaller than Common Black-headed Gull with shallowly forked tail and ***diagnostic tri-coloured upperwing-pattern***. In summer, adult has ***slate-grey hood***, in winter just blackish nape-patch; ***short black bill with yellow tip***. Juvenile separated from similar Black-legged Kittiwake ***by continuous brownish from crown and nape to mantle*** (also breast-sides), ***brownish forewing***, dusky bar on inner underwing (visible at close range), ***pale fleshy-grey legs*** and more forked tail (young Black-legged Kittiwake has black band at base of white hindneck, black band on inner wing, white underwing and black legs). Flight light; often swoops to water surface for food. **Habitat** Maritime, occasionally driven onshore. **Note** Vagrant Iran, UAE.

Black-legged Kittiwake *Rissa tridactyla* V

L: 40. W: 95. Slightly larger than Common Black-headed Gull. Adult resembles Common Gull but has ***all-black wing-tip without white spots***, darker grey back and inner wing but ***whitish, translucent outer wing and black legs***. First-winter recalls much smaller Little Gull, but head white (just small black spot behind eye) with ***bold, black band on hindneck, white secondaries and slightly forked tail***. First-summer birds often lack band on hindneck and sometimes lack bar on inner wing. **Habitat** Coastal and off-shore waters. **Note** Scarce in winter as hatched, also Aqaba, vagrant Iran, Oman, UAE.

1st-winter
ad winter
Common Black-headed Gull
1st-winter
ad winter
ad winter
ad summer
Brown-headed Gull
ad winter
ad summer
1st-winter
Slender-billed Gull
1st-winter
Sabine's Gull
ad
ad
1st-winter
Black-legged Kittiwake
ad winter

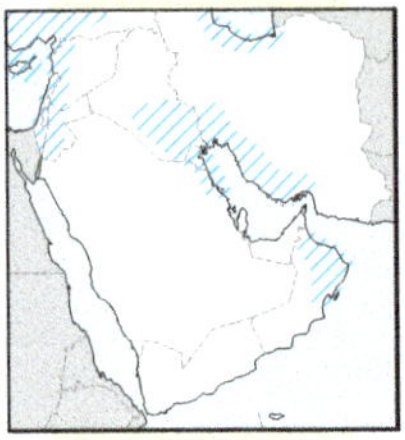

Common Gull *Larus canus* V

L: 43. W: 115. Recalls ***small Caspian Gull, but wings narrower and bill distinctly smaller, thinner and greenish-yellow without red spot***. Black wing-tip with prominent white spots, larger area than in Caspian. In winter, bill has narrow black band near tip; eye dark. Juvenile resembles Mediterranean Gull, but with greenish bill, mottled breast, brown markings on underwing and less contrasting upperwing pattern. ***In first-winter, back blue-grey***, upperwing pattern less contrasting and bill bicoloured. **Voice** Alarm call *klee-uu* emphasis on last high-pitched note. **Habitat** Coastal and inland waters; sometimes fields. **Note** Passage and winter hatched, rare in south; vagrant Bahrain, UAE. [Alt: Mew Gull]

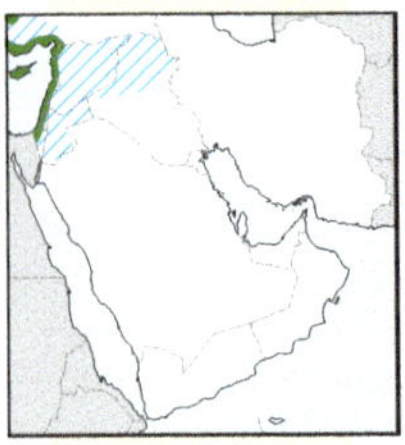

Yellow-legged Gull *Larus michahellis* No records

L: 52–58. W: 120–140. Adult similar to W. European Herring Gull ***L. argentatus***, with slightly darker grey mantle, ***yellow legs***, squarer head-shape with flattish crown, stouter bill and longer wings. ***Smaller white mirror on P10 (outermost primary), also usually P9. Greater area of black on outer primaries***. First-winter has paler head, with 'shadow' behind eye, pale underparts, and contrastingly pale back against dark wings in flight with translucent inner primaries. Immatures, with pinkish legs, possess ***neat tail-band on white tail***, often narrower on outer feathers. **Voice** Variable; deep *aar* or *wah* (much as *L. fuscus*); also high, disyllabic *kyee-arrr*. **Habitat** Coastal and inland waters, fields. **Note** Passage and winter hatched.

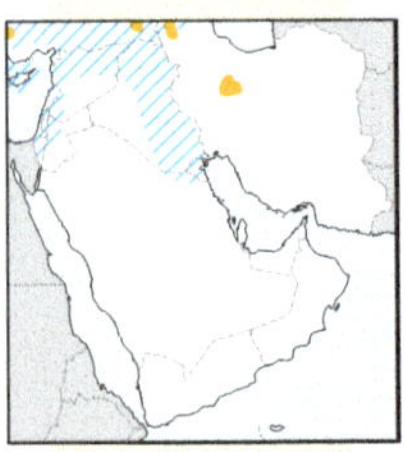

Armenian Gull *Larus armenicus* No records

L: 52–60. W: 120–140. Bill typically robust and blunt-ended with large red gonys-spot and variable, broad, black subterminal band (may be absent). Adult averages darker than Yellow-legged Gull, with orange-yellow legs (duller in winter); in winter, sparse streaking on hindneck. In flight, upperwing shows large, squared-off black primary area more extensive than Yellow-legged Gull (mirror on outermost primary P10, and sometimes small one on P9). Juvenile/imms. told by structure and call. **Voice** Distinctive, almost cat-like miaowing call. **Habitat** Coastal and inland waters. **Note** Passage and winter hatched.

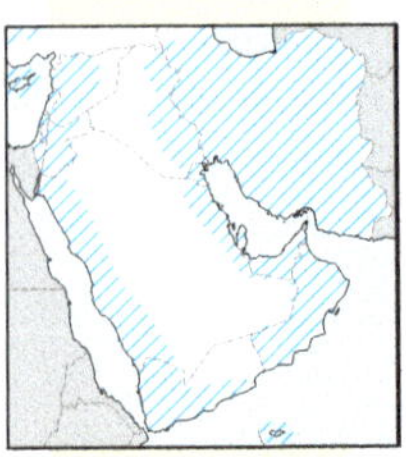

Caspian Gull *Larus cachinnans* wv, pm

L: 60–65. W: 125–150. Long-winged, long-legged, with ***long, parallel-sided bill and upright stance. Neat looking with characteristic blank or innocent facial expression, gently sloping forehead, rounded head and often a dark, bullet-hole eye***. Adult paler than all likely congeners (except Great Black-headed Gull). ***More white in wing-tip than congeners*** with large white tongue on inner web of P10 (outermost primary) and often with large all-white tip; some eastern populations have small amount of black within the large white wing-tip. Grey tongues protrude into more restricted black on primaries. Legs pale greyish-pink/straw; bill with a greenish tone to the base. Juvenile quickly moults to first-winter, a beautiful gull with a clean white head, nestled in 'shawl' around hindneck, and white underparts; pale tips to median and greater coverts form pale borders to finely patterned greater coverts; underwing normally white. Subtle pale line from eye to eye along nape/crown boundary, with 'shadow' behind eye. First-winter Heuglin's Gull can look similar to Caspian Gull. **Voice** Nasal and ringing, likened to a donkey braying. **Habitat** Coastal; inland waters. **Note** Passage and winter hatched; least numerous of larger gulls in much of Arabia.

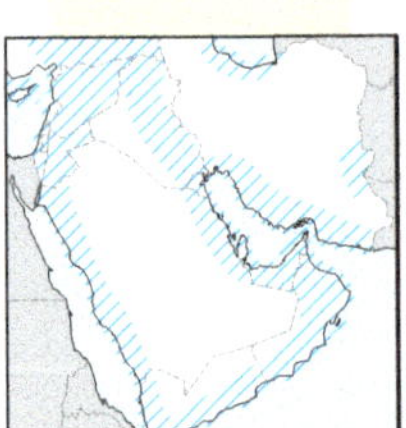

Great Black-headed Gull *Larus ichthyaetus* WV, pm

L: 68. W: 158. Very large; adult in summer unmistakable. In winter, has large dusky patch behind eye; ***white eye-crescents present in all ages. At rest appears deep-chested, with long sloping forehead accentuating length of bill***. Readily identified in flight even at extreme range by extensive pale-based primaries (upperwing thus tri-colour). Juvenile and first-winter from other young large gulls by ***unmarked white rump and tail with clear-cut band at tip***, unmarked white underparts (though juvenile has grey-mottled breast-band or patches at sides, lost in first-winter), white underwing with extensive black wing-tips and often black-tipped coverts forming underwing lines, ***pale mid-wing panel above and head pattern***. First-winter told from second-winter Caspian Gull by size, ***sharp tail-band***, head shape and pattern, dark-mottled hindneck, darker inner primaries and longer bill. Second-winter still shows fairly distinctive tail-band. **Voice** Loud and deep but hoarse, strangulated *kra-ah*. **Habitat** Coastal, mostly rare inland. **Note** Passage and winter hatched. [Alt: Pallas's Gull]

Common Gull
ad winter
ad
1st-winter
Yellow-legged Gull
ad summer
ad winter
1st-winter
Armenian Gull
ad summer
ad winter
1st-winter
Caspian Gull
ad summer
ad winter
1st-winter
ad winter
1st-year
ad summer
ad winter
Great Black-headed Gull

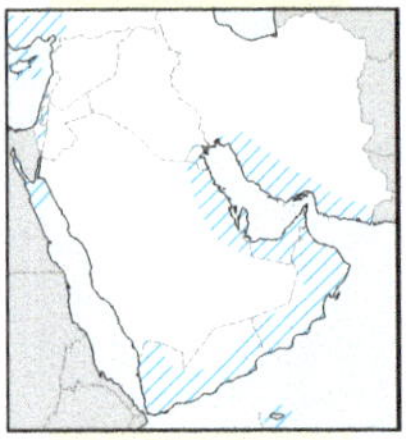

Heuglin's Gull *Larus heuglini* WV, PM

L: 58–65. W: 125–150. Large, with upright stance; recalling Steppe Gull but with fiercer face; ***one, rarely two white mirrors*** on wing-tip. Bold streaking on head and neck in winter to early April (to February in Steppe Gull). Juvenile can appear similar to Baltic Gull. First-winter/first-summer can appear Caspian Gull-like, but often has bicoloured bill, patterned greater coverts, more heavily marked scapulars, and less translucency on inner primaries. **Voice** Deep, nasal *gagaga*; usually silent in winter. **Habitat** Coast and lakes. **Note** Passage and winter hatched; rare in north in winter; locally common UAE. [Alt: Siberian Gull]

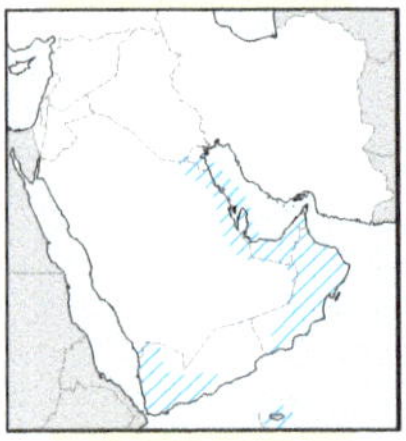

Steppe Gull *Larus barabensis* WV, PM

L: 55–65. W: 125–150. Slightly smaller than Caspian Gull with more rounded head, flattish forehead, medium/long bill, shorter legs and brighter bare parts; ***smallish, usually dark eye*** – birds that winter in Arabian Gulf often have pale eyes. Usually darker grey than Caspian, slightly paler than Heuglin's Gull. ***Grey tongues protrude into black wing-tip less than Caspian, but more than Heuglin's; mirrors on P9 and P10 (longest primaries). Underside of flight feathers grey against white coverts*** (unlike most Caspian but similar to Armenian, Yellow-legged and Heuglin's); legs generally deeper yellow in breeding condition, but some fleshy. Adult plumage and shape (except head) close to Armenian, with more slender, less blunt-tipped bill, and much less black in wing-tip. May show bill-band. Winter adult white-headed with weak streaks on hindneck. First-summer has mantle, scapulars and many coverts of adult type. **Voice** Gulping *yah-aah-aah-aah* and thin, high, drawn-out *peeeer* in winter. **Habitat** Coastal in winter. **Note** Known passage and winter range hatched; abundant UAE.

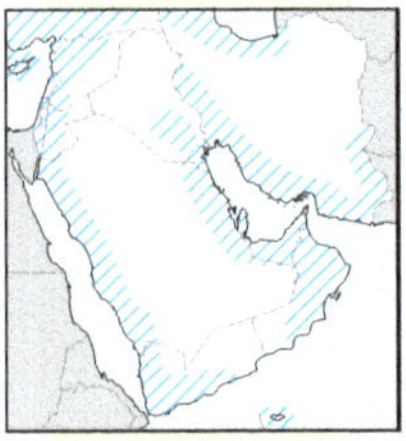

Baltic Gull *Larus fuscus fuscus* pm, wv

L: 53. W: 127. Smallest of the group, a slim, very long-winged gull. ***Adult with jet-black wings and mantle, though can wear greyer;*** rather sparse streaking on head and neck in winter; ***bright yellow legs and one, rarely two, small, white mirrors*** on wing-tip. In flight, wings noticeably long (also evident in settled bird) and slender. ***Little or no contrast in black tone in upperside of primaries***. Flying first-winter shows ***evenly coloured dark brown 'hand', secondaries and greater coverts without translucent inner primaries***, unlike Herring, Caspian, Armenian and Yellow-legged Gulls; tertials dark brown, narrowly fringed and tipped buffish-white; tail and rump white, tail-base boldly spotted or barred black and with broad, blackish terminal band. ***Underwing-coverts dark brown in first-winter and usually lack distinctive pattern*** (unlike Caspian and Armenian in which plain white). As immature plumage progresses, head and underparts turn whitish, and from spring of second calendar year blackish feathers are visible on shoulder, back and wing-coverts. **Voice** Calls deep and slightly nasal. **Habitat** Coastal; less frequently inland waters. **Note** Passage hatched; scarce or absent in winter. [Alt: Lesser Black-backed Gull]

Heuglin's Gull

Steppe Gull

Baltic Gull

PLATE 45: LARGE WHITE-HEADED GULLS – COMPARISON PLATE

The plate opposite shows the plumage progression for a typical 'large white-headed gull' (Yellow-legged Gull) from juvenile through first-winter and on to immediately pre-adult; also, for comparative purposes, all 'large white-headed gulls' occurring in the Middle East.

The table below provides comparisons of the adult plumage characteristics, bare part colouration and moult timings of the 'large white-headed gulls', as well as giving additional tips on identification. Care is needed in assessing the upperpart grey tone of adults as shade changes with light and angle.

Species	Status	Back Kodak greyscale	Leg colour	Bill	Wing-tip (best seen in flight)	Helpful tips
Herring (*argentatus*)	Vagrant	4.5–7.5 (palest grey)	pink		Least black except for Caspian. Mirror P10 usually merged with white tip. Mirror P9	Shorter-winged than the others, more dumpy
Caspian	Passage winter	4.5–6.5	pale greyish-pink/straw	Long, parallel-sided, often greenish at base.	Upper: narrow grey tongues into black, Mirror P10 merges with white tip, smaller mirror P9. Under: large white tongue on inner web of P10	Kindly face. Long, spindly legs. Long, parallel-sided wings. Dark bullet-hole eye.
Yellow-legged *michahellis*	Resident, passage winter	5–7	yellow	Strong billed.	More black than Herring and Caspian, less than Armenian. No tongues. Mirror P10 and often P9	Squarish head, strong bill.
Steppe	Passage winter	7–8.5	yellow	Often warm, bright yellow.	Between Caspian and Heuglin's, smaller grey tongues than Caspian, subterminal mirror P10, smaller P9. Smaller tongue on underside of P10.	Kindly face. Often dark, bullet-hole eye. Loses head-streaks by mid-Feb. c.f Heuglin's.
Armenian	Resident, passage winter	7–8.5	yellow	Robust, blunt-ended; often broad black sub-terminal band.	Extensive black wing-tip lacking grey tongues. Normally mirror on P10 only	Closest to Yellow-legged
Heuglin's	Passage winter	8–11	yellow		Some contrast with rest of wing, small mirror P10 and usually P9	Fiercer face. May keep head streaks until early April. c.f. Steppe.
Lesser Black-backed (*intermedius*)	Not recorded	8–13	yellow		Same as Heuglin's	Similar to Heuglin's.
Baltic	Passage winter	13–17 (darkest grey)	yellow		Little contrast with wing. Small mirror usually P10 only	Small, slim, long-winged. Truly black-backed

Note that in all these gulls males are larger than females with larger bills and a more fierce expression, and that gulls with yellow legs can, in winter or out of breeding condition, have pinkish legs. All species have a red gonys spot and all can show black near the bill tip.

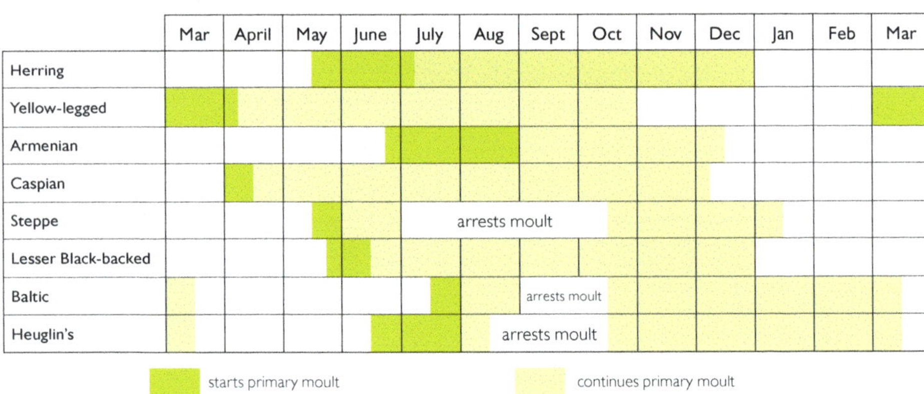

Adult primary moult progression interpreted from Olsen KM & Larsson H (2004) *Gulls of Europe, Asia and North America*, Christopher Helm, London.
Note: Northern breeding species moult later than southern species.

1st-winter
ad
Caspian Gull
Baltic Gull
4th-winter
1st-winter
1st-winter
Armenian Gull
ad
1st-winter
juv
ad
2nd-winter
2nd-winter
3rd-winter
Yellow-legged Gull
3rd-winter
ad
1st-winter
Heuglin's Gull
1st-winter
Steppe Gull
worn primaries

PLATE 46: LARGER TERNS

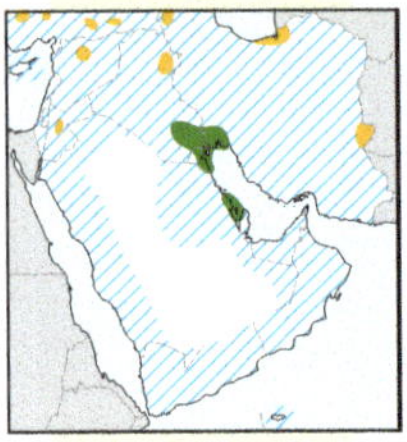

Gull-billed Tern *Gelochelidon nilotica* — WV, PM, sv

L: 38. W: 95. Medium-sized tern somewhat resembling Sandwich Tern ***but lacks crest and has shorter, thicker, gull-like all-black bill, shallower forked tail and whitish-grey rump and tail*** (white in Sandwich); lacks contrast shown by Sandwich Tern between dark outer and pale inner primaries above; outer primaries have distinct dark trailing edge. In winter, lacks black cap ***but has variable black 'mask', behind eye***. Juvenile is much less spotted on scapulars and innerwing-coverts than similar Sandwich, has white crown and ***black eye-patch, recalling winter adult Mediterranean Gull***; in first-winter head even whiter. Flight more leisurely and aerobatic than a Sandwich. When standing, note more robust build with ***long black legs***. Feeds over fields, marshes, dipping to surface. **Voice** Nasal *ger-wek* with stress on second syllable; alarm fast, agitated, nasal laughing notes *divdidit*; juvenile call soft *pre-eep*. **Habitat** Sheltered coasts, inland waters; nests colonially. **Note** Passage hatched; winters mainly Iran/south Arabia.

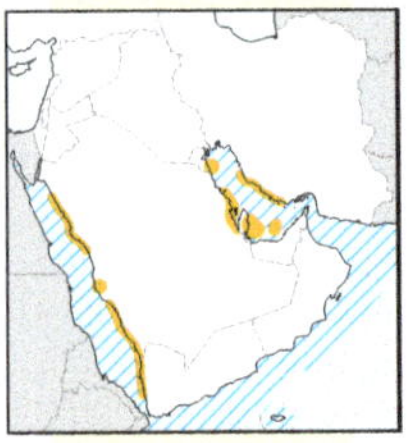

Lesser Crested Tern *Sterna bengalensis* — MB, PM, WV

L: 41. W: 92. Size as Sandwich Tern. ***Long, slim orange-yellow bill, upperparts pale ash-grey, with rump and tail paler grey***. Crested nape black; forehead in summer is white or black. In winter bill paler orange-yellow; solid black confined to nape. Juvenile has bill paler orange than adult; resembles juvenile Swift Tern but upperwing has paler dark bands and less pronounced pale mid-wing panel; dark markings on inner wing retained to first spring as are dark primaries above. ***Told from Sandwich Tern by bill colour*** and slightly darker upperparts. **Voice** Recalls Sandwich Tern's closely, but less grating and not so pronouncedly disyllabic, *krriik-krriik* or *kreet-kreet*; also *kir-eep* and *kee-kee-kee*; juvenile's call much as juvenile Sandwich Tern. **Habitat** Coastal; nests colonially on sandy or rocky shores or islands. **Note** Disperses to hatched areas.

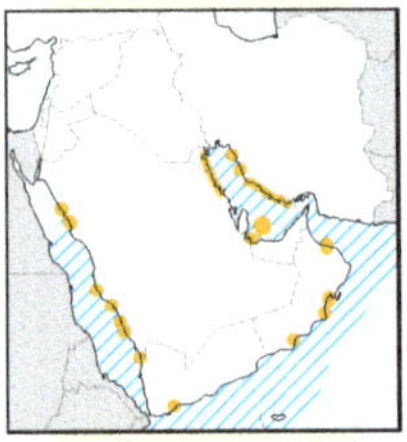

Swift Tern *Sterna bergii* — rb, WV, PM

L: 46. W: 105. Large, between Sandwich and Caspian Tern in size. Adult from smaller Lesser Crested Tern ***by longer, thicker, more drooping, waxy greenish-yellow bill and by darker grey upperparts***, with rump and tail paler grey than back; tail slightly longer, more forked, and wings broader than Lesser Crested. In summer, black crested cap always broken by white forehead-band (never black as in some Lesser Crested). In winter, bill paler greenish-yellow, or washed-out yellowish-orange; crown white with black confined to nape; tail and rump almost as dark as back. Juvenile has pale innerwing-panel above, framed by dark bar on forewing and dark covert- and secondary bars. Flies with shallow, deliberate wingbeats. **Voice** Deep, rough, *kee-rit*, or gravelly *craark*; also a high-pitched *kree-kree*. **Habitat** Coastal; nests colonially on sandy or rocky islands. **Note** Disperses to hatched areas.

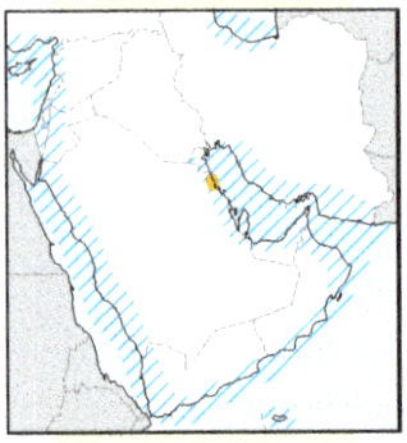

Sandwich Tern *Sterna sandvicensis* — WV, PM

L: 41. W: 92. Medium-sized with ***pale grey upperparts and wings (which may look almost white at sea), long slender black bill tipped yellow*** (yellow tip hard to see at distance), long narrow wings, medium-short forked tail and fairly long neck in flight, which is powerful with deep wingbeats. In summer black cap has ragged crest; in winter, black confined to nape extending forward to eye (like Lesser Crested Tern). Juvenile has all-black bill, dark-scaled mantle and forewing-coverts, but lacks the broad dark bars on inner wing of similar-aged Lesser Crested; first-winter has grey mantle and scapulars and head pattern like winter adult. **Voice** Loud, grating disyllabic *kerr-rick*; juvenile's call heard into winter, high-pitched *k-rill*. **Habitat** Coastal. **Note** Passage and winter hatched; many non-breeders summer.

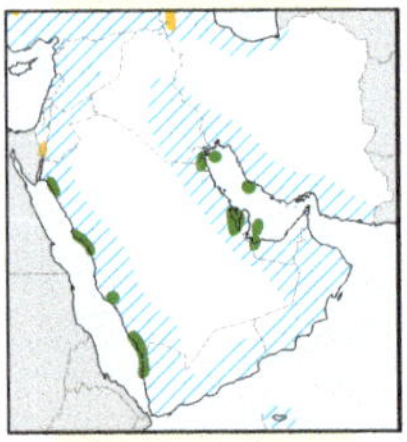

Caspian Tern *Hydroprogne caspia* — rb, WV, pm

L: 53. W: 135. Near size of a large gull, identified by ***large red bill, almost gull-like flight with slow steady wingbeats and by call; distinctive dark primaries below***. Juvenile has orangey bill, more extensive black cap to below eye (unlike adult), weak dark scales on mantle and wing-coverts, primaries dark both above and below; legs pale (black in adult). **Voice** Deep, loud, harsh and deliberate *aark*, recalling Grey Heron, or longer retching *kraa-jak*. **Habitat** Open coasts, lagoons; inland wetlands; nests in winter in Arabian Gulf, singly or colonially on small islands. **Note** Passage hatched, winters mainly Arabia.

juv
ad winter
juv
ad summer
Gull-billed Tern
juv
ad summer
Lesser Crested Tern
juv
ad winter
juv
ad summer
juv
ad winter
Swift Tern
ad winter
juv
ad summer
Sandwich Tern
juv
ad summer
ad winter
juv
Caspian Tern

Roseate Tern *Sterna dougallii* V

L: 38 (including tail streamers). W: 77. Similar to Common Tern but ***adult paler above; whiter below***, tinged pink in breeding season. ***Long, thin bill largely blackish*** in spring but in subspecies ***bangsi*** present in Middle East becomes red by July; ***tail-streamers very long and white***. Wing pattern diagnostic at all ages – ***dusky outer primaries lacking black line along trailing edge of outer primaries below***. At rest tail-streamers protrude far beyond tip of tail; legs longer than Common. Winter plumage similar to Common but retains long tail-streamers. Juvenile, which has short tail-streamers, told from Common/Arctic by underwing pattern (white trailing edge to outer primaries below); has darker forehead than Common/Arctic and more scaly upperparts and scapulars. Flight fast and direct with stiffer, more rapid wingbeats than Common. **Voice** Characteristic, soft guttural *cherr-wrick*. **Habitat** Coastal. **Note** Passage hatched; vagrant Bahrain, Saudi Arabia (Gulf), UAE.

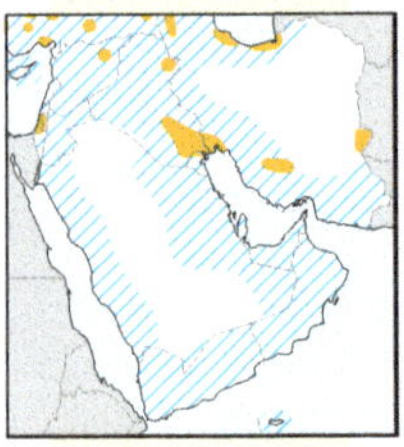

Common Tern *Sterna hirundo* PM, wv, sv

L: 35. W: 80. Closely resembles Arctic Tern, with wings slightly broader, tail-streamers shorter, ***bill and legs longer***, forehead flatter and wingbeats often more powerful. ***Dark outer primaries above abruptly cut off from grey inner primaries*** (uniform pale grey in Arctic); ***from below has blackish band on trailing edge of outer primaries and translucency confined to innermost primaries*** (all translucent in Arctic); ***bill dark orange-crimson, usually tipped black***; cheeks and throat whiter than in Arctic. Bill of adult in subspecies *minussensis* appears ***all black in summer***, (close to up to half of basal half and cutting edges deep red); also having ***white chin and throat but light grey breast and belly***, though grey paler than in breeding White-cheeked Tern, from which also differs in larger size, characteristic ***white rump, differing upper- and under-wing***. Juvenile best told from similar Arctic by flight, ***dark grey secondaries*** above (white in Arctic), broader, darker forewing-band above, translucency and pattern of primaries (similar to adult); centre of rump pale grey (clean white in Arctic). **Voice** Deeper than Arctic; sharp *kitt; kirri-kirri-kirri*, and a drawn-out *kreee-aeh*. **Habitat** Coastal and inland waters. **Note** Passage hatched; winters mainly south Arabia. Subspecies *minussensis* regular on passage Arabian Sea & Gulf, Red Sea and south Caspian Sea. (Subspecies not separable in winter plumage).

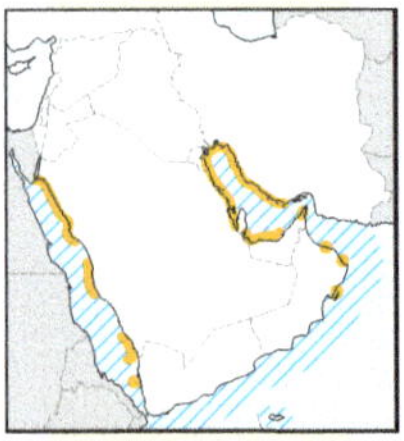

White-cheeked Tern *Sterna repressa* PM, wv, sv

L: 33. Slightly smaller than Common Tern (often mixes on passage/winter) with shorter wings, slightly shorter legs, and ***more slender bill, proportionately shorter than in Common Tern*** and usually slightly downcurved. Adult in summer is ***dark silver-grey above with grey underbody; underwing with whitish area in centre (also evident in juveniles)*** and ***primaries above appearing paler than inner wing (both visible, and diagnostic, at considerable range)***; ***white cheek-stripe*** recalls adult Whiskered Tern in summer (but tail-fork and bill longer); secondaries dull grey, with long, broad black line to tips of outer primaries below. In winter, remains ***dull grey above (slightly darker than Common), including rump and tail***; underparts white, mottled with dark grey in some; bill blackish. First-winter has broad blackish forewing-band and dark secondaries (like Common), but mid-wing dull greyish rather than greyish-white as in Common; underwing similar to adult, with rump and tail greyish (pale grey rump-centre only in Common). Settled juveniles/first-winter birds resemble marsh terns. **Voice** Often gentler than Common; loud *kee-err* or *ker-rit* with emphasis on short second syllable (on first note in Common) and single *kip* less sharp than Common. **Habitat** Coastal and maritime; nests colonially on bare islands. **Note** Passage hatched; scarce south Arabia in winter.

Arctic Tern *Sterna paradisaea* V

L.: 38. W: 80. Similar to Common Tern but rounder head, shorter thicker neck, longer tail-streamers and more elegant, buoyant flight. ***Bill dark red, usually without black tip; legs shorter. In flight all primaries appear translucent, revealing a thin, sharp black line to trailing edge of outer primaries***. In summer greyish underparts with whitish band below black cap. Juvenile/first-winter has ***white secondaries***, diffuse dark bar on leading forewing and white rump. **Voice** Similar to Common Tern, but harsher. **Habitat** Coastal; occasionally inland. **Note** Vagrant Kuwait, Oman, UAE.

Roseate Tern
juv
ad summer
juv
ad summer
Common Tern
1st-winter
ad summer
juv
ad summer
ad summer
minussensis
ad winter
White-cheeked Tern
ad winter
juv
ad summer
juv
ad summer
Arctic Tern
ad summer
1st-winter
ad summer
1st-winter

PLATE 48: PELAGIC TERNS

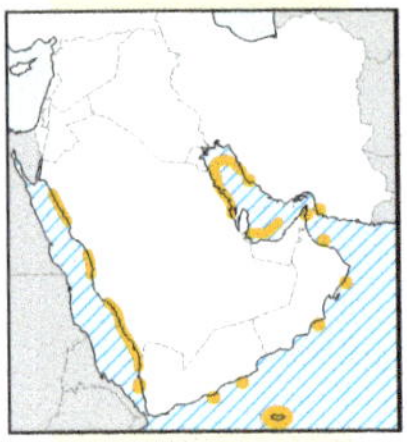

Bridled Tern *Onychoprion anaethetus* — MB, PM, wv

L: 37. W: 76. Medium-sized slender tern; Sooty Tern being the confusion species. ***Dark ashy upperparts*** (but often appearing blackish), ***narrow white forehead-band extending behind eye as narrow supercilium***, black cap contrasting with ***pale grey hindneck-collar*** and greyer mantle (often hard to see in flight); long outer feathers of deeply forked tail have white sides. Whitish underparts and underwing-coverts contrast with ***dark silver-grey central tail feathers and especially flight feathers***. Underparts can appear grey. Juvenile lacks distinct supercilium, has pale grey-brown back and wing-coverts, edged buffish. Flight graceful with slimmer wings than Sooty Tern. Does not dive but takes food from surface of water. May settle on sea, with tail pointed upward at 45°. **Voice** High-pitched *kee-yharr*, yelping *wep-wep*; also harsh grating *karr*; also a pleasing rolled *purrurr* or *prerrr*, day or night at sea or colony. **Habitat** Maritime; breeds colonially on rocky or sandy islands in crevices or under low vegetation. **Note** Passage hatched; winters Arabian Sea.

Sooty Tern *Onychoprion fuscata* — V

L: 44. W: 90. Larger, more stocky than Bridled Tern and with ***entire upper surface uniform blackish*** (paler mantle and hindneck in Bridled), ***broader but shorter white forehead-patch than Bridled not extending beyond eye***, and black loral streak reaching gape not base of upper mandible as in Bridled; white underparts and underwing-coverts ***contrast more with dark grey flight feathers, outer primaries lacking whitish wedge toward wing-tip***. Juvenile ***sooty-brown all over*** except for whitish belly and undertail-coverts; upperparts flecked whitish. Told from young Brown and Lesser Noddies by whitish vent and undertail-coverts, paler underwing and forked tail. First-summer like adult but throat and upper breast blackish. May settle on sea surface, as Bridled Tern. **Voice** Call diagnostic, a high-pitched *ker-wacki-wah*. **Habitat** Maritime; nests colonially in open on sandy, rocky or vegetated islands. **Note** Dispersal as hatched, but rare; vagrant Bahrain, Iran, UAE.

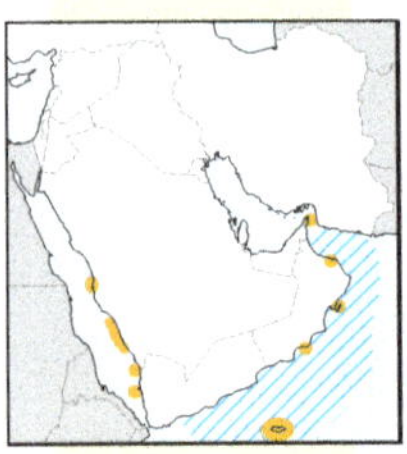

Brown Noddy *Anous stolidus* — sv

L: 43. W: 83. Size of Sandwich Tern with ***long, wedge-shaped tail (with shallow fork when spread) and long black bill. Dull chocolate-brown except for pale ash-grey crown, which grades into white forehead and contrasts sharply with black lores***; may appear white-capped in abraded birds in strong light; ***underwing-coverts contrast clearly with blackish-brown flight feathers*** (smaller Lesser Noddy has all-brown underwing). Juvenile has crown varying from grey-brown to whitish; immatures (up to three years old) are extremely abraded, appearing paler than adults. Separated from young Sooty Tern by darker underwing-coverts (can be whitish in Sooty) and shape of tail. Usually flies low over water with slower, more languid wingbeats than Lesser Noddy. Feeds by hovering or banking before swooping low to snatch prey from surface; often in mixed flocks of feeding terns and Persian Shearwaters. **Voice** At breeding site deep, guttural, corvid-like *kwok-kwok*, *karruk* or *krao*. **Habitat** Maritime; nests colonially on rocky islets and cliffs. **Note** Present in hatched area in summer and winter. [Alt: Common Noddy]

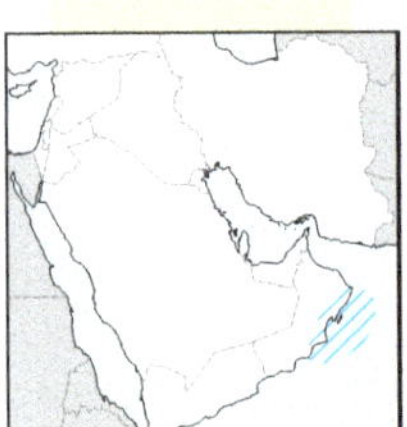

Lesser Noddy *Anous tenuirostris* — V

L: 32. W: 60. Very similar to Brown Noddy ***but smaller*** (size of Common Tern), shorter tailed and ***with proportionally longer, thinner bill***; lacks narrow black forehead band over bill and usually has pale ash-grey (not white) forehead and crown, ***lacking sharp demarcation with black lores; dark underwing without contrast between flight feathers and wing-coverts*** of Brown Noddy (but reflecting light may make underwing-coverts appear paler). Juvenile generally less pale-crowned than Brown Noddy. In flight, wings narrower, wingbeats faster, the 'jizz' lighter than the bulkier Brown Noddy, but separation difficult at distance. **Habitat** Maritime. **Note** Non-breeding summer visitor in small numbers Oman (as hatched on map); vagrant UAE, Yemen. [Alt: Sooty Noddy]

Bridled Tern
juv
ad
juv
Sooty Tern
juv
juv
ad
Brown Noddy
juv
ad
Lesser Noddy
ad
juv
ad

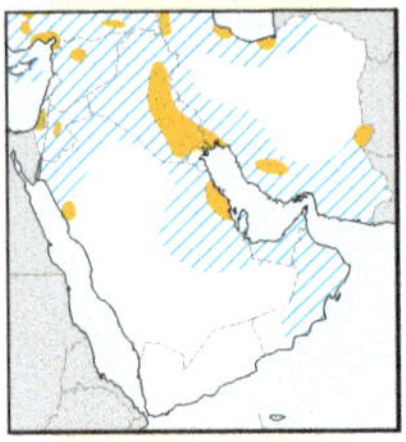

Little Tern *Sternula albifrons* PM, ?wv

L: 23. W: 53. Small with ***fast wingbeats***. Adult in summer pale grey above, extending in some onto centre of otherwise ***white rump***; ***usually two grey-black leading primaries (with shafts white)***; ***white forehead usually extending in point to rear of eye***; ***legs bright yellow to reddish-orange***. In winter bill black, legs dull grey or brown, usually with some yellow; black on head reduced to band around nape; rump and tail largely grey. Juvenile (dark bill with reddish base) has dark 'U'-shaped markings on pale grey mantle, scapulars and tertials, lost in first-winter when resembles winter adult; legs greyish-pink to yellow-brown; upperwing shows dark outer primaries becoming progressively paler inwards; leading forewing dark. **Voice** Excited, hoarse grating *kryik* or *pret-pret*. **Habitat** Coastal and inland waters; nests on beaches or sandbanks in rivers. **Note** Passage hatched; winter range obscured by confusion with Saunders's Tern; vagrant Yemen.

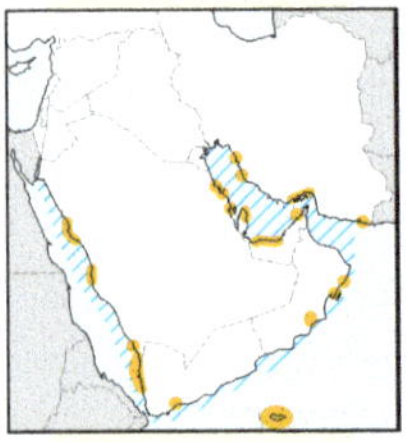

Saunders's Tern *Sternula (albifrons) saundersi* MB, PM, wv

L: 22. Separation from Little Tern problematic. In summer upperparts contrast markedly with ***blacker outer three primaries (dark-shafted)***; ***white forehead not usually reaching eye*** thus squarer; ***rump more extensively grey than in Little and concolorous with back; legs generally darker, reddish- or pinkish-brown***, sometimes some yellow on rear tarsus. In winter, adults and first-winter birds doubtfully separable (Saunders's has ***darker grey upperparts*** than similar-aged Little in Red Sea; apparently not so in Arabian Gulf). **Voice** Strident calls recall Little's closely, but often lacking same urgency; sharp *kip* or *wip*, and excited *tchijjick*. **Habitat** Coastal, rarely inland; nests on beaches. **Note** Winters Iranian and Arabian coasts.

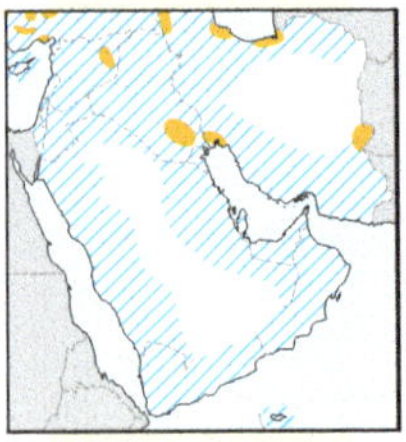

Whiskered Tern *Chlidonias hybrida* PM, WV, sv

L: 25. W: 73. Slightly larger than White-winged Tern with broader wings, more forked tail (adult), longer legs and ***heavier bill***. Flight buoyant but more ***Sterna***-like; dips or dives for food. Adult in summer has ***sooty-grey underparts contrasting with whitish cheek-stripe and underwing***; in winter separated from White-winged ***by flight action, larger bill, streaked crown and shape of ear-coverts patch***. Juvenile from similar White-winged by head pattern, practically no white collar on hindneck, paler 'saddle' and pronounced blackish and buffish markings in scapulars (contrasting with 'saddle'; scapulars dark in White-winged); upperwing slightly paler with virtually no dark bar on leading forewing; rump and tail usually concolorous pale grey (rump white in White-winged); sometimes has dark smudge at sides of breast, as some adults in winter, absent in White-winged. **Voice** Loud, hoarse *kreck*, almost a sneeze. **Habitat** Inland and coastal waters. **Note** Passage and winter hatched, rare in north in winter.

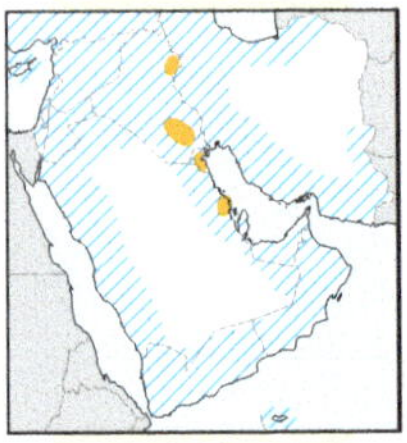

White-winged Tern *Chlidonias leucopterus* PM, wv

L: 22. W: 65. Summer adult distinctive; in winter from similar Black Tern ***by lack of dark patch at sides of breast***; whiter rump and tail (pale greyish in Black); a few show some black marks on underwing-coverts, making identification easier. ***Juvenile has more contrasting darker 'saddle' and paler upperwing than juvenile Black, and lacks dark breast-patch*** and white tips to rear scapulars. Rump and sides of tail whiter, with dark patch on ear-coverts extending farther downwards. Moults dark 'saddle' to pale grey in late autumn. **Voice** Harsh, dry *kesch* or *kruek*, less hoarse than Whiskered. **Habitat** Inland and coastal waters. **Note** Passage hatched, scarce in winter when mainly south Arabia. [Alt: White-winged Black Tern]

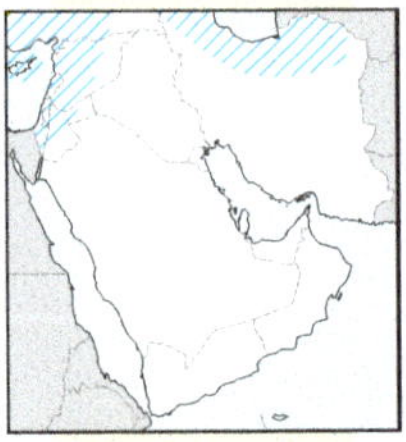

Black Tern *Chlidonias niger* V

L: 23. W: 66. Summer adult distinctive; in winter underparts white with ***dark patch on sides of breast***, more pronounced in juvenile; forehead and hindneck-collar white; upperparts greyish, darker grey and more uniform than White-winged. Juvenile has dark fore-mantle and leading upperwing-coverts; scapulars tipped whitish; sides of rump sometimes whitish, rest of rump and tail pale greyish. First-winter recalls adult in winter. **Voice** Short sharp *kjeh* and repeated *kit*, more squeaked than sneezed as in congeners. **Habitat** Inland and coastal waters. **Note** Passage hatched; scarce Iran; vagrant Iraq, Arabian Gulf and Arabian Sea.

Little Tern
juv
ad summer
ad
ad summer
ad
Saunders's Tern
Whiskered Tern
juv
ad winter
ad summer
White-winged Tern
juv
ad winter
ad summer
Black Tern
juv
ad winter
ad summer

Brown Skua *Stercorarius antarcticus* ?V

L: 63. ***Larger and darker than South Polar Skua***, with broad-based wings, ***heavier bill***, and lacking contrasting paler head and underbody, both species showing conspicuous white primary patches (above and below) even at long range. Adult very similar to dark morph of South Polar Skua, but upperparts less uniform with ***thick paler mottles and flecking*** (but only visible at close range). Juvenile warmer in tone without heavy mottling above. Some individuals cannot be safely told from South Polar Skua. Typically thicker billed and darker chocolate-brown than Great Skua, but latter not a reliable or constant field character. **Habitat** Maritime. **Note** Vagrant Oman; vagrant large skuas in Iran and UAE also considered to be this species. [Alt: Subantarctic Skua]

South Polar Skua *Stercorarius maccormicki* (not illustrated) No records

L: 53. W: 127. Closely resembles Brown Skua. Pale and intermediate morphs identifiable by ***creamy buff or grey-brown head, hindneck and underparts, contrasting with blackish underwing-coverts***. Dark morph often has pale forehead, hindneck with pale wash at sides and bill sometimes more bicoloured. During primary moult (July–August) immature may have ***conspicuous symmetrical gaps in each hand***. Pale 'noseband' at base of upper mandible shown by some birds. **Habitat** Maritime. **Note** Vagrant Oman, Yemen.

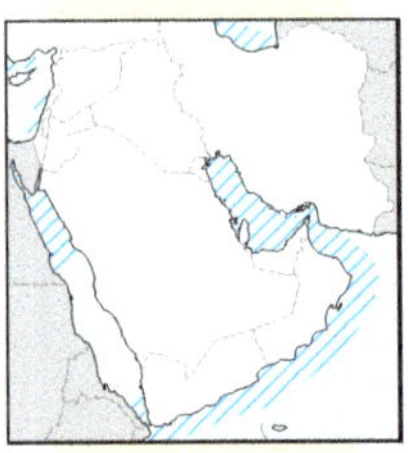

Pomarine Skua *Stercorarius pomarinus* PM, WV, sv

L: 51–56. W: 125. Size of large gull, ***heavily built*** and, in adult, ***elongated broad and twisted central tail feathers*** ('spoons') diagnostic, though these often lost. Two morphs: all dark (scarce) and a pale morph, which differs from Arctic Skua in more extensive black cap, darker flanks and blacker breast-band (sometimes absent) and vent. In winter, flanks and tail-coverts barred, elongated tail feathers often short and always blunt-ended. Juvenile best separated from similar Arctic Skua by ***more regular bars on flanks, vent, rump and tail-coverts***. Whitish base to greater primary coverts below often conspicuous, even at range (frequently absent in Arctic Skua). Bill proportionally larger, heavier at base and more distinctly bicoloured. With experience can be ***identified by larger size, slower wingbeats, broader wings, deeper breast and belly***; also, in juveniles, when present, short, blunt central tail feathers (pointed in Arctic Skua). Piratical attacks more direct, less agile than Arctic, frequently chasing large gulls. **Habitat** Maritime. **Note** Passage hatched; winters mostly Arabian Sea.

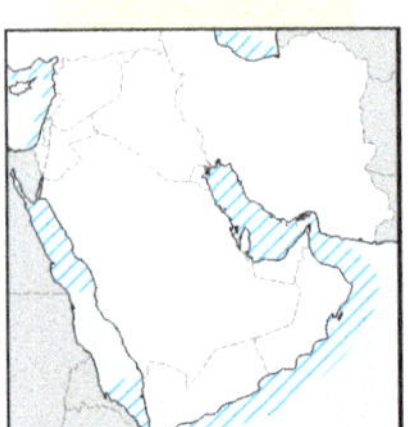

Arctic Skua *Stercorarius parasiticus* PM, WV, sv

L: 46. W: 117. Adult in summer has ***elongated, pointed central tail feathers*** (6–11cm). Two morphs occur (with frequent intermediates). Juvenile variable, from pale to very dark birds; ground-colour generally warmer, more rusty than young Pomarine Skua; juvenile told with experience from young Long-tailed Skua by ***broader wings, slightly shorter tail*** (equals width of wing-base), ***shorter, pointed central tail feathers*** (longer, blunt-ended in Long-tailed), ***thicker rear-body, less black on bill-tip and warmer brownish ground colour*** (colder, greyer in Long-tailed Skua), ***with rusty fringes and barring***. Normal flight steady, falcon-like, straight and fast; piratical attacks with sudden twists and turns, harassing birds the size of Sandwich Tern or Common Gull. **Habitat** Maritime. **Note** Passage hatched; winters mostly Arabian Sea.

Long-tailed Skua *Stercorarius longicaudus* V

L: 53. W: 111. Slim and lightly built; size of Black-headed Gull. Resembles pale and rather grey Arctic Skua but adult told by ***very long, flexible central tail feathers*** (16–24cm), ***lack of breast-band, white forebody gradually darkening towards rear, pale greyish upperparts contrasting with blackish flight feathers with pale shaft streaks only on outermost primaries*** and ***pale blue-grey legs***. Juvenile variable like young Arctic Skua, but has ***colder, greyer ground-colour and fringing*** (never rusty), ***more distinct barring on uppertail-coverts*** (except in darkest birds), ***thin white shaft-streak in leading primaries*** (hardly visible beyond 500m, unlike distinct white flash on most Arctic Skuas) and ***blunt-ended central tail feathers***. Primaries on settled bird plain (broad buffy tips on Arctic). At rest shows more rounded head and shorter, proportionally thicker ***bill with black tip to half the length***, unlike Arctic Skua. Also relatively deeper-chested with thinner rear-body. Flight more buoyant and tern-like than Arctic Skua; includes more circling and hovering; without making aggressive piratical attacks. **Habitat** Maritime. **Note** Vagrant Iran, Kuwait, Oman, Qatar, UAE.

Brown Skua
ad
ad
pale morph
dark morph
ad summer
Pomarine Skua
ad winter
juv
ad summer
juv
ad summer
juv
Arctic Skua
juv pale morph
juv intermediate morph
ad summer
juv dark morph
Long-tailed Skua

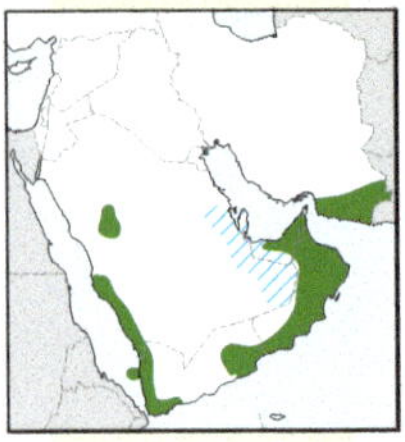

Chestnut-bellied Sandgrouse *Pterocles exustus* RB, WV

L: 32. W: 50. In fast flight shows ***diagnostic all-dark underwing continuous with dark chestnut belly***. Female shows black-spotted breast and narrowly vermiculated blackish upperparts; male has golden wings with some dark barring; both sexes show ***elongated tail and narrow black bar on lower breast***. Juvenile smaller, with reduced area of dark on belly, which can often be difficult to see. **Voice** Far-carrying, short, rather guttural but liquid *kwit-kwit-kwituroh-kwituroh-kwituroh* or *gattar-gattar*; also *ke-rep, kerep* with stress on last syllable. **Habitat** Sand and gravel semi-deserts, though often near agricultural land; also coastal dunes, beach-top gravels, inshore islands. Visits water to drink in large flocks in early to mid-morning. **Note** Nomadic, winter dispersal hatched; vagrant Kuwait.

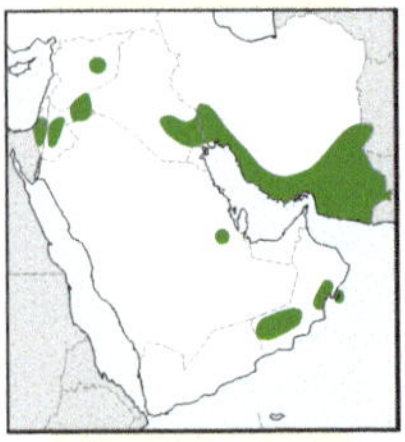

Spotted Sandgrouse *Pterocles senegallus* V

L: 32. W: 59. A pale sandgrouse ***with a long pointed tail***; in flight shows rather ***pale upperwing including primaries*** with indistinct dark rear border; underwing pale with contrasting dark flight feathers (especially secondaries) which separates from Chestnut-bellied Sandgrouse. Bold spots on upperparts and breast in female, confined to wing-coverts and shoulder in male. Flocks, often large, visit water-holes in morning; individuals occasionally join Chestnut-bellied flocks. **Voice** In flight, a disyllabic bubbling whistle, frequently repeated: *wi-dow*, first syllable higher. **Habitat** Mainly sandy deserts; also semi-deserts with sparse vegetation or scrub. **Note** Vagrant UAE.

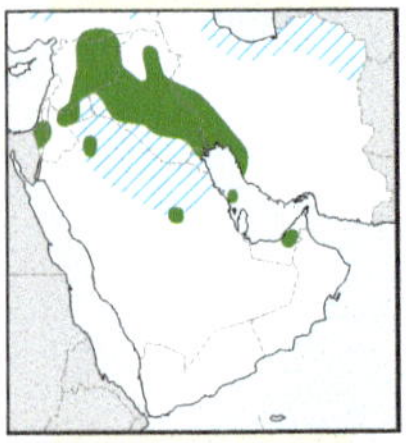

Pin-tailed Sandgrouse *Pterocles alchata* E/I

L: 35. W: 60. Large, plump sandgrouse, ***with long tapering tail***. In flight, ***white underparts contrast sharply with black primaries and chestnut-buff breast-band (male) bordered narrowly with black***; female has pale golden-yellow breast-band with ***three black bars***. Male has black throat and eye-stripe and spotted grey-green upperparts; female duller, mottled and barred above. Gregarious, often in large noisy flocks. **Voice** Call in flight a distinctive, repeated, slightly falling *arrrh, arrrh* or shorter *arrk-arrk-arrk-*. **Habitat** Dry plains and stony semi-deserts, fields. **Note** Nomadic or partial migrant as hatched; introduced UAE.

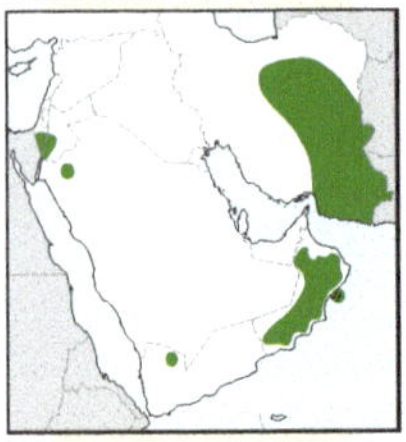

Crowned Sandgrouse *Pterocles coronatus* ?V

L: 28. W: 57. Rather pale with ***short tail***. Male has yellowish head and neck, and diagnostic ***black mask on forehead and around bill base***. Female sandy, finely vermiculated; head, neck and throat unmarked yellowish. In flight, ***black flight feathers contrast with sandy-grey upperparts and wing-coverts*** (unlike Spotted Sandgrouse); underwing-coverts white. Distinguished from Spotted Sandgrouse by short tail, which is white-tipped (noticeable when tail spread on landing) and distinctive upperwing pattern; Lichtenstein's Sandgrouse is more barred and lacks yellowish on head. Comes to water mainly in morning. Two subspecies occur in Arabia: *atratus* in sand desert and *saturatus* in the foothills of the Hajar mountains, Oman. **Voice** Frequently calls in flight on way to or from water: hard, accelerated, nasal *kaaa-kata-kata-kataah*. **Habitat** Stony and semi-deserts. **Note** Mainly resident. Historical reports, but no recent records UAE.

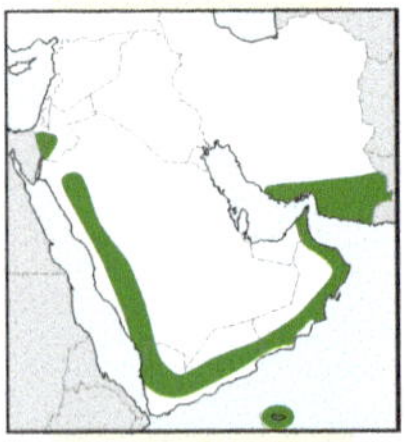

Lichtenstein's Sandgrouse *Pterocles lichtensteinii* RB

L: 25. W: 50. Small with ***short square-ended tail***. Yellowish-buff plumage ***finely vermiculated black and white all over***. Male has ***yellowish breast-patch framed by two black bars***; bill orange, ***white forehead with vertical black bars***. Female duller, lacking distinct markings on head and breast; lacks yellow throat of similar Crowned Sandgrouse. In flight, upperwings show black flight feathers and pale wing-coverts, underwing pale with slightly darker flight feathers. Singles or small parties drink at dusk or before dawn. Active by night, when calls frequently in flight. **Voice** Call, when coming to drink, a repeated disyllabic, melodic *whee-ak* with stress on first syllable; when flushed a harsh, whirring *arrk*; at night, in flight, a clear liquid *whit, wheet, wheeoo*. **Habitat** Rocky deserts, ***arid mountains***, wadis and hillsides with sparse scrub.

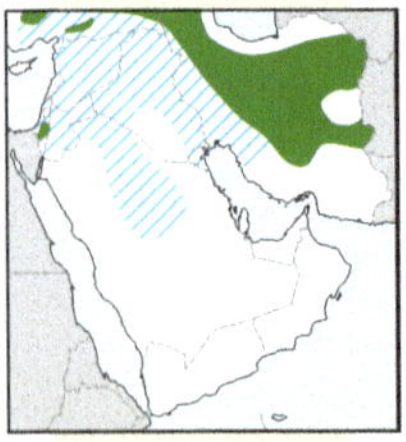

Black-bellied Sandgrouse *Pterocles orientalis* E/?V

L: 34. W: 72. A heavy sandgrouse. Male has plump ***black belly*** and ***prominent black-and-white underwing pattern***. Female has heavily spotted upperparts and breast, sharply demarcated from greyish lower breast and all-black belly. Far-carrying call highly distinctive. Chestnut-bellied Sandgrouse is smaller and slimmer with blackish underwings. **Voice** Flight call a characteristic ***rolling or bubbling*** durrrll, sometimes drawn out. **Habitat** Steppe, cultivation edge. **Note** Winter dispersal hatched, but rare Arabia; vagrant Bahrain.

Chestnut-bellied Sandgrouse
♂
♀
♂
♂
♀
Spotted Sandgrouse
♂
Pin-tailed Sandgrouse
♂
♀
♂
♂ atratus
Crowned Sandgrouse
♀
♂ atratus
♂ saturatus
♂
Lichtenstein's Sandgrouse
♀
♂
♂
♀
♂
Black-bellied Sandgrouse

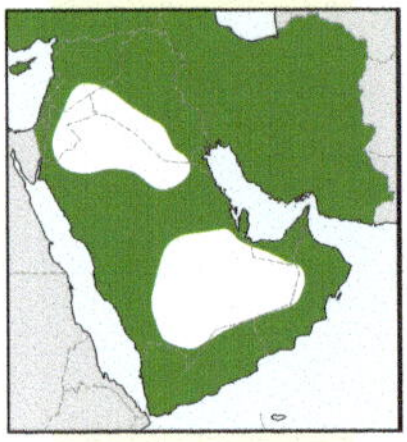

Rock Dove (including Feral Pigeon) *Columba livia* RB

L: 33. Pale blue-grey pigeon, recalling Stock Dove, with ***two broad black bands across secondaries above; rump white*** or grey (rump grey in most Arabian populations); underwing white with dark band at rear; black band on outer tail more pronounced than in Stock Dove. Rock Dove is the ancestor of the familiar Feral Pigeon, which can be blackish, whitish or even reddish; most show black bands on wing, white underwing and contrasting tail pattern. Gregarious, flight very fast and often aerobatic, gliding on lifted wings. **Voice** Cooing *kru-oo-u*, second syllable stressed and highest in pitch. **Habitat** Rocky wadis, sea-cliffs, mountains; nests in cave or rock-ledge. Feral Pigeon widespread, often in towns; nests on buildings and cliffs. **Note** No pure Rock Dove populations exist in UAE.

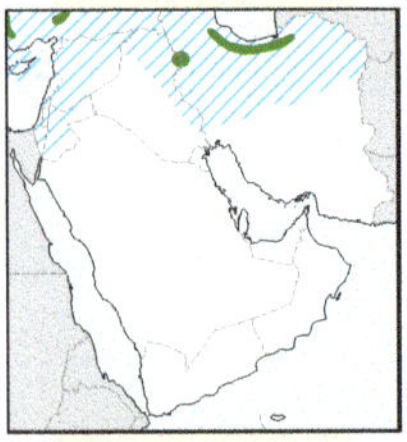

Stock Dove *Columba oenas* V

L: 33. Medium-sized and easily told from Common Woodpigeon by ***absence of white band on upperwing***, also smaller, more compact and shorter-tailed, with quicker wingbeats. From Rock Dove by less white underwing ***with darker flight feathers, pale ashy-grey wing-panel above and lack of bold black bars across secondaries***. Rump grey. Bill pale-tipped (dark in Rock Dove); iris dark. Often in small flocks. **Voice** Monotonous muffled, hollow 'cooing' with emphasis on first syllable, *uu-rur... uu-rur... uu-rur*; also similar single *ur... rer.* **Habitat** Open wooded areas with old trees; nests in hole in tree, rock or building. In fields on passage and winter. **Note** Winter hatched, rare in south; vagrant Bahrain, Kuwait, Oman, UAE.

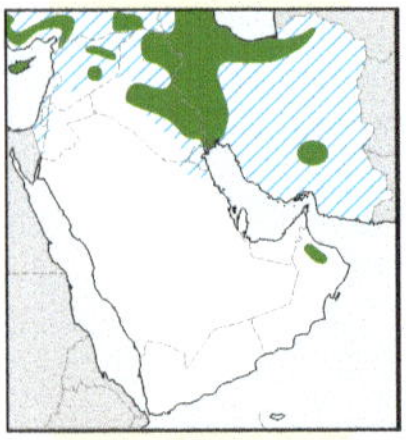

Common Woodpigeon *Columba palumbus* V, E

L: 41. Large stocky pigeon, told by size, ***bold white band on upperwing in flight*** (all ages) and ***white patches on sides of neck*** (absent in young). Flight slower and heavier than other pigeons, with tail proportionally longer. Gregarious outside breeding season. Breeding birds in Oman and SE Iran, *casiotis*, have smaller, buffy neck-patches. **Voice** Hoarse cooing, *cu-cooh-cu, coo-coo,* second note stressed and drawn out. **Habitat** Woods, parks, gardens, fields. **Note** Has bred Kuwait; winter range hatched; vagrant UAE (incl. one record of *casiotis*).

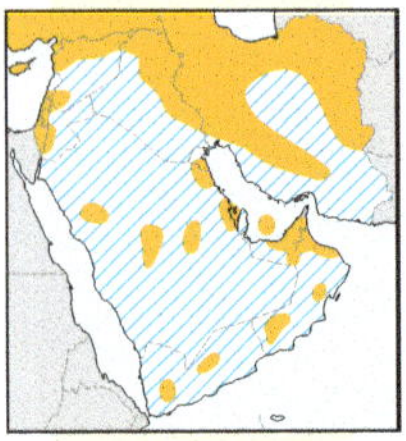

European Turtle Dove *Streptopelia turtur* MB, PM

L: 27. Small, fast-flying dove with ***shorter tail and more pointed wings than Eurasian Collared Dove***. Told by ***darker underwing, well-defined whitish belly-patch, rusty-edged, dark-spotted upperparts*** with blue-grey outer wing-panel; also ***contrasting pattern of uppertail with clear-cut white corners***, particularly when tail spread on landing. ***Sides of neck show black and white-streaked patch in adult***, absent in juvenile. From Laughing Dove by build, jerky wingbeats, scalloped upperparts and tail pattern. ***The subspecies* arenicola*, breeding Near East to Iran and Arabia, is paler, more washed-out grey-brown on mantle, wing-coverts (with less contrasting dark centres) and breast***. Often gregarious. **Voice** Soft, deep purring *roorrrr, roorrrr, roorrrr,* often persistent. **Habitat** Breeds in wooded country, sand desert with ghaf, oases; fields and livestock enclosures on passage. **Note** Passage hatched; rare in winter south Arabia.

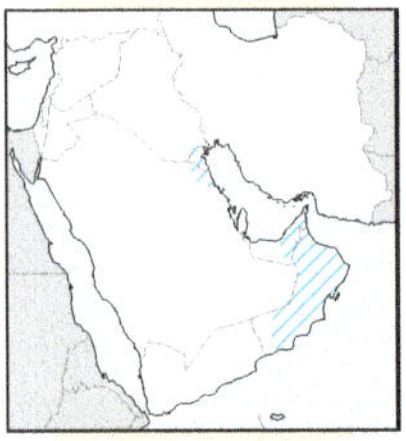

Rufous Turtle Dove *Streptopelia (orientalis) meena* pm

L: 33. Resembles European Turtle Dove but ***larger and heavier with broad-based wings, ill-defined dark centres to wing-coverts with narrow buff fringes*** (broader tan fringes in European Turtle Dove); ***forehead and crown pale grey, contrasting with browner rest of head***; neck-patch larger, ***streaked blue and black*** (usually white and black in European Turtle) and bare skin around eye ***rounded*** (larger area bare and lemon-shaped in European Turtle). Note also that Rufous Turtle Dove has broad buff fringes to primary coverts (narrow fringes in European Turtle Dove). Subspecies occurring in the region, *meena*, has whitish belly, undertail-coverts and distal part of tail. Juvenile lacks neck-patches, is paler brownish on body and wings, otherwise like adult. Flight heavier, straighter, with less jerky wingbeats than European Turtle Dove. **Voice** Alternating grating and clearer cooing notes, *gru-gror, co-co, gru-gror, co-co*. Migrants invariably silent. **Habitat** Open woodland, often near cultivation. **Note** Passage hatched, but rare; occasional in winter or even summer; vagrant Iran, Iraq, Saudi Arabia. [Alt: Oriental Turtle Dove]

Rock Dove
ad
Feral Pigeon (variations)
ad
Stock Dove
ad
juv
Common Woodpigeon
ad
arenicola
ads
European Turtle Dove
juv
ad
Rufous Turtle Dove

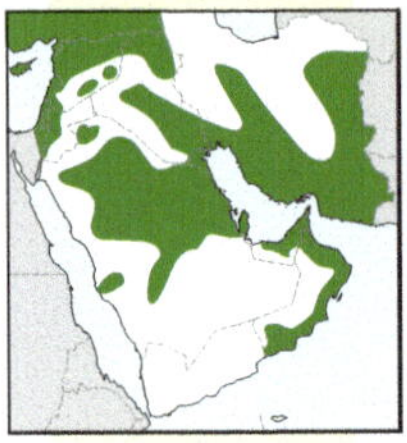

Eurasian Collared Dove *Streptopelia decaocto* RB

L: 31–34. Medium-small, sandy-brown dove with narrow black half-collar on hindneck (absent juvenile). Confusable with African Collared Dove, which see. ***From European Turtle Dove by unspotted sandy-brown forewing*** (dark spotting in Turtle), ***neck collar, whitish underwing*** (dark in European Turtle), plainer uppertail (bold pattern in European Turtle), ***absence of well-defined whitish belly-patch***, grey undertail, longer tail and less rapid flight with ***less jerky wingbeats of shorter, more rounded wings***. The domesticated Barbary Dove, *Streptopelia risoria*, is very like African and Eurasian Collareds but paler creamy, with clean white undertail-coverts, and lacks contrasting dark primaries above. **Voice** Loud, deep, trisyllabic *coo-cooh-co* with stress on middle note, which is also drawn-out and highest in pitch (Barbary Dove stresses first note). **Habitat** Towns, villages, parks, fields; often in flocks. **Note** Ongoing range expansion in Arabia; some winter dispersal.

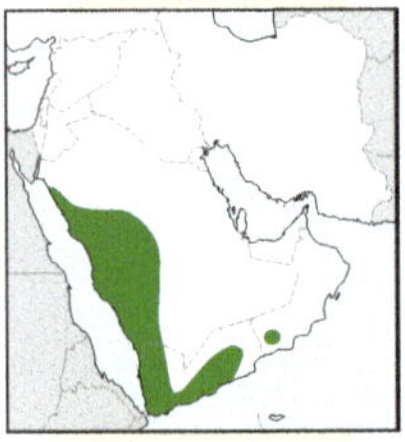

African Collared Dove *Streptopelia risoria* No records

L: 29. Similar colour and pattern on upperwing, tail-corners and underwing as slightly larger Eurasian Collared Dove. Best separated by ***white lower belly and undertail-coverts*** (dirty grey in Eurasian Collared), rather shorter tail, more obvious eye-ring and ***by voice. Dark border to entire trailing edge of pale underwing***. Young birds paler than adults, being whitish-grey on head and underparts (paler than Eurasian Collared). Gregarious, sometimes mixes with Eurasian Collared Doves. **Voice** Distinctive, ***high-pitched drawn-out note followed by a short pause then a series of broken, descending rolling notes, crooo, cro-cro-crococo or cruu… currruuu***; at distance sounds disyllabic *croo… cooorrr* (as feral Barbary Dove). **Habitat** Semi-desert and savanna with trees; also mangroves, parks. **Note** Range expanding; probably introduced Kuwait; vagrant Bahrain. Domesticated form, Barbary Dove, recorded Dubai and Abu Dhabi emirates.

Red Turtle Dove *Streptopelia tranquebarica* V

L: 23. Small, compact, short-tailed dove with black collar on hindneck. From larger Eurasian Collared Dove by combination of ***red-brown*** (male) ***or warm-brown*** (female) ***mantle and wing-coverts, darker ash-grey lower back and uppertail with more defined white corners***, browner breast (vinous-brown in male), ***whitish vent and undertail-coverts*** (grey in Eurasian Collared) and ***darker underwing*** (pale in Eurasian Collared). Laughing and European Turtle Doves have darkish underwing but both lack hindneck collar. Fast flight like European Turtle Dove. **Voice** Dry, rattling, rhythmic *ruk-a-duc-doo*, quickly repeated. **Habitat** Open wooded country. **Note** Vagrant Iran, Oman, UAE.

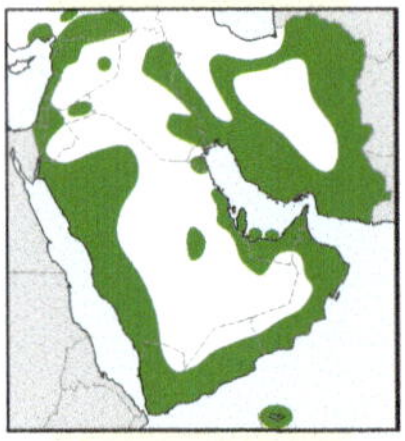

Laughing Dove *Spilopelia senegalensis* RB

L: 26. Small, dark red-brown to sandy grey-brown dove, with ***black-spotted patch on foreneck and upper breast, unspotted red-brown upperparts and large, blue-grey area in outer-coverts***. Uppertail less contrasting than in European Turtle Dove. Juvenile lacks patch on foreneck and is duller. Population in SW Arabia rather dark in plumage. Flight close to Eurasian Collared Dove but note short rounded wings and long tail. Gregarious; usually abundant. **Voice** Usually five syllables, subdued cooing with third and fourth notes slightly longer and higher in pitch, *do, do, dooh, dooh, do*. **Habitat** Towns, villages, gardens, oases and agricultural areas. **Note** Recent extensive range expansion in region. [Alt: Palm Dove]

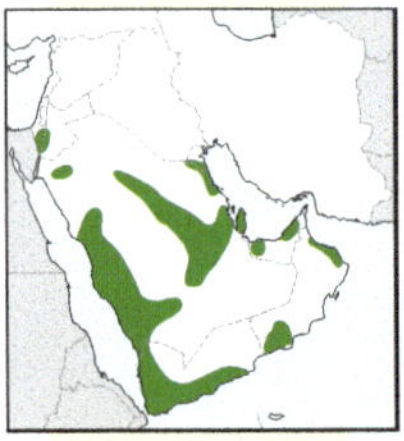

Namaqua Dove *Oena capensi* PM, WV, rb sv

L: 29 (including 9cm tail). ***Very small, slim dove***, blue-grey or grey-brown with ***long, pointed black central tail feathers***, recalling large Budgerigar. In flight, ***black primaries show large red-brown patch***, rump with two transverse black bands. Male has black face and upper breast, which are brownish-grey in female. Juvenile barred black and buff on crown, throat, wing-coverts and back. Flight very fast and direct. Unobtrusive; spends much time on the ground. **Voice** Mournful *hu-hu, hu-hu*; also a deep coo. **Habitat** Savanna and semi-desert with thorn-bush or scrub. **Note** Range expanding in Arabia; vagrant Iran, Iraq.

Eurasian Collared Dove
ad
African Collared Dove
ad
ad
Red Turtle Dove
Laughing Dove
ad
ad
display
dark individual
♀
♂
Namaqua Dove

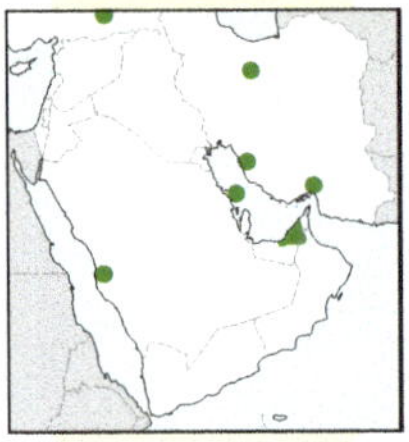

Alexandrine Parakeet *Psittacula eupatria* RB (E/I)

L: 53–58. Resembles ***distinctly larger version*** of Rose-ringed Parakeet, with clearly heavier bill and ***large red patch on shoulders at all ages*** (sometimes partly hidden when perched), which absent in Rose-ringed Parakeet. Rose-pink neck ring of adult male easily seen; other sex and age differences as in Rose-ringed. **Voice** Range of notes uttered, most loud and hoarse, commonest a deep macaw-like screaming note *kerrrck* with stress on first syllable, repeated 2–3 times. **Habitat** Parks, gardens and plantations; nests in hole in tree or building. **Note** Not native; escaped birds now naturalised.

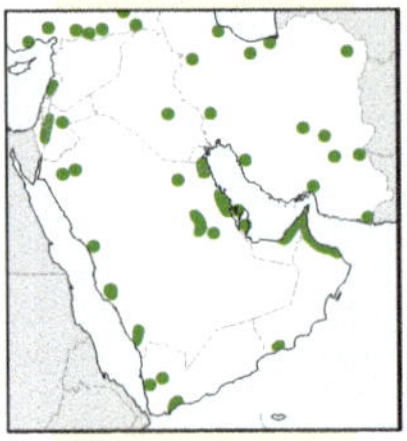

Rose-ringed Parakeet *Psittacula krameri* RB (E/I)

L: 42. Large green parakeet with ***long, graduated, pointed tail and red bill***. In male, ***black throat continues round neck as a narrow rosy ring***, absent in female; also absent in juvenile, which has green throat and horn-coloured bill. Flight swift, fast and direct but flocks often change direction rather suddenly. **Voice** Noisy; loud screaming *kee-ak*, rather falcon-like and piercing. **Habitat** Gardens and open wooded; also near cultivation; nests, often colonially, in holes in tree or wall. **Note** Colonies outside Iran probably originate from escapes.

Monk Parakeet *Myiopsitta monachus* E/I

L: 29. Small with pointed tail; sexes similar. Green above with ***greyish forehead, cheeks, chin and throat, breast grey-brown vaguely barred***; belly ochre, bill yellow-horn. Often in flocks. **Voice** Loud staccato rasping *krraaaak*. **Habitat** Urban parks and gardens; social nester; builds bulky stick nest in tree or on pole. **Note** Not native (from South America); escaped birds present Dubai.

Nanday Parakeet *Nandayus nenday* E/I

L: 35. Medium-sized parakeet with blue flight feathers; green above, paler below, tinged blue on breast with thighs orangey and ***head mostly black***. Tail long and pointed. **Note** Not native (from South America); escaped birds reported Dubai. [Alt: Black-hooded Parakeet]

Plum-headed Parakeet *Psittacula cyanocephala* E/I

L: 36. Slim, fast-flying parakeet; tail blue-green with whitish tip. ***Male has plum-red head; female has grey head with yellow upper breast and rear neck collar***. Both sexes have yellow-horn upper and dark lower mandibles, as does plainer, green-headed juvenile. **Voice** A shrill *too-ik, too-ik*; male chatters in flight, often ending with distinctive, hurried ringing *der-wink*. **Habitat** Wooded parks and gardens. **Note** A few free-flying Dubai and Abu Dhabi; may have bred but not yet naturalised.

Blossom-headed Parakeet *Psittacula roseata* E/I

L: 36. Very similar to Plum-headed Parakeet; tail blue-green with yellow tip. Head ***pinky in male, all light grey in female, which lacks yellow on breast, and has light green not yellow collar***. Bill of male as Plum-headed, but ***female and juvenile have both mandibles pale***. **Voice** Soft, musical *twee-too… twee… too… too… twee-too*; other calls much as Plum-headed. **Habitat** Wooded parks and gardens. **Note** Free-flying individuals regularly noted Abu Dhabi and Dubai.

Alexandrine Parakeet
♀
♂
Rose-ringed Parakeet
♂
♀
Monk Parakeet
ad
Nanday Parakeet
ad
♂
♀
Plum-headed Parakeet
♂
Blossom-headed Parakeet
♀

PLATE 55: CUCKOOS, KOEL & SCOPS OWLS

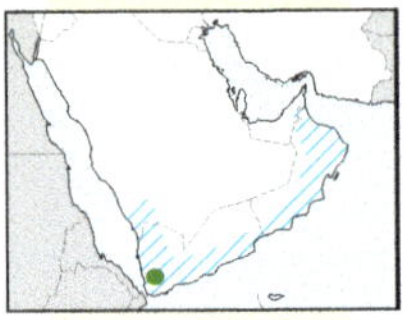

Pied Cuckoo *Oxylophus jacobinus* V

L: 33. Size of Common Cuckoo; distinctive ***black-and-white plumage, crest, long graduated tail and conspicuous white wing-patches in flight***. A rarer black morph occurs (but which also shows white in wing). Juvenile is sooty above with dirty-white underparts and smaller crest. **Voice** Loud, metallic *piu-piu-pee-pee-piu, pee-pee-piu*. **Habitat** Scrub, woodland edge. **Note** Scarce summer/passage visitor as hatched; vagrant Iran, UAE. [Alt: Jacobin Cuckoo]

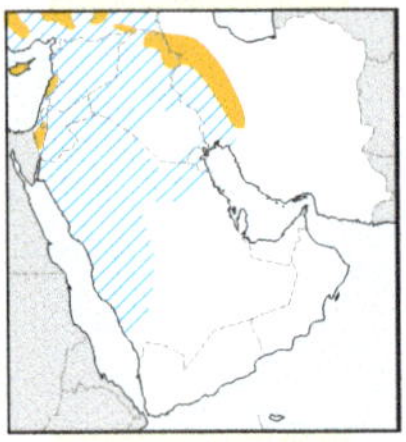

Great Spotted Cuckoo *Clamator glandarius* V

L: 40. Large cuckoo with ***slight crest, very long tail, grey upperparts and white-spotted coverts; underparts white***, with throat buffy. Juvenile is dark brown above with blackish head, white-spotted coverts and shows ***distinctive chestnut primaries (above and below)***. Often on ground; flight rather fast, with shallow wingbeats. **Voice** Harsh, chattering *chil, chil, chil, chil*. **Habitat** Woodland, ocultivation with bushes and trees. **Note** Passage hatched (rarer in autumn); vagrant Bahrain, Oman, Qatar, UAE, Yemen.

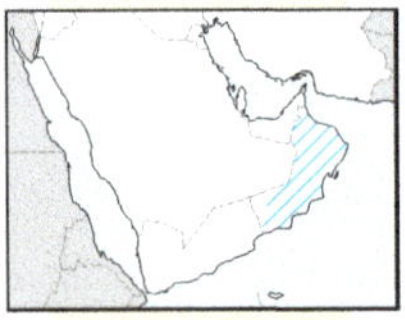

Asian Koel *Eudynamys scolopaceus* V

L: 43. Large with stout bill, short wings and rather long tail. ***Male black with bluish gloss and bright yellow-green bill***. Female has ***drab brown upperparts, thinly streaked and spotted white***; throat heavily streaked dark brown, rest of underparts densely barred dark brown and buff; dark tail with many thin whitish bars. Immature resembles female. Eye garnet red. When not alarmed has characteristic hunched stance. Flight direct. **Voice** Seldom heard in region; loud whistled *kooyl*; also a rapid ascending series of bubbling notes *kwow-kwow-kwow-kwow*. **Habitat** Scrub, gardens, woodland, fruit trees. **Note** Rare winter visitor to hatched area; vagrant Iran, Kuwait, UAE.

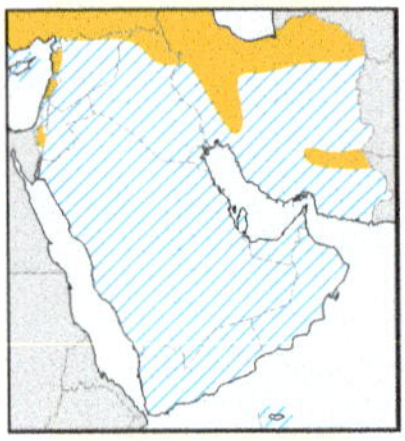

Common Cuckoo *Cuculus canorus* PM, ?mb

L: 33. In flight, has resemblance to Common Kestrel, but head held slightly raised and ***wings not lifted above level of body***. Grey above with barring on underparts; grey underwing has pale line through centre. Female similar, with warm wash on breast. Rufous morph, in which female and juvenile have brownish-red plumage with noticeable barring on upperparts and breast, relatively common in Middle East. Grey morph juvenile is dark brownish-grey, with pale fringes above and whitish spot on nape. **Voice** Far-carrying song ***unmistakable, kwuk-koo*** or *kwuk, kwuk-koo*, first note highest; female has distinctive descending, long bubbling call. **Habitat** Open country, trees and bushes, ***open*** woodland. **Note** May breed north UAE/Oman; passage hatched.

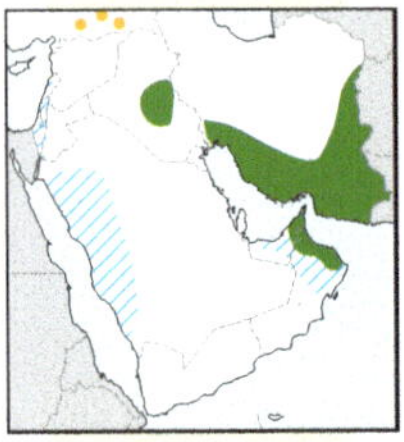

Pallid Scops Owl *Otus brucei* RB, pm, wv

L: 20–21. Typical scops owl, sandy-grey or grey above without white spots on crown or hindneck, usually with ***clear-cut pencil-fine black streaks below*** (grey morph of Eurasian Scops Owl has black streaks crossed by vermiculations and paler blotches below, submerged by darker ground-colour). Braces creamy to cinnamon-ginger. ***Tail extends marginally beyond wing-tips; tail-bands usually diffuse, especially toward tip. Feathering on legs extends to base of toes, middle toe especially. Juvenile completely barred below*** (unlike juvenile Eurasian Scops). **Voice** Diagnostic, ***soft and dove-like***, carrying only short distance (unlike Eurasian Scops): hollow, resonant, low-pitched *whoop* or *whoo* repeated regularly about eight times in five seconds (like distant water-pump); also longer *whooo* repeated irregularly at 3–5-second intervals or *ooo-ooo... ooo-ooo*. **Habitat** Arid hills, semi-desert, wadis with trees, parks, palm groves. Nests in hole. **Note** Passage and winter hatched; vagrant Bahrain, Kuwait, Qatar, Yemen. [Alt: Striated Scops Owl, Bruce's Scops Owl]

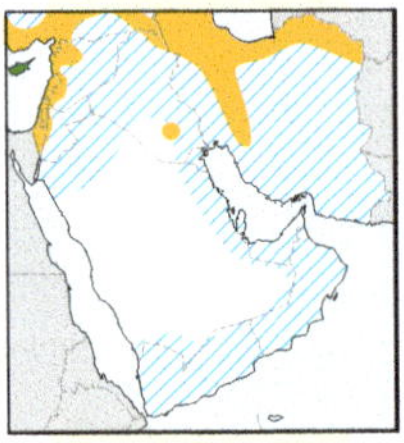

Eurasian Scops Owl *Otus scops* PM

L: 18–20. Typical scops owl with small ear-tufts; ***best distinguished by voice*** though migrant birds usually silent. Mothy grey-brown, or rufous-brown (rarer morph), with paler face and dark surround to yellow eyes. ***Wing-tips reach tail-tip; whitish tail-bands bordered narrowly black and usually clear-cut*** (but often cloaked by closed wings). Upperparts streaked and vermiculated; braces off-white, buff or creamy, sometimes rusty on outer web, with parallel ***deep rust or rufous line outside braces*** (if not hidden); underparts streaked, barred and vermiculated, ***interspersed with white blotches*** (lacking in Pallid Scops). Feathering on legs stops square-cut short of toe bases. Nocturnal **Voice** Clear, soft ***whistle repeated rhythmically*** and monotonously, *pwoo, pwoo* at 2–3-second intervals. **Habitat** Trees, groves. **Note** Passage hatched; occasional in winter.

Great Spotted Cuckoo
juv
pale morph
Pied Cuckoo
ad
dark morph
juv
Asian Koel
♀
♂
Common Cuckoo
♀
Pallid Scops Owl
♂
rufous morph
♀
rufous morph
grey morph
Eurasian Scops Owl

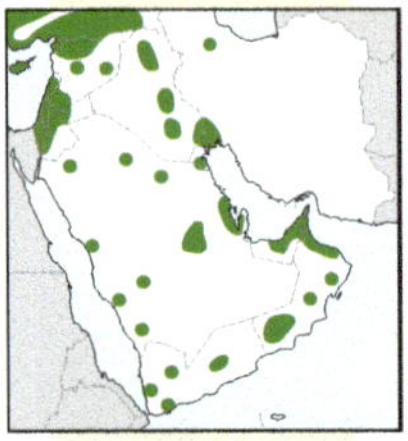

Barn Owl *Tyto alba* rb, wv

L: 35. W: 89. Strikingly pale, with ***creamy-buff upperparts, white underparts and heart-shaped face with black eyes***. Mainly crepuscular or nocturnal in region. **Voice** Territorial call of male a clear, vibrant, chilling shriek of about two seconds; alarm call in flight a shrill shriek, when disturbed will hiss; young beg with a drawn-out hissing. **Habitat** Open country with trees, semi-deserts, edges of woods, often near human habitation. Nests in hole in tree, building, ruins, crag, cave or nestbox. **Note** Much winter dispersal.

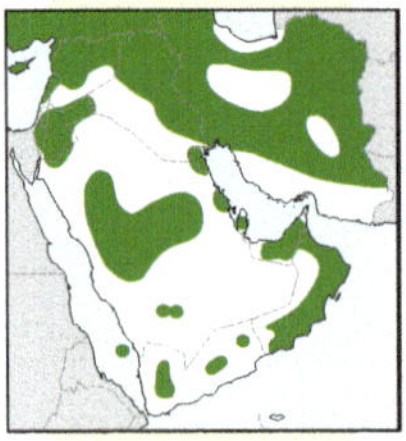

Little Owl *Athene noctua* RB (includes Lilith Owl)

L: 22. Small with ***round, flat-crowned head and long legs***. Subspecies vary in upperpart colour. Pale Arabian birds usually assigned to the subspecies ***saharae***, but birds in eastern Arabia often cold chocolate-brown may represent another subspecies. All subspecies have ***crown and nape distinctly white-spotted***, with white-blotched upperparts and wings, ***underparts boldly streaked***; eyes yellow, framed white. Flight deeply undulating, alternating rapid flapping with closing of wings. Largely crepuscular, but often sits in the open in daytime on rock, building or telegraph pole. If agitated, may bob (pogo) in an upright posture. **Voice** Territorial call a drawn-out and wailing *koooah*, also a short *kiu*; alarm call a sharp series of dog-like yapping notes *kip-kip-kip*.... **Habitat** Open country with trees, stony wasteland, wadis, rocky semi-deserts, sand desert with outcrops, cultivated areas. Nests in hole in tree, in rocks, buildings and burrows. **Note** Map includes Lilith Owl.

Lilith Owl *Athene (noctua) lilith*

L: 21–22. Possibly a separate species. ***Pale sandy-grey***, the palest birds with ***sparse, very pale buff streaks below***. Slightly smaller and slimmer than Little Owl, with ***smaller talons; rim of facial disc usually less distinct***. **Habitat** Much as Little Owl, but away from habitation and usually also trees. **Voice** Song of male somewhat drawn-out and slightly hoarse *gwuuh* or *gwuah*, less nasal and without the upward inflection typical of Little Owl; alarm calls less yapping. **Note** Not mapped separately. Range overlaps with Little Owl; reportedly resident S. Turkey, Cyprus, W. Iran and Arabia.

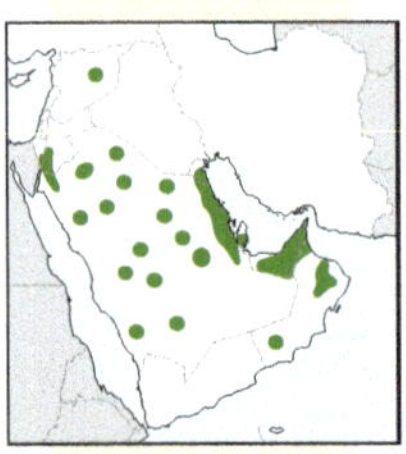

Pharaoh Eagle Owl *Bubo ascalaphus* RB

L: 50. A large owl, ***cinnamon or ginger in tone***, with most of ***lower breast and belly distinctly barred***. Eyes orange-yellow. Often sits out in daytime. **Voice** Loud; a ***booming* booor** and variety of other, mainly deep calls. **Habitat** Mountains, steppe, frequenting cliffs, crags and rocky outcrops; sand desert with large trees or bushes. Nests on ledge, down well, under roots or in old tree nest of raven or buzzard. **Note** Resident with some dispersal post-breeding and wandering in winter; vagrant Bahrain. [Alt: Desert Eagle Owl]

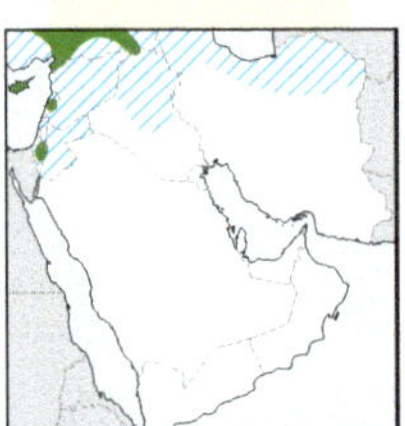

Long-eared Owl *Asio otus* V

L: 36. W: 95. At roost, muted brown and buff with distinctive ***long ear-tufts*** (though invisible when flattened); ***facial disc noticeably warm buff with striking white divide and orange eyes***. Flight jerky; fairly stiff wingbeats with glides on level wings (sometimes slightly raised). Separated from Short-eared Owl mainly by face pattern, ***entirely streaked underparts*** and wing pattern in flight (see Short-eared Owl). Nocturnal. **Voice** Vagrants silent. **Habitat** Woodland. **Note** Winter hatched; vagrant Kuwait, Oman, Qatar, Saudi Arabia, UAE.

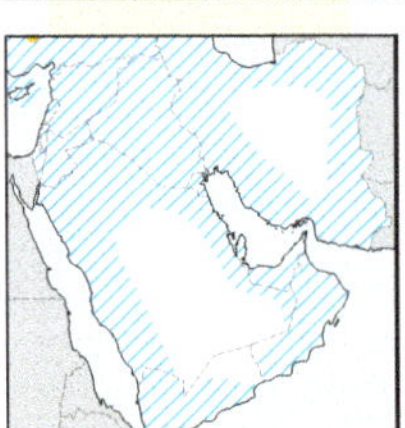

Short-eared Owl *Asio flammeus* wv

L: 38. W: 102. Often diurnal, flying with slow, ***elegant and high wingbeats*** on long, slender wings, ***raised during glides in shallow 'V'*** (shorter wings with shallower, faster wingbeats in Long-eared); sits on ground. Further differs from Long-eared Owl in strongly buff-spotted upperparts, paler greyish facial disc with striking black surround to ***glaring yellow (not orange) eyes, ear tufts short***. In flight, separated from Long-eared by ***yellow-buff base of primaries*** (buff-orange in Long-eared), ***more contrasting bars on flight feathers and tail, white trailing edge to upperwing***, black tips to underwing (not diffusely barred, as in Long-eared Owl) and ***dark streaking on underparts mostly confined to breast and contrasting with paler belly***. **Voice** Silent in winter. **Habitat** Open country, often marshy. **Note** Winter and passage hatched; rare south Arabia; vagrant Qatar.

Barn Owl
Little Owl
Pharaoh Eagle Owl
Lilith Owl
Long-eared Owl
Short-eared Owl

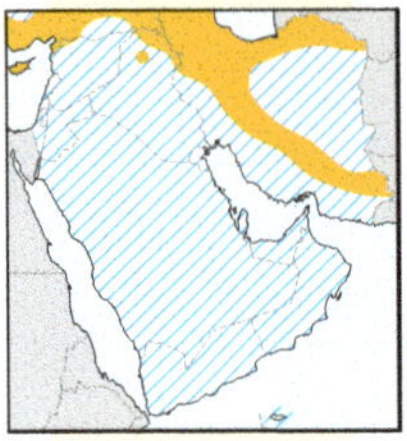

European Nightjar *Caprimulgus europaeus* PM

L: 26. The most widely encountered nightjar in the region; not resident. Distinguishing features (breeding subspecies *meridionalis* and passage migrant *europaeus*) are the ***dark-streaked grey crown contrasting with browner cheeks and throat, grey upperparts with broad bands of dark brown and buff-white on the scapulars and rows of whitish to buffy spotting on the coverts***. Male has white spots in wing and tail. Subspecies *unwini* (breeding in Iran and Iraq) is distinctly paler, with more sandy-grey upperparts, white-and-buff spotted coverts, and longer, whiter lower throat-patches; male shows larger white spots in primaries. (Migrant *plumipes*, assumed to occur, paler sandy above, much as Egyptian Nightjar, but retains prominent white cheek-stripe and has greyish flight feathers). Most often seen in flight at dawn and dusk; will hunt moths attracted to street lighting or car headlights. Frequently encountered in day perched longways along branch (also sits on ground) and if flushed usually flies only a short distance; note rather slow, soft wingbeats with long glides on stiffly held wings (as other nightjars), recalling cuckoo or falcon and similarly causing small birds in vicinity to alarm. Often sits on tracks at night. In courtship display wing-claps in flight, often also giving distinctive *kru-ipp* call. **Voice** Song at night (only rarely during daylight) long rising and falling churr, which can go on for many minutes, alternating on two pitches and highly ventriloquial. Commonly calls again after dusk or pre-dawn, a distinctive, loud throaty *kru-ipp*, or falling, discontented guttural *kworr-kworr*. **Habitat** Edges of woods and heaths, steppes with sparse vegetation; any open areas on migration, often in or near trees or shrubs. **Note** Passage hatched.

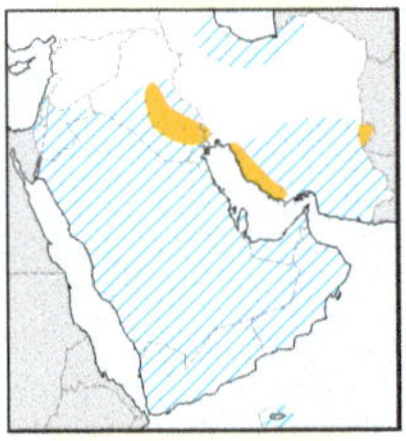

Egyptian Nightjar *Caprimulgus aegyptius* WV, PM

L: 25. The palest nightjar of the region, Egyptian Nightjar having ***pale sandy-grey base-colour, broad rows of inconspicuous buff tips to wing-coverts and white patch on side of neck (often very hard to see)***. Crown plain, finely dark flecked on rear. ***In flight, very pale underwing, and above, dark flight feathers contrasting with pale upperwing-coverts. Both sexes lack white patch in wing***, but male shows pale creamy spots on outer tail feathers below. Juvenile has ill-defined spotting, not barring, on underparts. From European Nightjar by pale sandy plumage, contrast between flight feathers and paler coverts and lack of white spots in less pointed wings. Invariably only ever sits on ground; often on tracks at night. **Voice** Song a regular, rapidly repeated *kowrr-kowrr-kowrr* slowing towards end. **Habitat** Semi-deserts often with palms or scrub. **Note** Passage hatched, often rare; winters south Arabia. Song has been reported in spring from Acacia plain in Ras al Khaimah, UAE.

Sykes's Nightjar *Caprimulgus mahrattensis* V

L: 22. ***Rather plain and greyish***, features that help to distinguish it when perched from the other nightjars encountered in region. Told from Egyptian Nightjar by ***white primary spots*** and, in male, large white tips to undertail (Egyptian Nightjar lacks white spots on primaries and has only diffuse pale tips to undertail) and, if seen well, ***feathered tarsi***. Told from European Nightjar by plain plumage (lacking European Nightjar's contrasting greyish or buffish crown, dark cheeks and two broad pale bands on scapulars and coverts). **Voice** Readily told by voice, ***a long frog-like purr***, after sunset and pre-dawn. Soft *cluck-cluck* when flushed by day. **Habitat** Deserts and semi-deserts, sandy or stony terrain; abandoned cultivation, often with tamarisks. **Note** Vagrant UAE.

European Nightjar

Egyptian Nightjar

Sykes's Nightjar

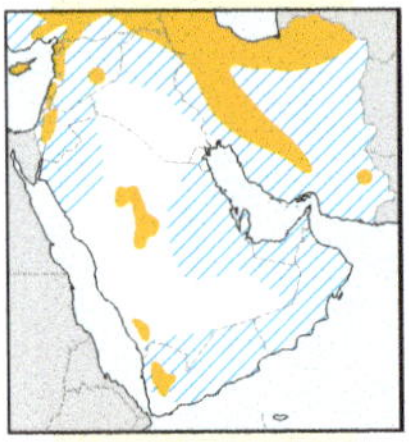

Alpine Swift *Tachymarptis melba* — V

L: 21. W: 57. Shape and behaviour as Common and Pallid Swifts, but ***much larger with markedly slower and deeper, scything wingbeats; white underparts broken by dark breast-band; vent and undertail-coverts brown***, as upperparts. **Voice** Loud, dry chattering trill unlike that of Common Swift, *trit-it-it-it-itititit-it-it-it*, accelerating then decelerating, rising and falling. **Habitat** Mountain ridges, steep rock faces, also sea-cliffs. Nests, usually colonially, in natural crevices on cliff face, occasionally in ancient building. **Note** Mainly summer visitor; passage hatched, rare SE Arabia; winter records Iran, UAE (where vagrant), Yemen.

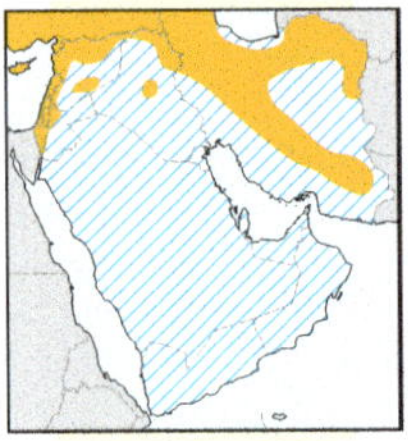

Common Swift *Apus apus* — PM

L: 16. W: 45. ***Almost uniform sooty in fresh plumage***, but brownish and more contrasting in worn plumage and in eastern birds, though much depending on light conditions. ***Whitish throat-patch round, but variable, sometimes appearing absent; lacks contrast above but underparts show inner flight feathers clearly paler than body and wing-coverts***; tail-fork deep. Easily confused with Pallid Swift (which see). Wing moult of adults is entirely in winter quarters (unlike Pallid Swift). From other swifts by size, structure and lack of white rump-patch. **Voice** High-pitched ringing *shreee*, variable in duration and pitch; hard to tell from Pallid Swift. Mostly silent on migration. **Habitat** Aerial; congregates in areas with suitable nesting sites or abundant food. Nests in buildings, under eaves, occasionally in cliffs. **Note** Passage hatched.

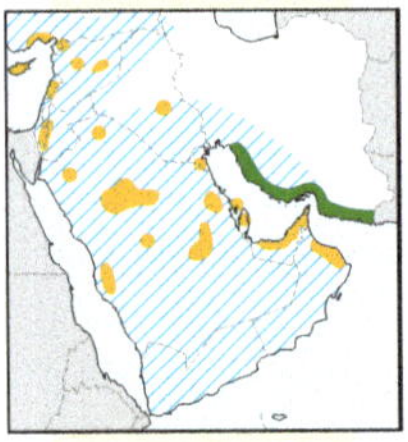

Pallid Swift *Apus pallidus* — MB, PM

L: 16. W: 44. Slightly bulkier than Common Swift but all brownish with broader, less tapering outer wing and sometimes blunter wing-tips, tail-fork variable but slightly shallower (adults) with less pointed tips. Brown tones variable, depending on light conditions and wear and often difficult to separate from Common Swift but ***head looks broad and flat with larger, more triangular, whitish throat-patch; forehead and lores paler, contrasting with dark eye-patch. Faint scaling is visible on underparts*** when seen passing close (rarely seen in Common Swift); ***upperparts show darker 'saddle'*** slightly contrasting with paler inner flight feathers, head and rump. Underwing shows darker outer primaries than inner wing, much as in Common Swift. Adults commence wing moult in breeding areas in summer/autumn, unlike Common Swifts, which often appear badly worn. **Voice** Similar to Common Swift, but deeper, hoarser (tinnier, less ringing; individual notes less clear-cut) and slightly disyllabic, *sree-er*, which falls off or ends abruptly. **Habitat** As Common Swift; nests in historic buildings, new towers, cliffs or craggy outcrops. **Note** Migrant breeder; apparently resident Iran; arrives Gulf colonies November (until May/June); passage hatched.

Fork-tailed Swift *Apus pacificus* — V

L: 18–19. Similar to smaller White-rumped Swift (which see). Larger than Common Swift with ***white rump, more deeply forked tail***, longer wings, ***pale scaling to underparts*** and slimmer build with ***attenuated (Arctic Skua-like) rear end***. Also ***more protruding head*** and larger white chin-patch. Partially leucistic Common Swifts may possess partly white rump, but lack underpart scaling. Pallid Swift, scaled underneath, easily eliminated by all-brown plumage. **Voice** Unlikely to be heard in region; coarser and harsher than Common Swift, closer to that of Pallid Swift in dropping off at end. **Habitat** Open country. **Note** Vagrant UAE (from Asia). [Alt: Pacific Swift]

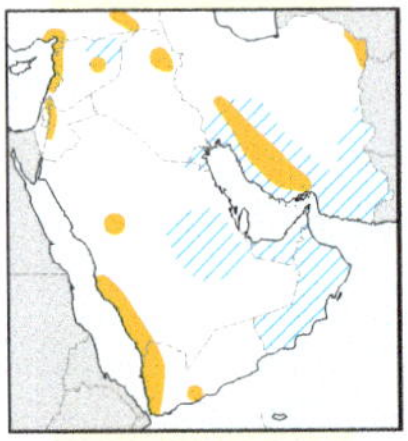

Little Swift *Apus affinis* — pm, wv

L: 12. W: 34. Smallest swift in the region, distinctive stout silhouette with ***short, square-ended tail*** (round when spread); prominent ***deep, white rump-band 'wrapped-around' onto rear flanks*** (some white visible at all angles). At longer distance can recall house martin, but dark underparts and stiff wingbeats separate. Somewhat fluttering flight, alternating with short glides. **Voice** Fast, high-pitched, rippling trill *dillillillillill*, regularly rising and falling in pitch, much higher-pitched and faster than Common Swift. **Habitat** Over grassland or near water, often with other swifts, also gorges, towns and cities (nests colonially on ceiling of open building, under rock overhangs, cave roofs). **Note** Passage hatched, but rare; vagrant Bahrain.

Alpine Swift
Common Swift
Pallid Swift
Little Swift
Fork-tailed Swift

PLATE 59: ROLLERS & KINGFISHERS

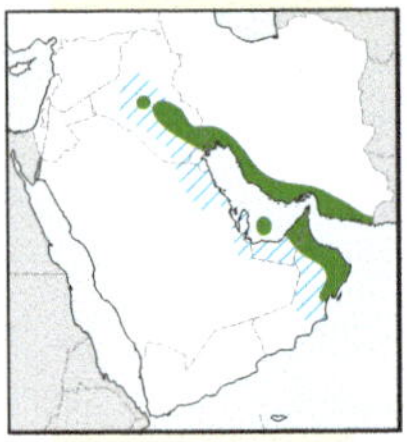

Indian Roller *Coracias benghalensis* RB

L: 30. Stocky and multicoloured; told at all ages in flight ***by large, pale turquoise-blue primary patch*** (above and below) ***and pale turquoise-blue rectangles in sides of tail base*** (seen when tail spread). Wing-tip clearly blunter than in European Roller. When perched, ***lightly white-streaked neck, throat and breast are vinous-cinnamon*** (turquoise-blue in European Roller) ***and cap dark turquoise-green***; mantle earth-brown (pale chestnut in European). Aerobatic, In pursuit of flying insects and in sky-diving display. **Voice** Similar to European Roller, but with more barking *rak*; agitated sneezed *chew-chew-chew* in display or towards intruder in territory. **Habitat** Open cultivated country with scattered trees, plantations, parks, gardens; usually below 1,000m. Nests in hole in tree or wall. **Note** Mainly resident, autumn dispersal hatched, but rare; vagrant Qatar, Yemen.

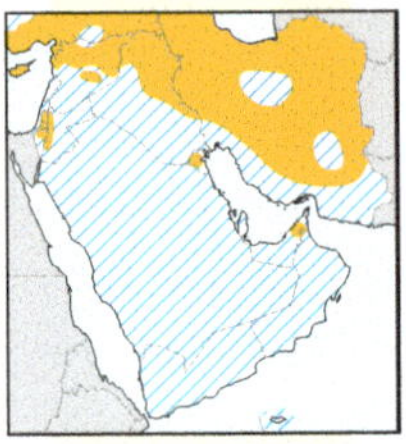

European Roller *Coracias garrulus* PM

L: 30. Stocky; ***turquoise-blue body and most wing-coverts contrasting with blackish flight feathers, chestnut back and deep blue leading forewing above***. Colours often faded in autumn (and paler-hued in subspecies *semenowi*, breeding in Iran and Iraq). Juvenile duller and browner with lightly streaked neck and breast. Often sits on prominent perch (wires, poles, dead branches) taking prey on ground. In display flight 'tumbles' from side to side in downward dive (not unlike Northern Lapwing). **Voice** Sonorous, hoarse *rack-rack* (recalling Eurasian Magpie); also in display or when agitated a loud piercing *keer-keer-keer* and repeated dry gravelly grating note. **Habitat** Open country with large trees, rarely to 2,000m. **Note** Passage hatched; formerly bred UAE (as mapped).

White-throated Kingfisher *Halcyon smyrnensis* V

L: 26. Large, brightly coloured kingfisher with ***enormous red bill, dark chocolate-brown head and belly, large white throat and breast 'bib', brilliant turquoise-blue upperparts*** and black forewing. Flight fast and straight, showing conspicuous white primary patches. Often perches on wires (but can sit hidden) looking for prey on ground. Rather noisy. **Voice** Loud raucous yelping *kril-kril-kril-kril*; also tittering descending song. **Habitat** Dry woodland glades and palm groves, as well as tree-lined lakes, rivers or other wetlands. **Note** Some winter dispersal, including to Saudi Arabia; vagrant Qatar, UAE.

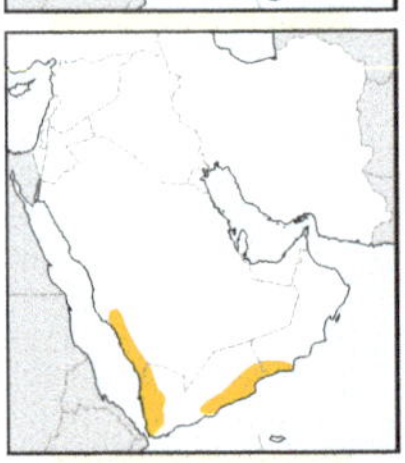

Grey-headed Kingfisher *Halcyon leucocephala* V

L: 20. Fairly large, with large red bill, ***buffish-grey head and neck, off-white throat and upper breast; bluish-black wing-coverts and back bluish, with chestnut belly***. White patch in primaries visible in flight. In first autumn, slightly darker head, dark scalloped pectoral band and black-tipped bill. Often sits on dead branch, poles, wires or other prominent perch. Not dependent on water. Flight sluggish and undulating. **Voice** Weak chattering *ji, ji, ji-jeee*. **Habitat** Wadis with trees (with or without water). **Note** Vagrant UAE.

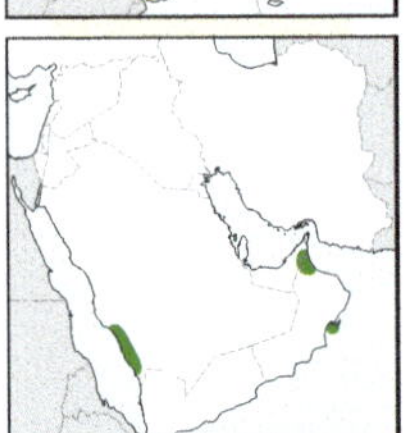

Collared Kingfisher *Todirhamphus chloris* RB

L: 24. Large kingfisher easily told by ***turquoise upperparts, all-white underparts and white collar***, bordered above by long black eye-stripe and white supercilium to nape; base of bill horn, upper mandible greyish. Juvenile has dusky barring on breast. Flight appears rather weak. Noisy, particularly at dawn, intermittently at other times. Often perches low in mangroves on look-out for crabs, on which it mostly feeds; when perched on ground tail usually held cocked. **Voice** Distinctive kookaburra-like calls; a fast series of loud, ringing notes, each ascending, until end *chei-chei-chei-chei*. **Habitat** Mangroves. Nests in hole in tree or bank. **Note** Endangered endemic subspecies *kalbaensis* restricted to Khor Kalba, Sharjah and nearby sites on Batinah of Oman.

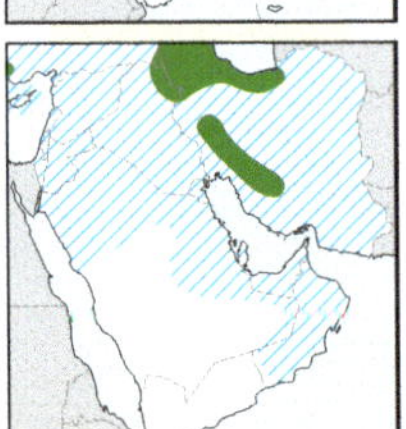

Common Kingfisher *Alcedo atthis* WV, PM

L: 17. Small kingfisher with ***brilliant blue and green upperparts, reddish-orange underparts*** (often bleached in autumn), but buffish-white throat and neck-patch and long, blackish bill (though female has red on most of lower mandible). In flight, which is swift, direct and low over water, luminous back and tail obvious. Fairly shy, often inconspicuous when perched, sitting on overhanging branch for long periods before diving for fish; often hovers. **Voice** High-pitched, thin, piercing *cheee* or *tzeee*, mostly in flight. **Habitat** Rivers, streams, canals, lakes; in winter also coasts. **Note** Passage and winter hatched; vagrant Yemen.

ad
ad
Indian Roller
European Roller
ad
ad
White-throated Kingfisher
Grey-headed Kingfisher
ad
ad ♀
Collared Kingfisher
Common Kingfisher

Pied Kingfisher *Ceryle rudis* — V

L: 25. Large and unmistakable; ***the only black-and-white kingfisher in region, frequently seen hovering well above water,*** plunging for fish (but also fishes from perch). White underparts have two more or less complete black breast-bands in male, one in female. Black eye-mask, white supercilium, short crest at nape; white-sided tail has black band at tip. Juvenile has greyish breast-band. Sometimes in small groups. Rather vocal. **Voice** Loud, noisy chattering *chirrick, chirrick, chirrick*. **Habitat** Rivers, lakes, ponds, mangroves and open coasts. Nests in hole in bank. **Note** Winter hatched; vagrant Oman, UAE.

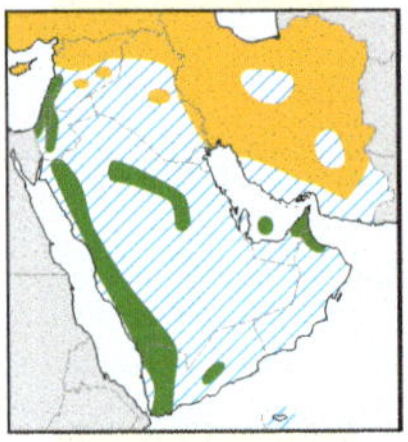

Eurasian Hoopoe *Upupa epops* — RB, mb, PM, wv

L: 28. ***Distinctive pinkish-buff with bold black-and-white bars on wings and tail, long black-tipped crest, usually depressed*** (raised on landing), ***long decurved bill; flight with flaps of broad wings***. Spends much time on the ground. **Voice** Male's song distinctive, ***repeated hollow poo-poo-poo***; also a dry *terrr* when agitated and strange, thin squeaking and hissing notes when courting. **Habitat** Woodland, olive and palm groves, parks, gardens, oases; open and wooded areas in winter. Nests in hole in tree or ruin. **Note** Passage and winter hatched (but absent from north in winter).

White-throated Bee-eater *Merops albicollis* — V

L: 30. More elegant than European Bee-eater, ***easily distinguished by black-and-white head pattern, white underparts with black collar around throat and very long central tail-streamers;*** also has bluish-green upperparts and blue tail. In flight shows ochre upperwing and coppery underwing, both with black trailing edge. Often remains below tree canopy and easily overlooked. **Voice** Higher-pitched and softer than European Bee-eater, *prrrp, prrrp, pruik*. **Habitat** Hills, plains and wadis with bushes and trees, agricultural land. **Note** Vagrant Oman, UAE.

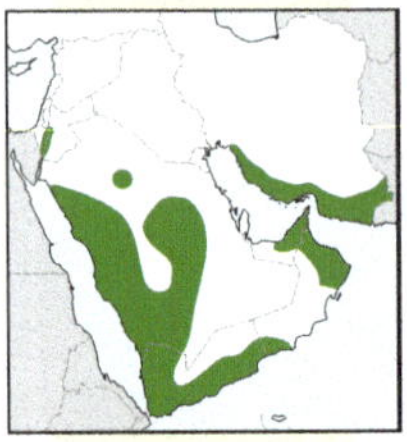

Green Bee-eater *Merops orientalis* — RB

L: 24. W: 30. ***Small, mainly green bee-eater*** with black eye-stripe and elongated central tail feathers. Arabian birds (*cyanophrys/muscatensis*) have blue on supercilium and throat, rather diffuse dark breast-band (often absent) and shortish tail-streamers. Crown and nape shining coppery-gold in fresh-plumaged adults. S. Iranian subspecies, *beludschicus*, has blue confined to chin and below eye-stripe, narrow breast-band and longer tail-streamers. Juvenile duller and lacking tail-streamers. Usually in pairs. **Voice** In flight, a high-pitched *trreet-trreet* or *prrrit*; often burbles excitedly. **Habitat** Open country with trees, semi-desert, wadis, cultivations, parks, gardens. Nests in tunnel in bank or hole excavated in ground. **Note** Some dispersal and seasonal movements, vagrant Bahrain, Qatar.

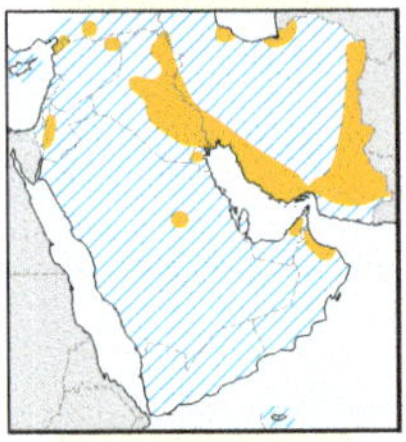

Blue-cheeked Bee-eater *Merops persicus* — mb, PM

L: 30. W: 48. Larger than European Bee-eater and ***distinctly green or turquoise-green*** with long central tail-streamers and strongly ***rusty-red underwings*** framed dark along trailing edge. Juvenile duller and lacks long tail projections. From young European Bee-eater by entirely green plumage, including crown and underparts, and rusty-red underwings. Gregarious; hunts insects in flight. Vocal and audible at long range. **Voice** Very similar to European Bee-eater, but higher pitched and ***certainly hoarser (throatier)***, the notes sometimes being disyllabic, *prrllip-prrllip* or *prl-rip*. **Habitat** Dry open country with scattered trees; often on overhead wires. Almost anywhere on diurnal passage. Nests colonially in holes excavated in sandy ground. **Note** Passage hatched.

European Bee-eater *Merops apiaster* — PM

L: 28. W: 46. Easily distinguished by ***chestnut crown and back, bright yellow throat contrasting with turquoise-blue underparts*** and, in adult, ***chestnut upperwing-coverts***. In flight shows paler, pinkier underwings than other bee-eaters in the region. Juvenile has greenish upperparts, but shows chestnut crown and yellow throat (both paler); central tail feathers mere spikes or lacking. Migrates in vocal flocks, often high overhead, sometimes audible but remaining unseen. Hunts insects in flight. **Voice** Similar to Blue-cheeked Bee-eater but softer, lower pitched and more liquid, a far-carrying *prruup*, usually not disyllabic. **Habitat** Open bushy country with scattered trees, riversides and woodland glades; often on overhead wires. Nests colonially in holes excavated in sand banks, riversides and roadside cuttings. **Note** Passage hatched; formerly bred UAE (as mapped).

ad ♂
Pied Kingfisher
ad
Eurasian Hoopoe
ad
White-throated Bee-eater
ad
Green Bee-eater
ad
Blue-cheeked Bee-eater
juv
ad
ad
juv
European Bee-eater

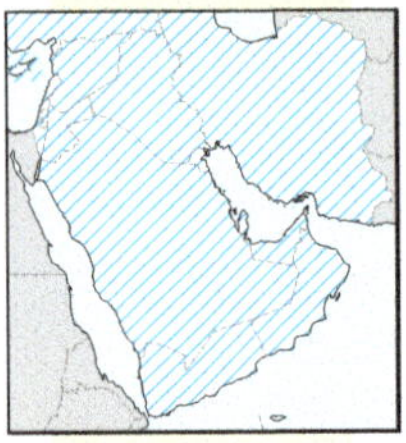

Eurasian Wryneck *Jynx torquilla* PM, wv

L: 16. Inconspicuous slim, atypical, brown woodpecker with finely vermiculated cryptic plumage. Flight direct, shallowly undulating, reminiscent of large, long-tailed warbler or female Red-backed Shrike. Distinctive features are long tail, ***brown scaling on off-white underparts, dark eye-stripe, black central crown streak onto mantle, buff mottled wings and finely barred yellowish throat***. Often forages on ground. **Voice** Song a loud, monotonous, plaintive *vee-vee-vee-vee* recalling small, distant falcon. **Habitat** Open woodland, orchards, parks; nests in hole in tree. Any cover on migration. **Note** May breed NW Iran/NE Iraq; passage hatched, some winter Arabia.

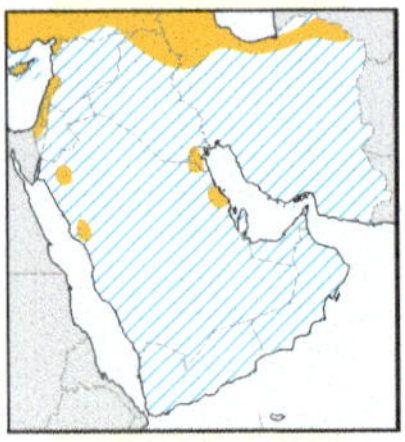

Eurasian Golden Oriole *Oriolus oriolus* PM

L: 24. Male unmistakable; female and immature male greenish above with olive-brown wings and tail, rump yellowish-green, underparts yellowish-white, indistinctly dark streaked. Appears relatively short-tailed in flight, with woodpecker-like undulations, but strong and fast; often changes direction, tilting and angling wings accordingly. Heard more often than seen. Subspecies *kundoo*, 'Indian Golden Oriole', possibly a separate species, may visit (collected Iran; reported Oman); longer billed with black extending behind eye, male with tertials and inner secondaries broadly tipped yellow; female and immature with extensively yellowish washed underparts and finer streaks than Eurasian. **Voice** Song loud ***yodelling* tjoh-wlee-kleeooh**, the last note (often given alone) emphasised and descending, with tone of Eurasian Blackbird; alarm hoarse, mewing Jay-like *kra-eik*. **Habitat** Parks, gardens, broad-leaved woodland. **Note** Passage hatched; occasional in winter Oman, UAE.

Ashy Drongo *Dicrurus leucophaeus* V

L: 29. Similar to Black Drongo but plumage greyer and ***bill proportionally longer and slimmer; lacks white rictal spot***. Adult glossy dark slaty-grey above; dark grey below; ear-coverts contrastingly darker depending on light. First-winter birds, palest on belly, have matt smoky-grey underparts ***without white fringes***, latter also absent above (white blotches or fringes present above and below in first-winter Black Drongo); undertail-coverts white-tipped, eye fairly bright red (garnet in young Black Drongo). Crown flattish, nape slightly peaked (more rounded in Black Drongo). ***Hunts from within or under canopy***, less often from exposed open perch. Migrant subspecies *longicaudatus* recorded. **Voice** Alarm an abrupt, short, harsh, dry rattle. **Habitat** Woodland, parkland. **Note** Vagrant Kuwait, UAE.

Black Drongo *Dicrurus macrocercus* V

L: 28–30. Adult glossy blue-black; long, deeply forked tail curves outwards towards end; ***small white rictal spot in all ages, bill heavy***. First-winter duller with white tips on mantle, flanks, rump and especially belly feathers. Sits upright on conspicuous perch, making aerobatic aerial sallies in pursuit of insects. **Voice** Harsh, throaty *schweep-schweep*, also high-pitched whistles and clicks; vagrants usually silent. **Habitat** Trees, scrub, cultivation, often near habitation and water. **Note** Formerly bred SE Iran; vagrant Iran, Oman, UAE.

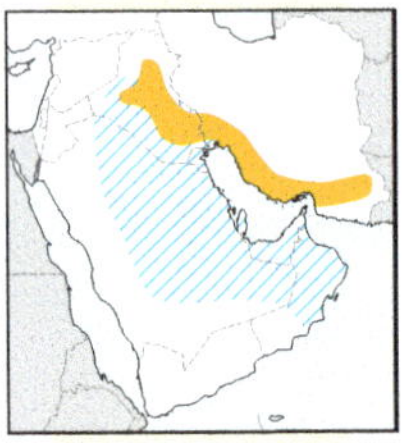

Hypocolius *Hypocolius ampelinus* PM, WV

L: 23. Recalls slim Southern Grey Shrike but with longer tail. ***Sleek, soft blue-grey male, with black-tipped tail and black eye-mask joining over nape*** (obscure in young male); primaries black (barely showing at rest) with ***pure white tips, prominent in flight***. Female and immature featureless mousey grey-brown, lacking black on head and with only diffuse dark tip to tail. May raise slight crest. Can be tame; often remains still, lurking in cover. Will fly high with steep climb, often for a distance, and flocks will circle for several minutes; note long tail, short wings and rather rapid wingbeats. **Voice** Mellow flight call, a ***liquid* tre-tur-tur**, notes running together and the last two lower pitched. When perched, ***a descending* whee-oo**, like Eurasian Wigeon. **Habitat** Fruit-bearing trees, scrub, palm groves on passage and in winter. **Note** Winter hatched (some birds resident in breeding range); rare Oman. [Alt: Grey Hypocolius]

Eurasian Golden Oriole
Eurasian Wryneck
♀
♂
Ashy Drongo
Black Drongo
Hypocolius
♂
♀

Woodchat Shrike *Lanius senator* PM

L: 18. Perches prominently; ***rufous-brown crown and nape and large white shoulder-patch on black upperparts*** diagnostic. Female duller than male, with more white-buff at base of bill, and rump only slightly greyer than mantle. Juvenile/first-winter resembles juvenile Red-backed Shrike but has ***paler grey-brown upperparts with creamy-white markings on scapulars, wing-coverts and rump, and base of primaries*** (pale wing-patch noticeable in flight). **Voice** Song varied whistles and trills, often mimics; calls *schak-schak* and grating rattles. **Habitat** Open bushy country, parks. **Note** Passage hatched, rarely winters Arabia and Iran.

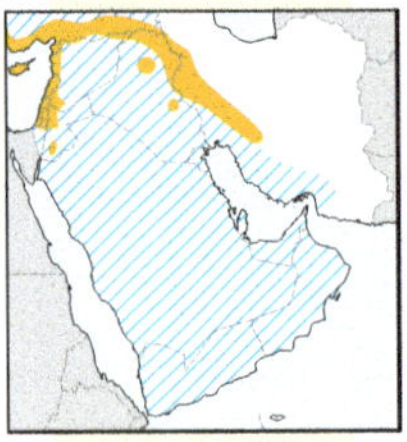

Masked Shrike *Lanius nubicus* PM, wv

L: 18. Male with ***black upperparts and mask, white face, orange wash on flanks*** (female similar but duller) and distinctive white shoulder-patch. ***White base of primaries and outer tail feathers obvious in flight***. Juvenile similar to juvenile Woodchat but with finer bill, longer tail, paler forehead, colder, ***greyer ground colour to upperparts and rump concolorous with back***. First-winter lacks orange flanks. Perches mid-level in tree or bush with ***long tail pumped slowly downwards or slightly cocked and waved up and down***. Sometimes secretive. **Voice** Harsh, scolding, dry *krrrrrr*. Song rather quiet, a tuneless jumble of notes. **Habitat** Open woodland, cultivation, parks, scrub. **Note** Passage hatched; occasional in winter south Arabia.

Brown Shrike *Lanius cristatus* V

L: 18–19. Dark rufous-brown to rusty olive-brown upperparts; tertials pale fringed, ***crown, mantle and tail concolorous; tail graduated*** sometimes faintly barred. Subspecies *luscionensis* has grey crown and nape; intergrades with nominate *cristatus* occur. Breast creamy- to yellowish-buff, throat white, flanks rufous-buff. Bill usually appears deep, heavy. Immature told from similar immature Red-backed Shrike by ***absence of white at base of primaries*** (rarely small amount present), ***shorter primary projection (only 4–5 primary tips showing beyond tertials), tail shape and underparts colour;*** bill pale-based with dark tip. Often furtive in cover. **Voice** Loud, excitable staccato ratcheting. **Habitat** Scrub, woodland edge. **Note** Vagrant Oman, UAE.

Red-backed Shrike *Lanius collurio* PM

L: 18. Adult male unmistakable; female and first-autumn birds warm brown above, tail dark brown, underparts whitish with dense crescentic barring on breast and flanks, also above in immatures. Barring, colour of rump, tail and upperparts separates from Daurian and Turkestan Shrikes. Immature similar to Woodchat Shrike (which see). ***Primary projection long***, thus eliminating Brown Shrike. **Voice** Song quiet and musical warbling with mimicry. Call a short *shack*, alarm hoarse, hard *keck-keck-keck*. **Habitat** Scrub, thickets, lightly wooded areas. **Note** Passage hatched.

Daurian Shrike *Lanius isabellinus* PM, WV

L: 16–18. Pale sandy to sandy-grey above, whitish to creamy- or orange-buff below, with burnt orange flanks. ***Rump and tail foxy- to orange-red***, to dull brown in some females, square or only slightly rounded. Immatures have ***upperparts except mantle finely barred*** (crescents on upperparts of Red-backed Shrike have ***dark and pale fringes adjacent***); faintly barred on flanks. In females white at base of primaries often reduced, rarely absent. Adult from Turkestan Shrike by ***paler upperparts, creamy to warm orange-buff supercilium*** (***never white***), ***pale lores and creamier breast and belly***. **Voice** Short, hard dry staccato ratcheting. **Habitat** Open wooded or scrubby areas, cultivations, parks. **Note** Passage and winter hatched (absent north in winter). [Alt: Isabelline Shrike]

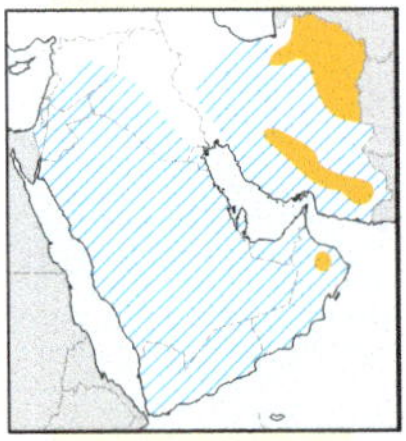

Turkestan Shrike *Lanius (isabellinus) phoenicuroides* PM, wv

L: 16–18. Sandy grey-brown to rich warm earth-brown above, some almost as rich as Brown Shrike, with contrasting foxy-red rump and tail, latter often ruddier-chestnut than in Daurian Shrike. Spring male from Daurian by ***rufous crown, white supercilium, darker upperparts and dark lores, also whiter underparts*** with flanks rusty; female and immature similarly darker above but probably not all immatures safely distinguishable. Intergrades occur; form *karelini* sandy or sandy-grey above with pale rufous to concolorous sandy-grey crown. From similar Brown Shrike by tail shape, slightly longer primary projection (six primaries exposed beyond tertials, but beware in winter moult) and white at base of primaries (sometimes not visible at rest). **Voice** Calls as Daurian Shrike; also mimetic. **Habitat** As Daurian. **Note** Passage hatched; rare or absent in winter.

Woodchat Shrike
♂
1st-winter
Masked Shrike
♂
1st-winter
Brown Shrike
1st-winter
♂
♀
Red-backed Shrike
1st-winter
♂
Daurian Shrike
♂
♀
phoenicuroides
1st-winter
♂
Turkestan Shrike
♂
karelini

Bay-backed Shrike *Lanius vittatus* V

L: 18. Plump, long-tailed and large-headed shrike; adult has ***broad black forehead-band reaching forecrown, greyish rump, whitish upper tail-coverts, white sides to tail and large white mirror at base of primaries, forming conspicuous bar in flight*** (larger Long-tailed Shrike has graduated tail with buff outer feathers, narrow black forehead-band, bright orange-buff rump and upper tail-coverts and minute wing-mirror); female duller than male. Juvenile has rufous tail and faint wing-mirror, told from young Daurian and Turkestan Shrikes by rufous-edged greater coverts and grey lower rump. First-winter birds superficially like adult but have wavy barring on flanks and upperparts and the black forehead-band may be absent. **Voice** Calls grating; song quiet and pleasant, imitating other species. **Habitat** Open, often rocky country with scattered trees and scrub. **Note** Vagrant Oman (has bred), Qatar, UAE (has bred).

Lesser Grey Shrike *Lanius minor* PM

L: 20. Resembles Southern Grey Shrike but smaller with ***stouter bill, proportionally longer wings with long primary-projection*** and shorter, less graduated tail; ***broader white wing-panel*** confined to primaries. ***Adult shows extensive black forehead*** without white supercilium, bluish-grey upperparts lacking white on shoulders (sometimes faintly) ***and pinkish-white underparts***. Juvenile/first-winter lack black forehead, have paler bill, brown-grey upperparts, finely barred darker, and paler tips to wing-coverts and flight feathers; underparts creamy-white sometimes faintly barred. **Voice** Chattering, varied whistles, trills and mimicry. **Habitat** Open cultivated country with scattered trees and bushes. **Note** Passage hatched.

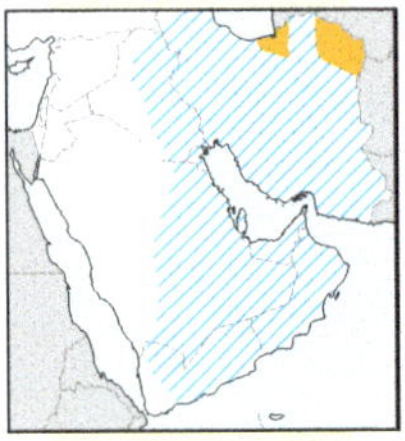

Steppe Grey Shrike *Lanius (meridionalis) pallidirostris* WV, PM

L: 25. Much as Southern Grey Shrike with very pale grey upperparts, white underparts usually blushed pinky-buff on breast and flanks, ***pale (not black) lores, broad white wing-bar not extending onto secondaries***; in winter ***bill horn-coloured or grey***, tipped dark (but all black in summer). **Voice** Demonstrative raucous screeches, mechanical throaty whirrs and clicks strung together. **Habitat** Steppe and semi-desert scrub. In winter ***sparsely vegetated habitats***, mostly sand desert and semi-desert scrub, acacia savanna. **Note** Passage and winter hatched.

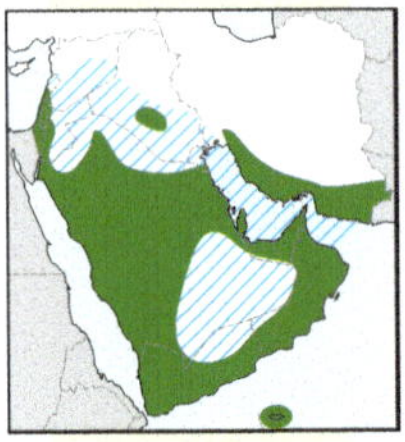

Southern Grey Shrike *Lanius meridionalis* RB, PM, WV

L: 25. Resembles Steppe Grey Shrike closely, but adult ***lacks white on secondaries***. Juvenile pale-billed, without black over bill base, unbarred above (unlike juvenile Lesser Grey); told from wintering adult Steppe Grey by ochre-buff on breast and wing-bars. Several subspecies occur in region, *aucheri* (resident UAE) dark grey above with grey wash on flanks, dark lores, broad black line over bill and relatively small amount of white in wing. **Voice** Song a mixture of quiet, melodious ramblings, raucous notes and mimicry. Calls includes harsh *sheck sheck* often extended into chatter. **Habitat** Open wooded and scrubby areas, rarely such sparse habitats preferred by Steppe Grey. **Note** Winter hatched.

Long-tailed Shrike *Lanius schach* V

L: 24. Size of Southern Grey Shrike, but ***graduated tail with buff (not white) outer feathers, and rump and upper tail-coverts bright orange-buff***. Subspecies *erythronotus* depicted. White mirror at base of primaries small and inconspicuous (sometimes absent). First-winter told from smaller Bay-backed Shrike by rufous-buff rump and uppertail-coverts; lacks chestnut on back of latter. **Voice** Vagrants usually silent. **Habitat** Open scrubland and cultivated regions. **Note** Old records NE Iran; vagrant Kuwait, Oman, UAE.

Bay-backed Shrike
♂
1st-winter
Lesser Grey Shrike
ad
juv
Steppe Grey Shrike
Southern Grey Shrike
aucheri
ad
1st-winter
♂
Long-tailed Shrike
1st-winter

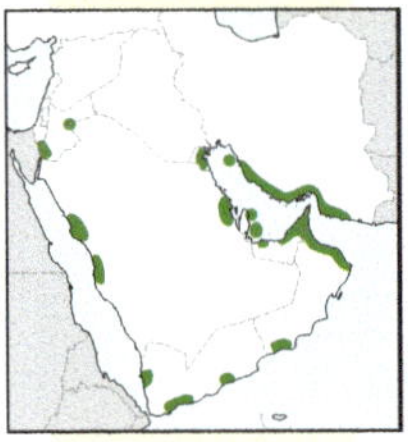

House Crow *Corvus splendens* V

L: 43. Readily told by ***steep forehead, domed crown, grey nape, neck and breast clearly demarcated from black face***, but merging with rest of black plumage. Some individuals (adults of different subspecies, apparently including introduced/stowaway Sinhalese subspecies *protegatus*), also some, probably young birds, show little grey on head or breast. Most Arabian Gulf populations (subspecies *zugmayeri*) consistently with much grey. Wings broad, tail long and with relatively light and nimble flight (can recall chough) with slow or fast wingbeats. Bold, noisy and gregarious, often in very large numbers, especially at roosts. **Voice** High or low-pitched; harsh *grehr* or *waaa waaa waaa*; often also higher *aah-aah*, recalling archetypal crow heard in spaghetti Westerns. **Habitat** Ports, coastal towns and villages. Nests semi-colonially in trees or manmade structure. **Note** A pest species in region [Alt: Indian House Crow]

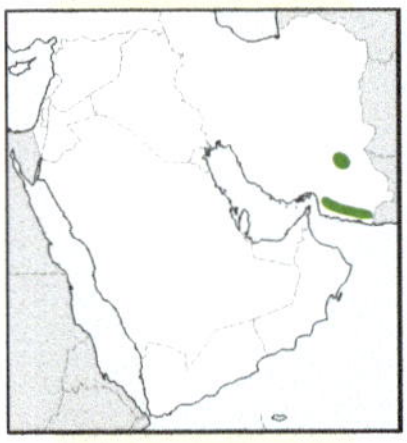

Large-billed Crow *Corvus macrorhynchos* E/?V

L: 46. A large crow with ***long, slightly wedge-shaped tail, head with steep forehead and long, heavy bill with strongly decurved culmen***. Smaller and sleeker than Northern Raven, but more domed crown, lack of shaggy throat-feathers as well as size should prevent confusion, although flight outline resembles that in soaring Northern Raven, and may also soar high overhead. **Voice** A fairly deep, throaty *hroarr-hroarr*, unlike calls of Northern Raven. **Habitat** Wooded countryside, also edges of towns and villages. **Note** No recent records from area mapped in Iran; vagrant or escape UAE. [Alt: Jungle Crow]

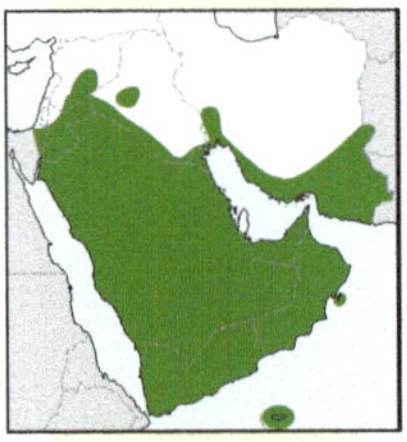

Brown-necked Raven *Corvus ruficollis* RB

L: 50. Smaller than Northern Raven with ***proportionally longer, slimmer wings, longer head*** and ***slimmer bill*** (often held drooping in flight), and ***bronzy-brown sheen on nape and neck*** (can be difficult to see); longish, wedge-shaped tail often shows central feathers protruding beyond tail outline, whereas Northern Raven more evenly wedge-shaped. ***At rest wings reach to or beyond tail-tip***, in Northern Raven they usually fall well short. Juvenile lacks brown on neck. In pairs or flocks, roosts communally post-breeding. **Voice** Croaking *raark*, less deep than Northern Raven. **Habitat** Deserts, semi-deserts, arid mountains, often near remote habitation, camps, livestock enclosures, villages.

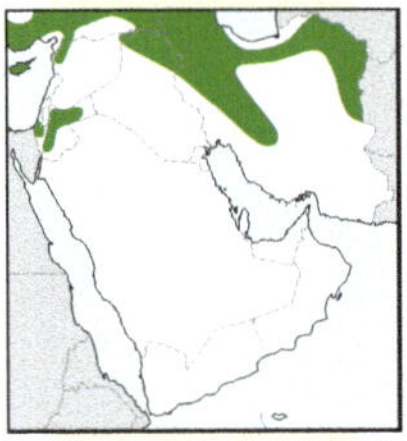

Northern Raven *Corvus corax* V

L: 64. A large, powerful corvid similar to Brown-necked Raven (which barely overlaps in range) but somewhat larger with broader wings, heavier head and bill and ***often showing shaggy feathers on throat***. For further distinctions see Brown-necked Raven. Often soars on flat wings and in breeding season performs tumbling and rolling display nights. **Voice** Loud, far-carrying, ***deep*** **krroak, krroak**, a quieter *kroak* and hollow *klong*. **Habitat** Mainly mountain areas, except when wandering. **Note** Some winter dispersal; vagrant UAE.

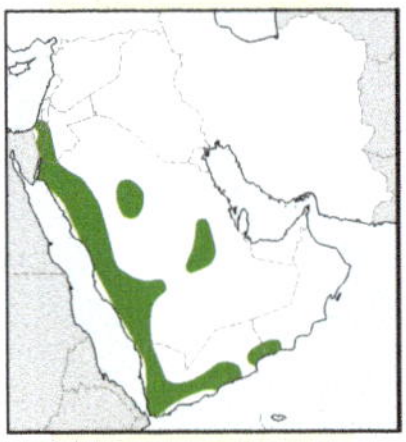

Fan-tailed Raven *Corvus rhipidurus* V

L: 47. Readily told from other crows (though only Brown-necked Raven and House Crow occur in the same range) by ***very short tail and bulging trailing edge, giving it unmistakable, almost 'flying backwards' (vulturine) flight silhouette***. Strong bill is shorter and heavier than in Brown-necked Raven. When soaring overhead black coverts contrast with slightly paler flight feathers and the greyish feet may be visible. ***At rest wings extend well beyond tail-tip***. Often congregates in large groups, which will soar, raptor-like, in thermals or updrafts. **Voice** High-pitched, rather gull-like croak. **Habitat** A wide variety of habitats from sea level to over 3,000m; often close to human habitation. Nests on ledge or hole in rock face. **Note** Some winter dispersal; vagrant UAE.

House Crow
Large-billed Crow
Brown-necked Raven
Northern Raven
Fan-tailed Raven

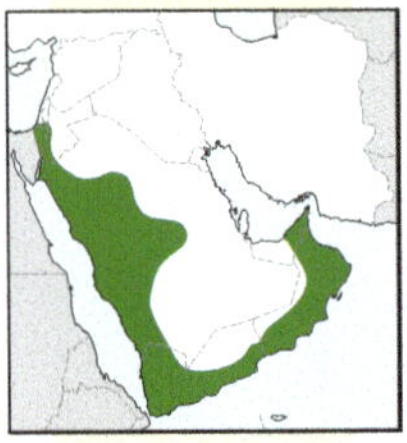

Arabian Babbler *Turdoides squamiceps* RB

L: 26. Greyish-brown, lightly streaked, often with head appearing 'moth-eaten'; faint dark mottling on throat and breast; bill blackish with paler base, legs brownish to dark grey. (Birds of Yemen lowlands and some highland areas have variable off-white face with whitish eye-surround and bill varying from orange-red to yellowish-orange). Most often in close-knit groups, on ground, in bush or tree. **Voice** Typical calls squeaky or piping, including high piping *piu-piu-piu-piu-piu*, decelerating towards end. **Habitat** Dry scrubby areas, wadis, arid hills and open wooded savanna, especially acacia, from sea level to 2,400m; also irrigated plantations, shelterbelts. **Note** Recent range expansion in Arabian peninsula.

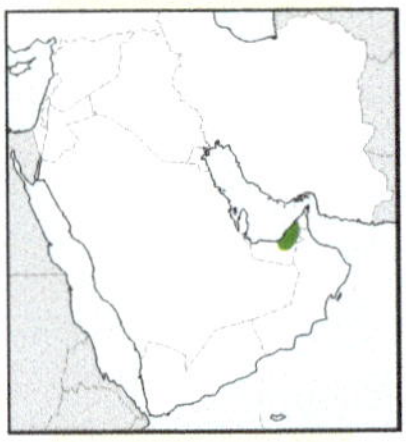

Red-whiskered Bulbul *Pycnonotus jocosus* E/I

L: 20. Told by sooty-black crown with ***tall pointed crest, white cheeks separated from white throat by black line, white underparts*** and red undertail-coverts; blackish tail has noticeable white tips below. Small red patch behind eye noticeable when close. Juvenile lacks this red patch and undertail-coverts are rufous-orange. Somewhat shy, remaining under canopy. **Voice** Vocal and varied. More musical than Red-vented Bulbul, with a distinctive single loud call *kleeoo*, unlike any note of congeners. **Habitat** Woodland, plantations, large gardens. **Note** Not native; populations originate from escapes, possibly not self-sustaining in UAE and may die out.

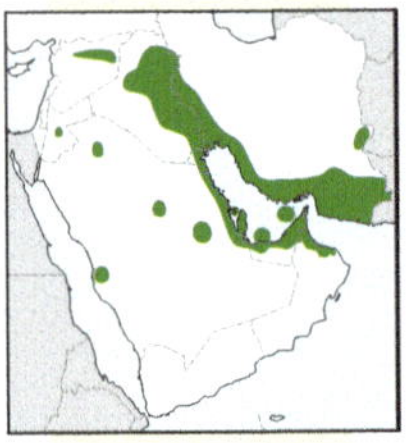

White-eared Bulbul *Pycnonotus (leucogenys) leucotis* RB (E/I)

L: 18. Easily told by black head and throat, and ***large white cheek-patch***, often shows slight crest; undertail-coverts yellow, longish tail with noticeable white tip. Juvenile has browner head than adult. Rump can be pale, as in Red-vented Bulbul, with which hybridises readily, progeny having variable intermediate characteristics including indistinct cheek-patch, dark spotting on upper breast and orange undertail-coverts. Often makes flycatching sallies; gregarious. **Voice** Lively bubbling jumble of notes, with simple repetitive tune; more musical than Red-vented, *too-tiddly-ooo; twee-ooo-wee-ooo* and longer bubbling variations *doo-widdly-iddly-wick*. **Habitat** Woodland, parks and gardens, urban and rural settings near human habitation. **Note** Native, but now widely introduced in Arabia. [Formerly White-cheeked Bulbul *P. leucogenys*]

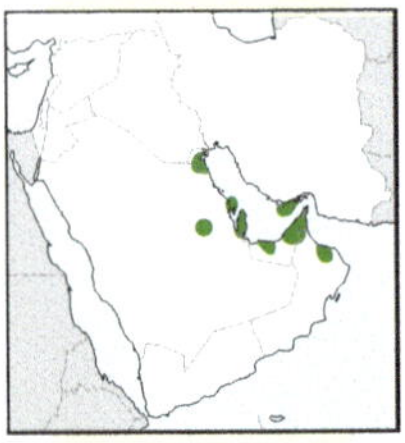

Red-vented Bulbul *Pycnonotus cafer* RB (E/I)

L: 22.5. Sooty-brown with black head, slight crest, and ***fine pale scalloping on upperparts and breast; undertail-coverts red***. In flight, reveals ***off-white rump*** (White-eared Bulbul's rump rarely as pale) and white tip to long blackish tail. Often makes flycatching sallies; gregarious. **Voice** Noisy; calls being bubbly or a burbling chatter and fairly loud, including typical *pick-yow-you* or *pee-who*. **Habitat** As White-eared Bulbul, but intolerant of drier native wooded habitats. **Note** Not native; breeding populations originate from escapes.

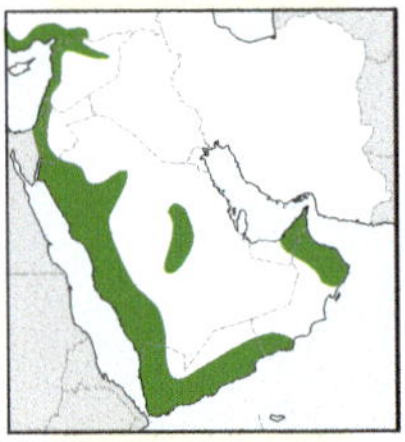

White-spectacled Bulbul *Pycnonotus xanthopygos* RB

L: 19. Size of small slim thrush, often noisy, with rather floppy flight action. Drab with ***sooty-black head*** shading into grey-brown upperparts and paler greyish underparts; with an ***obvious white eye-ring***. Tail rather long, dark brown ***without white tip; undertail-coverts yellow***. Crown feathers often slightly raised. Sociable, can occur in large groups. **Voice** Fluty and fitful, but rather loud and obvious *bli-bli-bli-bli* or *bul-bul-bul-bul-bul*. Calls include a loud, rather harsh *pwitch* and *trratsh*. **Habitat** Fruiting trees, gardens, palm groves, wadis with cover. **Note** Locally common native. [Formerly Yellow-vented Bulbul]

Arabian Babbler
White-eared Bulbul
Red-whiskered Bulbul
ad
ad
Red-vented Bulbul
ad
ad
White-spectacled Bulbul

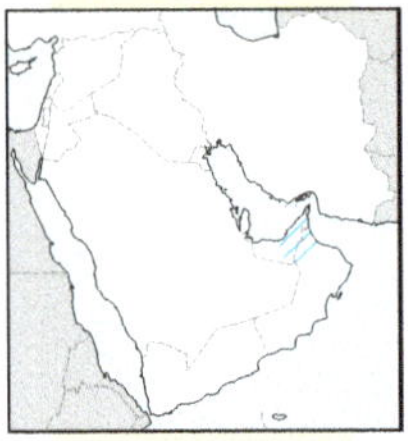

Brown-throated Martin *Riparia paludicola* pm, wv

L: 12. Small, ***mouse-brown above*** with greyish-white underparts ***merging into darker greyish-brown upper breast, throat and head***; lacks breast-band of Sand Martin, has shorter tail accentuating broader-based appearance to wings, thus recalling larger Pale Crag Martin; differs from that species in ***more fluttering flight, lack of white tail-spots and by white vent/undertail***. See also Pale Martin. Flight and behaviour similar to Sand Martin. **Voice** Soft *dree-dree* or *dree-err, dree-err*, not as dry or buzzy as Sand Martin. **Habitat** Rivers, lakes, grasslands. **Note** Rare UAE as mapped; vagrant Iran, Oman, Saudi Arabia, Yemen. In Iran and SE Arabia assumed (but none confirmed) to be of Asian subspecies *chinensis* (Grey-throated Martin), but those noted SW Arabia/Red Sea more probably of African subspecies *minor*. [Alt: Plain Martin]

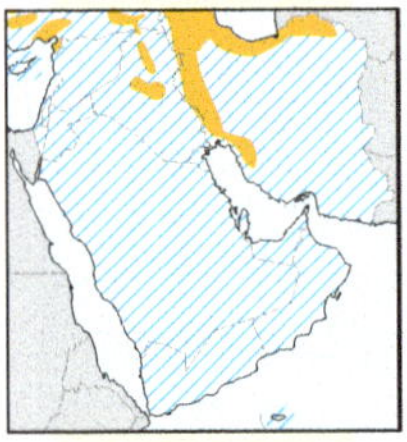

Sand Martin *Riparia riparia* PM, WV

L: 12. Small, dull ***brown and white martin*** recalling Brown-throated Martin; distinguished by white underparts with well-marked ***brown breast-band*** (though paler and indistinct in juveniles) and slightly longer and deeper-forked tail. From Pale Crag Martin by fluttering flight action, breast-band, white lower underparts and lack of white tail-spots. Juvenile has scaly, buffier upperparts and poorly defined breast-band (see also Pale Martin). Often perches on wires with other hirundines. **Voice** Usual call in flight a vowel-less, rasping repeated *tschr* or *zrrrr-zrrrr*, persistent and slightly buzzy in groups, louder ***chirr*** given in alarm. **Habitat** Open country with wetlands. Widespread on migration, though usually over water or grasslands. Nests colonially in tunnel excavated in sandy banks. **Note** Passage hatched; some winter south Arabia.

Pale Martin *Riparia (riparia) diluta* WV, pm

L: 11.5. Marginally smaller than Sand Martin (with which may consort, although respective flocks typically remain separate), being pale ***mousey grey-brown above, with breast-band ill-defined or absent***, and ***less deeply forked, almost notched, tail***. Also has slightly paler underwing; off-white throat, with ***ear-coverts smudged dingy-brown and their lower edge not clear-cut***. Note young Sand Martin often has poorly defined breast-band, and some individuals perhaps not safely separable. **Voice** Usually silent in winter but quiet, sweet chatter sometimes audible; not heard to give 'buzzing' characteristic of Sand Martin. **Habitat** As Sand Martin, over water or irrigated grasslands. **Note** Winter/passage hatched (variable numbers annually); vagrant Oman.

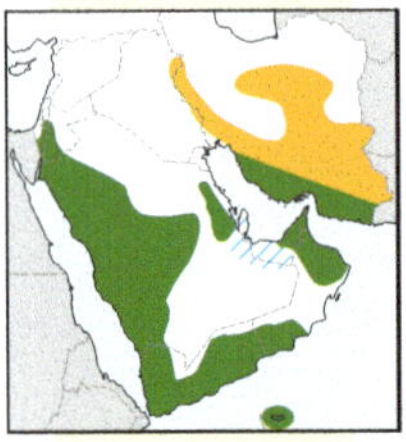

Pale Crag Martin *Ptyonoprogne (fuligula) obsoleta* RB

L: 12.5. Similar to larger Eurasian Crag Martin but usually discernibly ***smaller and paler***, more grey-brown; upperparts, especially ***back and rump, appear slightly greyer than wings; underparts off-white including chin*** (lacking dark spots of Eurasian Crag Martin) merging into pale mouse-grey undertail-coverts; ***less contrasting head pattern*** though ear-coverts sometimes darker than crown; white spots in spread tail distinctly visible when flying overhead or from above; underwing pale grey with ***brownish-grey coverts contrasting much less than in Eurasian Crag Martin*** (overhead, against light, may look confusingly close to Eurasian Crag). **Voice** Dry, rather quiet twittering and repeated single *drrrrt*. **Habitat** Hilly country with gorges and ravines, but often also lowlands near habitation in sand desert, towns and cities, where will nest on low- or high-rise buildings. Often over wetlands near breeding areas. Nest as Eurasian Crag Martin. **Note** Some dispersal occurs; vagrant Kuwait.

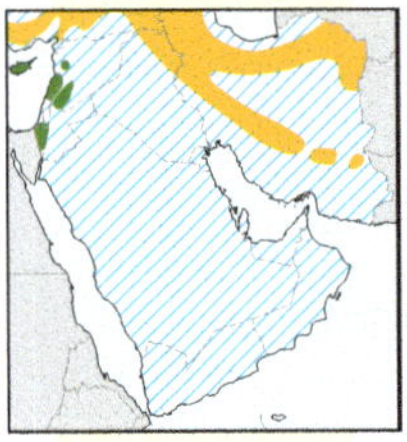

Eurasian Crag Martin *Ptyonoprogne rupestris* pm, wv

L: 14.5. Large broad-winged, brown martin, almost square-tailed, similar to smaller and typically paler Pale Crag Martin but ***larger, heavier-built, with darker upperparts lacking greyish tinge to rump; cheeks fairly dark, usually contrasting with pale throat and underparts; fine spots on chin and throat*** (can be hard to see). Underparts buffish-grey becoming darker towards undertail; white spots in spread tail as in Pale Crag Martin but ***underwing-coverts distinctly darker than flight feathers***. Flight more swooping and diving than in most other hirundines. **Voice** Song a quiet twittering; in flight a short *chip* or *chirr*. **Habitat** Mountain gorges and rocky inland and coastal cliffs. Builds open half-cup-shaped nest in caves, on cliff or under eaves of building. On passage open country, hills, fields, sometimes over wetlands. **Note** Passage hatched, but rare in south; some winter Arabia.

Brown-throated Martin
Sand Martin
Pale Martin
Pale Crag Martin
Eurasian Crag Martin

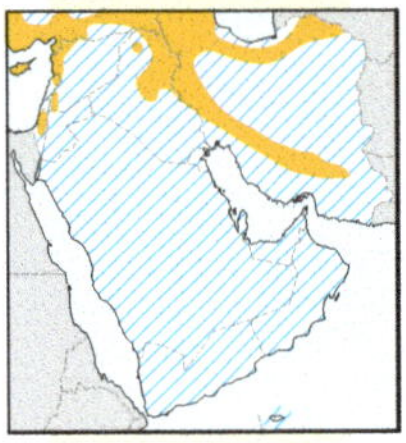

Common House Martin *Delichon urbicum* PM, wv

L: 12.5. Smaller than Barn Swallow; easily recognised by ***bluish-black upperparts with striking white rump and all-white underparts***; short, forked tail with white undertail-coverts and rather dark underwings. Juvenile duller with brownish wash on head and breast-sides. Flight more fluttering than Barn Swallow with long glides often high in air. Frequently perches on wires. Gregarious, migrating in flocks and nesting in noisy colonies. **Voice** Commonest flight call a short dry warbling *prrlit*, often sounding conversational. Gives repeated high trilled *scheeer* in alarm. **Habitat** Mountains, hills, villages. Mud nest built under eaves of house, on cliff or in cave. **Note** Passage hatched, some winter Arabia.

Asian House Martin *Delichon (urbicum) dasypus* V

L: 12. Appearance much as Common House Martin but at ***close range note that blue-black extends below bill onto chin***. Also marginally smaller with dusky underparts, ***darker underwing-coverts, smaller off-white rump and dark uppertail-coverts, with more shallowly forked (notched) tail. Dusky-centred undertail-coverts*** sometimes visible. **Voice** Call similar to Common House Martin but buzzy with urgency of, and recalling, Sand Martin. **Habitat** Records from region are of single birds over irrigated grassland and man-made lakes in winter. **Note** Vagrant UAE (from Asia).

Streak-throated Swallow *Petrochelidon fluvicola* V

L: 11. Diminutive and front-heavy; dusky throat and upper breast seen at close quarters is strongly darkly streaked; thighs with dark flecks. Crown dingy-brown (immature) to ginger-chestnut (adult), rump contrastingly brownish. White streaks on mantle in adult. Tail only slightly forked. Viewed overhead can resemble Brown-throated Martin, but more compact and thickset. Often joins flocks of other hirundines. **Voice** Not heard from vagrant individuals. **Habitat** Lakes, ponds and open country. **Note** Vagrant Oman, UAE (from Asia). [Alt: Indian Cliff Swallow]

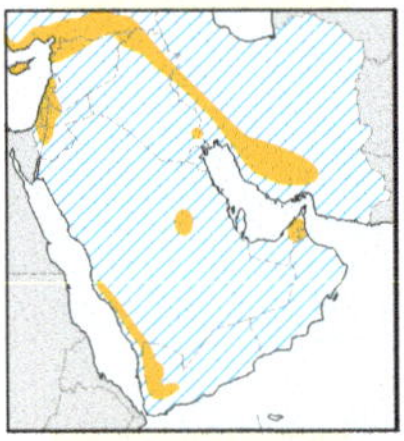

Red-rumped Swallow *Cecropis daurica* PM, wv

L: 17 (including tail-streamers). Closely resembles Barn Swallow but easily told by ***rufous collar and broad pale rufous rump***; underparts, including underwing-coverts, buffish-white, faintly streaked (but variably so, streaks obvious in some, almost absent in others), ***undertail-coverts striking black***; ***lacks breast-band and white in tail***. Juvenile browner with paler collar and rump, and shorter outer tail feathers. Flight slow and graceful, frequently gliding for longer periods than Barn Swallow, tail flared or closed. **Voice** Usually rather quiet. Song shorter and quieter than Barn Swallow; in flight a short soft nasal *tweit* reminiscent of House Sparrow. **Habitat** Inland and sea-cliffs, cultivated areas; in flat country frequents bridges and buildings. ***Builds flask-shaped mud nest with spout-shaped entrance*** in caves, under overhang, bridge or on building. **Note** Has bred UAE; passage hatched, some winter south Arabia.

Barn Swallow *Hirundo rustica* PM, WV

L: 16 (including tail-streamers). Easily distinguished by ***bluish-black upperparts, long tail-streamers,*** (small ***white patches visible in tail*** when spread), ***chestnut forehead and throat, and solid dark breast-band***; underparts buffish-white including underwing-coverts. Some subspecies have underparts reddish-buff. Juvenile lacks tail-streamers, has brownish breast-band and pale rusty forehead. Flight is strong and elegant with much banking and turning, often hunting rather low. Often perches on wires with other hirundines. **Voice** Song a melodious twittering; contact call *witt-witt*, alarm call disyllabic *tsi-wit*. **Habitat** Open cultivated country with settlements; over any area on migration. Nests on ledge in building. **Note** Passage and winter hatched, but absent from north in winter.

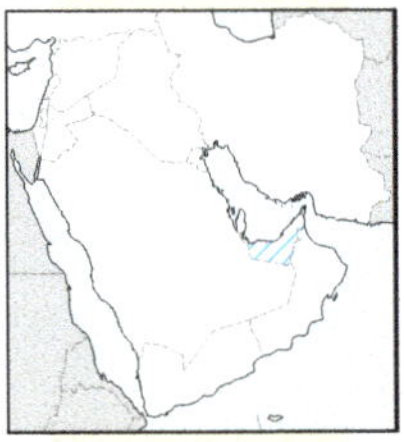

Wire-tailed Swallow *Hirundo smithii* wv

L: 16 (including tail-streamers). Slightly smaller than Barn Swallow, without breast-band. Clean-coloured with rich ***blue upperparts and eye-mask***, white tail-patches as in Barn Swallow, ***rufous-chestnut crown*** and rather square-ended tail with long, fine outer tail feathers (shorter in female and often difficult to see in flight); ***underparts gleaming white including underwing-coverts*** with bluish flanks and broken vent-bar. Juvenile Wire-tailed Swallow lacks tail-streamers, has almost square tail, dull bluish wings, brown back and crown with darker eye-mask; underparts as adult. Flight similar to Barn Swallow, but with ***more compact, triangular outline*** discernible. Often joins other hirundines. **Voice** Call a sharp *tchik*. **Habitat** Along rivers, over lakes, grasslands. **Note** Winter hatched, but rare; vagrant Oman.

Common House Martin
Asian House Martin
Streak-throated Swallow
Red-rumped Swallow
Barn Swallow
juv
ad
ad ♂
Wire-tailed Swallow

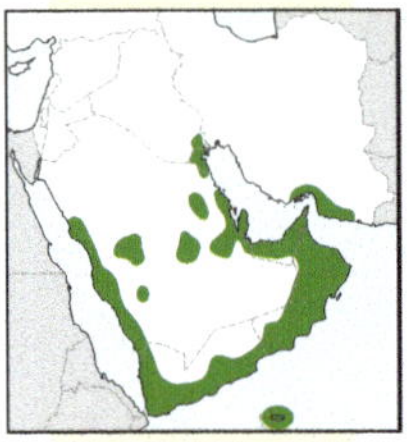

Black-crowned Sparrow-Lark *Eremopterix nigriceps* RB

L: 12. Small stocky lark with ***stout, deep-based, pale blue-grey bill with curved culmen***. Male unmistakable. Female has ***unstreaked pale sandy-grey upperparts***, faintly streaked crown, pale face and hindneck, buffish underparts, faintly streaked across upper breast and ***darkish underwing-coverts***. From Bar-tailed Lark by heavier bill and unmarked tail-tip. Flight bouncing; feeds on bare ground. **Voice** Song, often uttered in circling display flight, a repetition of 2–4 loud, sweet notes *chee-dee-vee* or *pooo, pee-voo-pee*. Flight calls bubbling twitter, dry *rrrp* or soft *tchep*. **Habitat** Semi-desert, sandy or stony plains with low scrub, edges of cultivation, salt-flats, coastal dunes. **Note** Nomadic.

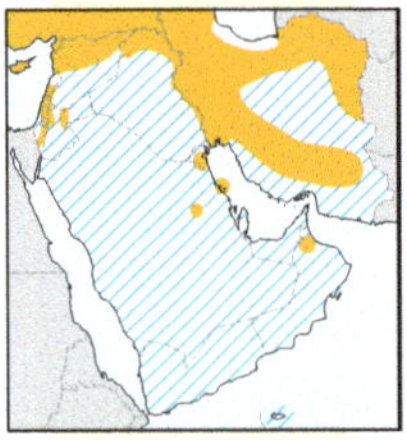

Greater Short-toed Lark *Calandrella brachydactyla* PM, ?mb

L: 14. Small, with streaked upperparts, buffish-white ***generally unstreaked underparts with variable black patch at sides of neck*** (adult); relatively stout, pointed pale bill, ***long tertials almost covering wing-tip***; median coverts boldly patterned. Streaked crown greyish, tinged rufous in some males. From Lesser Short-toed Lark by ***black neck-patches (in some just thinly streaked smudge), long tertials and longer bill***. Flight undulating; flocks densely and flies low. **Voice** Typical flight call a sparrow-like *tjirp*, and *drelit*. Song unmusical, repetitive short bursts (includes mimicry), in circling flight. **Habitat** Steppe, semi-desert, cultivated plains. **Note** Passage and winter hatched, absent from north in winter.

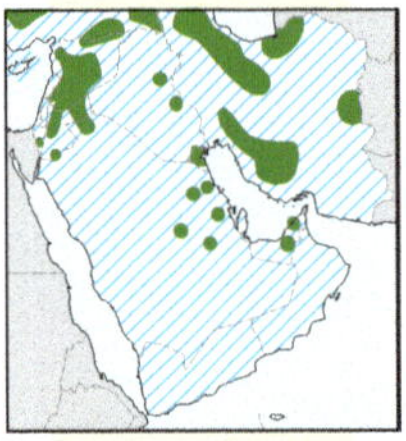

Lesser Short-toed Lark *Calandrella rufescens* mb, PM, WV

L: 13. Resembles Greater Short-toed Lark; but stockier, ***with shorter, stubbier bill, bulging jowls, more distinctly streaked breast, lacking black patches at sides, and with wing-tip clearly exposed beyond tertials*** (beware Greater Short-toed with worn tertials). Upperparts sandy-grey to rufous-brown; crown occasionally tinged rufous; supercilium generally less noticeable than in Greater Short-toed, but patterning below eye more pronounced. Flocks in winter and on passage. **Voice** Flight call abrupt, ***fast dry staccato prrrrt or prrr-rrr-rrr***; also Eurasian Skylark-like *drrie* and quick *dreeup*. Song, in spiralling flight with unbroken deliberate wingbeats, varied, melodious and heavily mimetic. **Habitat** Steppe, saltflats, stony desert, cultivation. **Note** Passage and winter hatched.

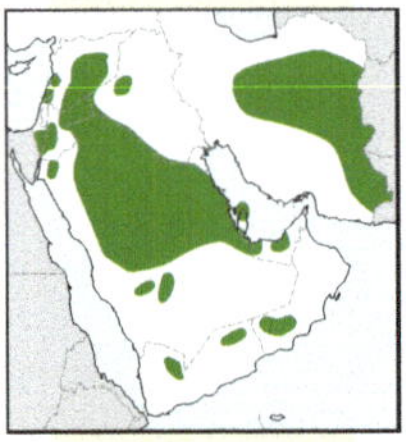

Bar-tailed Lark *Ammomanes cincturus* rb

L: 13. Resembles Desert Lark but smaller, with more upright stance, rounder head, ***smaller bill, clear-cut black band to tip of reddish-brown tail***, and rufous-buff ***wings having blackish tips***. In subspecies *arenicola* (Near East/Arabia) unstreaked upperparts pale sandy-rufous. Runs faster than Desert Lark with more abrupt stops. From Dunn's Lark ***by much smaller bill, different tail pattern, unstreaked upperparts and crown, and blackish wing-tip projecting beyond tertials;*** also habitat. **Voice** Song can be mistaken for Black-crowned Sparrow-Lark, like swinging pub sign creaking in wind, 2–3 notes *tlee-tloo-hee*; also *dee-dee-doo*. Flight call purring, soft hoarse *twer*; also *see-oo*. **Habitat** Flat or undulating desert with scattered vegetation and gravelly or stony rises. **Note** Some winter dispersal.

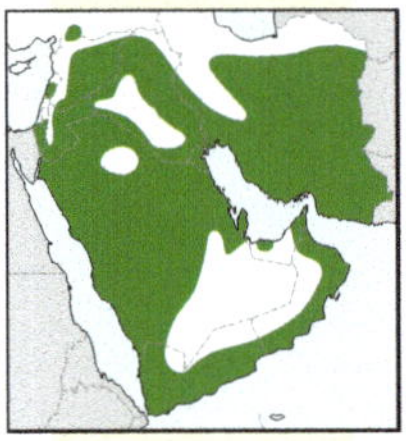

Desert Lark *Ammomanes deserti* RB

L: 15. Short-tailed with ***broad, rounded wings and slow, floppy, undulating flight. Unstreaked above*** except for vague mottling on mantle in some; underparts buffish or greyish-white, sometimes unmarked, but often diffusely streaked on breast; rufous flight feathers and underwing in some birds. ***Tertials fall clearly short of wing-tip. Longish, stout pointed bill***, yellowish-horn with dark culmen, gently tapering. Subspecies in Hajar mountains, *taimuri*, sandy-grey to grey-brown above; *Insularis* on coast in western Abu Dhabi rather paler. **Voice** Typically short, soft and melodious *dee-leeut*. Song includes phrases of call, given from ground or in descending glide. **Habitat** Arid hills, stony or rocky slopes with sparse vegetation.

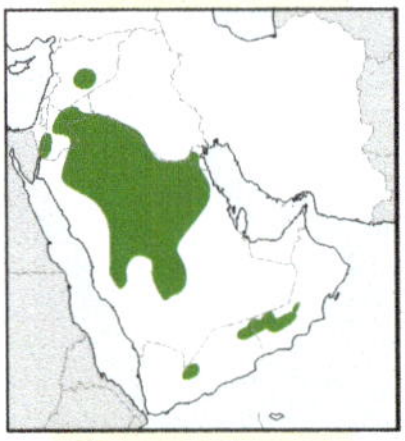

Dunn's Lark *Eremalauda dunni* V

L: 14. Small, sandy rufous-brown with dark streaks on crown and, vaguely, on mantle; black sides to sandy-brown tail (white sides in Greater and Lesser Short-toed Larks), and large pinkish bill with pronounced curve near tip. Long tertials almost reach wing-tip. Broad whitish eye-ring bordered below by dark line, with a dark moustache and line behind eye. Broad, rounded wings and relatively short tail in flight. Appears large-headed, with upright stance; runs fast for short distances with sudden stops. **Voice** Flight call drawn-out, soft *wazz* or *ziup*; also thin liquid *prrrp*. **Habitat** Flat sandy or stony semi-desert with low scrub and grass. **Note** Vagrant UAE.

Black-crowned Sparrow-Lark
♂
♀
Greater Short-toed Lark
longipennis
ad
brachydactyla
ad
Lesser Short-toed Lark
ad fresh
ad worn
Bar-tailed Lark
ad
Desert Lark
ad
Dunn's Lark
ad

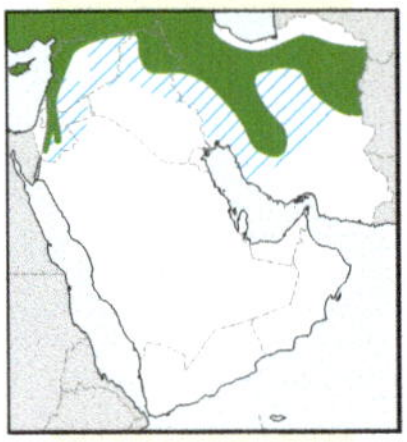

Calandra Lark *Melanocorypha calandra* V

L: 20. Large, heavy-billed lark with relatively short tail. In flight ***shows blackish underwing with conspicuous white trailing edge and white sides to tail***. On the ground ***shows swollen yellowish-horn bill, black patches at sides of lower throat*** (of variable shape, inconspicuous in some autumn birds) and variable whitish supercilium. Smaller Bimaculated Lark also has black neck-patches, but has paler underwing without conspicuous white trailing edge, lacks white at sides of tail, which instead has white tip. See also female Black Lark. Undulating flight low with deliberate 'wader-like' wingbeats. Flocks outside breeding season. **Voice** Flight call harsh, rolling *terrelet*; also Eurasian Skylark-like note. **Habitat** Open cultivated plains, grass and cereal fields, steppe and wastelands. **Note** Partial migrant, summer visitor to some breeding areas; winter hatched; vagrant Bahrain, Kuwait, Saudi Arabia, UAE.

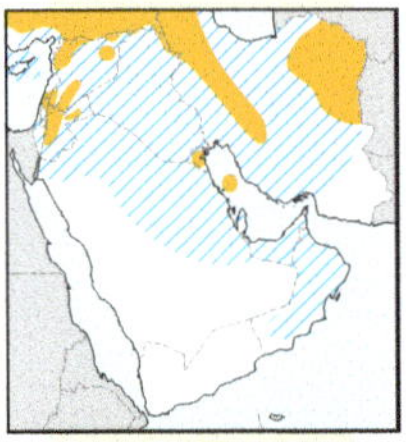

Bimaculated Lark *Melanocorypha bimaculata* WV, pm

L: 16. Resembles small, shorter-tailed Calandra Lark; in flight has ***dull grey-brown underwing without clear white trailing edge, and white-tipped tail*** with outer feathers buff-brown. On ground ***upperparts show more prominent scaling*** with head pattern more contrasting than in Calandra, with rusty cheeks, pronounced ***long white supercilium and dark lores*** giving capped appearance. **Voice** Flight call recalls Calandra, *trrelit*, with rather gravelly scrunching; also a Short-toed Lark-like *dre-lit*. Song includes drawn-out rolling call-note, delivered from ground or air. **Habitat** Thinly vegetated hills or marginal stony cultivation to 2,400m; in winter to sea level in agricultural areas, often near penned livestock camps. **Note** Partial migrant; passage hatched; some winter Arabia.

Crested Lark *Galerida cristata* RB

L: 17. Rather stocky, short-tailed lark with ***long spiky crest***. Upperparts sandy-grey or rusty, diffusely streaked darker on hindneck and mantle; breast more heavily streaked. Flight flappy, ***broad wings showing rusty-buff underwings; short tail blackish-brown with cinnamon sides. Lacks white trailing edge to wing***. Juvenile heavily pale-spotted above. **Voice** Clear *du-ee*, also varying fluty ee or *uu* sounds. Song sweet and plaintive with phrases of 4–6 repeated notes; slower, clearer and shorter than Eurasian Skylark, often includes mimicry; from exposed perch or high in air. **Habitat** Grassy or arid country, cultivated plains and semi-deserts; often near habitation, tracks and roadsides.

Eurasian Skylark *Alauda arvensis* WV, PM

L: 18. Medium-sized with earth-grey to sandy upperparts, streaked dark with warm brown edges to tertials and coverts in fresh plumage; underparts buffish-white heavily streaked on breast; can show a small crest (much smaller than Crested Lark); head markings rather indistinct. Juvenile has upperparts spotted dark with scaly ochre markings. ***In flight, shows distinct whitish trailing edge to wings*** (except in juveniles) and broad triangular ***tail with white sides***. **Voice** When flushed and in flight gives a variable rolled *chrriup*; or *trruwee*. **Habitat** High and low grasslands, cultivated fields; in winter more widely in open areas. Avoids deserts. **Note** Passage and winter hatched.

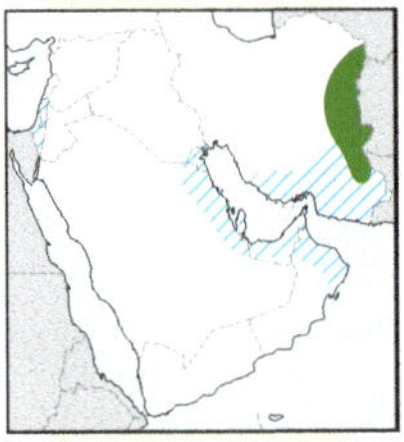

Oriental Skylark *Alauda gulgula* wv, pm

L: 16. Similar to Eurasian Skylark, but smaller with obviously ***shorter tail***, comparatively longer, more pointed bill, and very short primary projection; also slightly rusty tinge to ear-coverts and fringes of flight feathers. In flight, size of Woodlark with similarly short tail; ***broad wings have inconspicuous buffish trailing edge*** (white in Eurasian Skylark). Also from Eurasian Skylark by call, duller head marks, and tail having buffish-white outer feathers (bleaching to white). **Voice** Distinctive flight calls are soft *pyhp* or *twip* recalling Ortolan Bunting, and a ***characteristic hard buzzing* bzzeebz *or* baz-baz**. **Habitat** Open grassy and cultivated lowlands, but also grassy hills. **Note** Passage and winter hatched, but rare; probably overlooked. [Alt: Small Skylark]

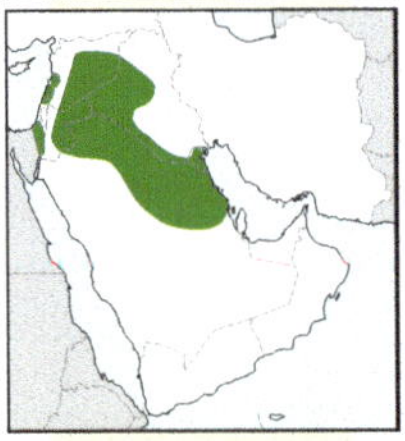

Temminck's Lark *Eremophila bilopha* V

L: 14. Pale sandy upperparts contrast with black and white facial pattern and black breast-band. Juvenile, which lacks black markings on head, throat and breast, uniform sandy-cinnamon above and easily mistaken for other lark species. **Voice** Song and calls squeaky jingle. **Habitat** Open, flat, stony or sandy desert with sparse grassy vegetation. **Note** Vagrant UAE, Yemen.

Calandra Lark
ad
ad
Bimaculated Lark
ad
Crested Lark
Eurasian Skylark
ad
Oriental Skylark
ad
Temminck's Lark
ad

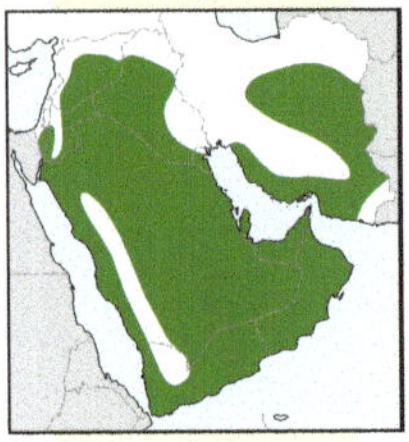

Greater Hoopoe-Lark *Alaemon alaudipes* RB

L: 18.5. Large, slender sandy-buff lark, with ***long tail, long decurved bill and variable black band across mostly white inner wing***. Underparts whitish, often spotted black on breast. Juvenile with few breast-spots and shorter, less decurved bill. Runs speedily with sudden stops in upright position; also creeps away slowly. In characteristic songflight, male ascends vertically a few metres, twists over and spirals to bush or ground with outstretched wings. Solitary or in pairs, not mixing with other larks. **Voice** Song, mostly at sunrise, melodious and melancholy, starting slowly, accelerating and ascending, then dropping and dying away slowly *dee-dee-dee-dee-dee, dee, de-de-de-de-de-dee-dee*; also distinctive *weerrp* or *jeerrp* on territory. **Habitat** Sandy desert, semi-desert, coastal dunes.

Purple Sunbird *Cinnyris asiaticus* RB

L: 10. The only sunbird in the UAE. Metallic bluish-black male sometimes has a narrow red-brown breast-band; mousey-brown female has underparts washed yellow. Male in eclipse (Sept–Dec) like female but underparts yellower with dark line down centre of throat and breast. Short-tailed, hummingbird-like appearance, feeding largely on nectar from flowers, with corresponding seasonal movements. **Voice** Excited song repeated 2–6 times, *cheewit-cheewit*, with Willow Warbler-like cadence. Male's sub-song a low twitter; calls *dzit-dzit* and pronounced *tsweet*. **Habitat** Gardens, cultivation, tamarisks along rivers, thorn scrub, dry forest, stony desert with flowering trees and shrubs (acacia, Sodom's apple), mangroves. Hanging nest pear-shaped, suspended in tree or bush. **Note** Vagrant Kuwait.

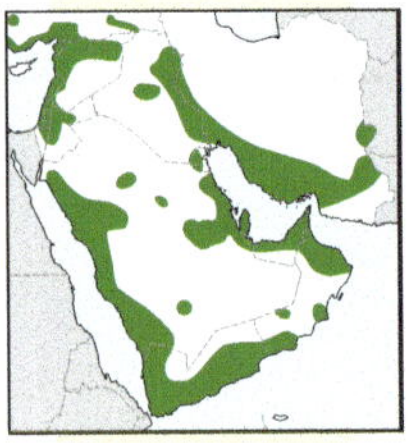

Graceful Prinia *Prinia gracilis* RB

L: 11. Small, active, variable (across subspecies) grey-buff warbler with ***long graduated tail***, frequently cocked and slightly fanned; narrowly streaked on crown, mantle and back; roundish head and ***pale face with prominent eye, but without supercilium and eye-stripe*** (unlike Scrub Warbler); tail dark above, paler below with ***black-and-white tips***; underparts off-white, unstreaked (streaked in Scrub Warbler); bill fine, black in breeding male, brown in female. Skulks in cover, but usually confiding. Flight weak, the long tail being obvious. **Voice** Rather noisy. Song monotonous, winding *drrir-drrir-drrir-drrir-drrir* or fast *di-der di-der di-der di-der*; chittering calls include an abrupt *chlip; churr-churr*, metallic winding *srrrrrrt*, and hard *tsiit* or *chig* in alarm. **Habitat** Scrub and low vegetation; cultivated areas, gardens, wasteground in villages and towns. Domed or cupped nest in low bush or thick grass.

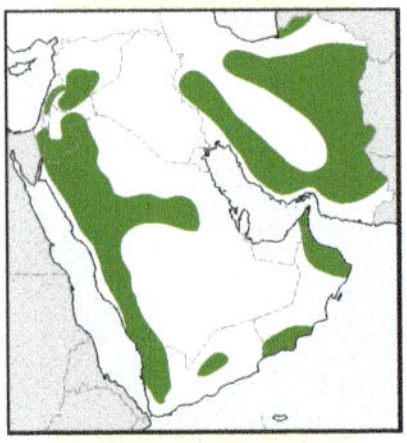

Scrub Warbler *Scotocerca inquieta* RB

L: 11. Recalls Graceful Prinia. Furtive; often in pairs or small groups and on ground. ***Long, scarcely graduated tail cocked and constantly manoeuvred***; differs from Graceful Prinia in having blackish tail, tipped white to underside of outer feathers. ***Distinct white supercilium, dark eye-stripe*** and flat crown; ***breast finely streaked***; bill yellow-horn and legs rather long. **Voice** Song thin *di-di-di-di-di*, descending *di-di-di-de-de*, also dry *dzit, dzit* followed by warbling *toodle toodle toodle*. Calls include *drzip, dri-dirrirri*, loud rolling *tlyip-tlyip-tlyip*, sometimes fast and descending, also scolding *prrt*; alarm-note, sharp piping *pip*, often repeated. Contact call characteristic, disyllabic, clear *dee-düü*, latter note lower. **Habitat** Rather barren, stony hillsides, semi-deserts and sandy plains with low scrub, up to 2,600m. Builds domed nest in low bush.

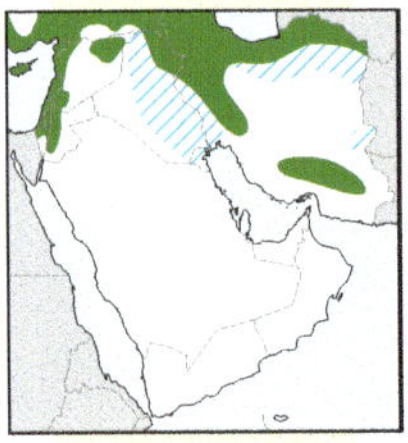

Cetti's Warbler *Cettia cetti* V

L: 13.5. Short-winged, ***round-tailed***, rather robust-looking, ***warm brown warbler***; keeps well-concealed in thickets and more often heard than seen; ***song distinctive***. Off-white below, tinged rufous on flanks and breast-sides; ***vent and undertail-coverts brown with feathers fringed white***; uniform chestnut-brown tail often jerked or held cocked; rather thin but ***distinct white supercilium***. **Voice** Abruptly delivered song from well-concealed perch a ***sudden, loud, explosive outburst*** **plit… plitiplitipliti… plhi-(pliti)**. Alarm call hard, forceful *tlitt*, or rapid, rattle, *tlitt-tlitt-tlitt*. **Habitat** Dense vegetation, bushy thickets, mostly near streams, ditches and reedbeds. **Note** Winter dispersal hatched; vagrant Oman, UAE.

Greater Hoopoe-Lark
ad
♀
♂ eclipse
♂
Purple Sunbird
ad
Graceful Prinia
Scrub Warbler
ads
ad
Cetti's Warbler

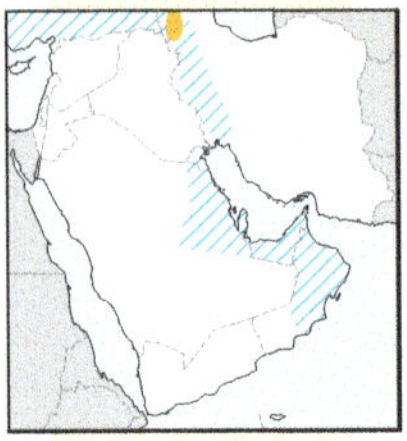

Common Grasshopper Warbler *Locustella naevia* PM

L: 12.5. Small, rather dark, skulking warbler with obviously rounded tail when flushed, often from underfoot! Flies to cover, in short, low, jerky flight. ***Upperparts olive-brown, heavily streaked*** (but appearing obscure in flight) ***with faint supercilium***; underparts dirty-white or yellowish with streaking on rear flanks and undertail-coverts, sometimes diffusely streaked on breast, juveniles especially. **Voice** Migrants typically silent. Sings, often at night, a dry reeling (likened to reel on fishing rod), longer, more sibilant than Savi's Warbler; infrequently gives a short *chik* call. **Habitat** Thick moist vegetation, dense bushes, thickets near streams, reedbeds; on passage, also grassland, herbaceous cover. **Note** Passage mapped, but scarce or overlooked.

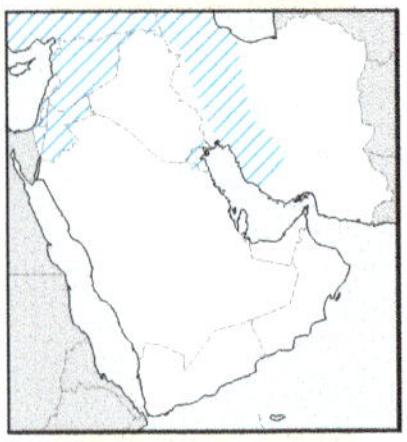

River Warbler *Locustella fluviatilis* V

L: 13. Secretive olive-brown warbler, with ***unstreaked upperparts, diffusely streaked breast and long undertail-coverts with brownish feathers tipped white***. Confusable with Savi's Warbler but is darker and greyer, being colder, darker olive-brown above, with much more obvious darker spotting on breast (diffuse in Savi's); generally keeps low in vegetation, but will walk on ground with slightly cocked tail. **Voice** Migrants silent in region. **Habitat** Thick cover, usually near water; grasslands on passage. **Note** Passage hatched but rare; vagrant Bahrain, Iraq, Oman, Saudi Arabia, UAE, Yemen.

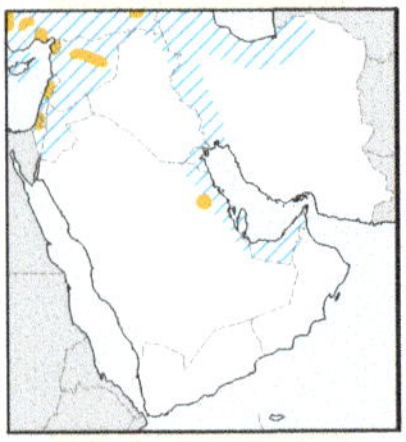

Savi's Warbler *Locustella luscinioides* pm

L: 14. A large *Locustella*. Plumage warmer brown than River Warbler; eastern subspecies ***fusca***, which occurs, is grey cast above, but diffusely marked on breast and also has the long undertail-coverts tipped pale, only less distinctly so than in River Warbler. Often on ground, walking slowly and stealthily with horizontal stance and jerky movements, the graduated tail raised; often bobs tail (a character shown by all *Locustella* warblers). **Voice** Song, uttered from reed stem or bush, by day or night, a monotonous, fast reeling. Call (at intruder or in alarm) also distinctive, a loud, hard, twinky *pitch* repeated 2–3 times. **Habitat** Reedbeds, swamps, fields, scrub and rank grass. **Note** Passage hatched; vagrant Oman.

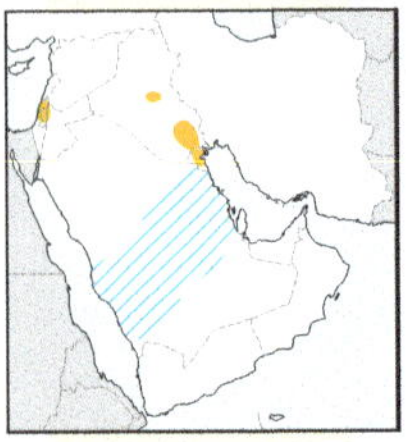

Basra Reed Warbler *Acrocephalus griseldis* V

L: 15. Larger than Eurasian Reed Warbler; told from larger Great Reed Warbler by ***narrower bill***, short but ***more conspicuous supercilium (especially above and behind eye)***, greyer, less-warm upperparts, ***whiter underparts (especially flanks and undertail)***, shorter, more square-cut tail and shorter, ***greyer legs***. Moults during migration at stop-over sites in Arabia. **Voice** Song between Eurasian Reed and Great Reed, with the same rhythm of Eurasian Reed but quieter and without the harsh notes of Great Reed. Call-note, a harsh *chaarr*. **Habitat** Reedbeds; on migration also bushes and other cover. **Note** May breed Iran; has bred Kuwait; regular known passage hatched. Vagrant UAE.

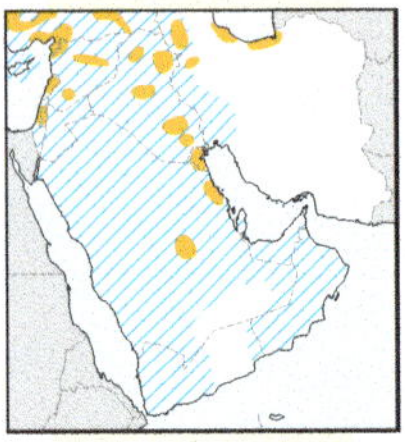

Great Reed Warbler *Acrocephalus arundinaceus* PM

L: 19. Eastern subspecies *zarudnyi* occurs on passage in UAE; compared with nominate subspecies is more olive, less rufous, especially on rump, and whiter below, with a pale supercilium, dusky lores and light brown streaks on the white throat. Contrasting buffier rump shows in flight. Compared with Indian Reed has ***longer primary projection with 7–8 primary tips showing***, proportionally shorter tail and shorter, slightly stouter bill. Moults in winter quarters. **Voice** Loud and powerful song with repetitive character; common phrase *trr-trr, karra-karra-karra, kreee-kreee-kreee*. **Habitat** Reedbeds; on passage also drier habitats. **Note** Passage hatched.

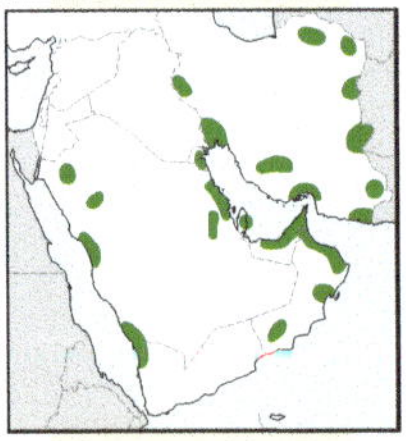

Indian Reed Warbler *Acrocephalus (stentoreus) brunnescens* RB, pm, wv

L: 18. A subspecies of Clamorous Reed Warbler *A. stentoreus*, or perhaps a separate species. Greyish-olive in fresh plumage, whitish below with buff flanks, with long tapering pointed bill. Primary projection shorter than in Great Reed. Moults post-breeding. **Voice** Loud, ***strident song***, similar in tempo but more melodious than Great Reed Warbler, *witch-a-witch-a witch, chew-chew-chew-chew, skatchy, skatchy, skatchy, vachoo vachoo vachoo*, frequently including the phrase *rod-o-petch-iss*. Call a loud abrupt *tjuck* or rolled *churr*. **Habitat** Breeds in mangroves and reedbeds (including inland); also in scrub, woodland, date gardens post-breeding and in winter. **Note** Resident Iran and Arabia, dispersive or augmented by (usually biometrically larger) migrants in non-breeding season; range gradually expanding north and west. [Alt: Clamorous Reed Warbler]

juv
ad
Common Grasshopper Warbler
ad
River Warbler
ad
Basra Reed Warbler
ad
fusca
Savi's Warbler
ad
Great Reed Warbler
ad
Indian Reed Warbler

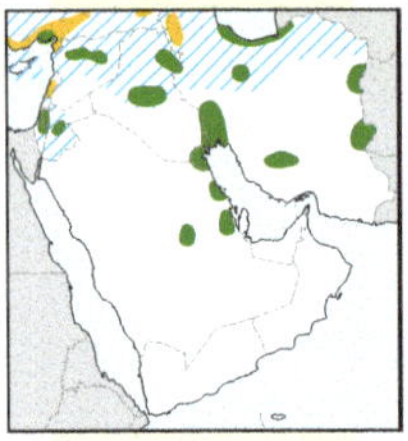

Moustached Warbler *Acrocephalus melanopogon* V

L: 13. Similar to Sedge Warbler but has broad, ***clear-cut white supercilium, ending squarely on side of nape and dividing blackish crown and greyish ear-coverts***; ***more uniform, rufous-brown upperparts*** (less obviously streaked than in Sedge), with ***whiter throat*** and warmer brown flanks. Eastern subspecies *mimicus* occurs, has shortish primary projection. Keeps low in vegetation or on ground where unobtrusive, ***often cocking and flicking tail nervously***, also when singing. **Voice** Song with intermittent ***characteristic Common Nightingale-like*** lu-lu-lu. Call a loud *trr-trr*, soft, short *tcht*, longer *trr-trrrrr* and a hard *tack*. **Habitat** Reedbeds, swampy thickets. **Note** Passage hatched; vagrant Bahrain, Oman, UAE.

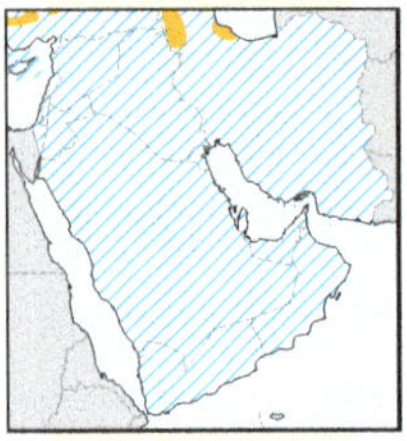

Sedge Warbler *Acrocephalus schoenobaenus* PM

L: 13. Streaked, with buffish-white supercilium; confusable with Moustached Warbler and told ***by buffier, less square-cut supercilium, slightly paler crown, paler ear-coverts and buffier, more streaked, upperparts, which merge into warmer-coloured rump, and distinctly longer primary projection***. First-winter Sedge Warbler is yellower than adult, shows paler centre to crown and fine spotting on breast. **Voice** Song fast, often rising and falling; often starts song with rapid *trr* notes. Mimetic. Call a hard *chek* and fast, churring *trrr*. **Habitat** Reedbeds, swampy thickets; drier habitats on passage. **Note** Passage hatched.

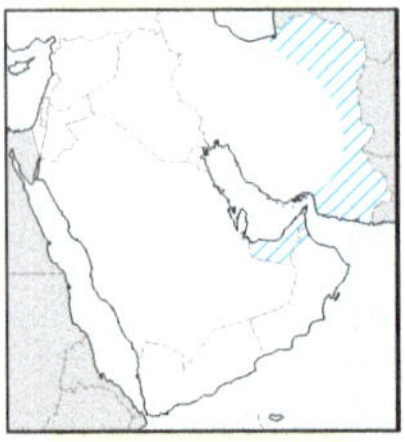

Blyth's Reed Warbler *Acrocephalus dumetorum* pm

L: 12.5. ***Similar to Caspian Reed and Marsh Warblers***. Identified by combination of ***shorter, more rounded wings and short primary projection with six primary-tips visible*** (7–8 visible in Reed and Marsh Warblers), ***uniform upperparts with little or no rump contrast, plainer tertials***, near-concolorous alula, short supercilium, which often bulges in front of eye, and dark grey bill with flesh-coloured base to lower mandible. Sometimes shows flicking and fanning movements of tail. **Voice** Song musical and highly imitative; slower, more hesitant than Marsh Warbler; often has high-pitched *lo-ly-lia* and utters *tjeck-tjeck* between phrases. Contact call a soft *thik* or *chck*. **Habitat** Bushy vegetation. **Note** Passage hatched; vagrant Bahrain, Kuwait, Oman, Saudi Arabia.

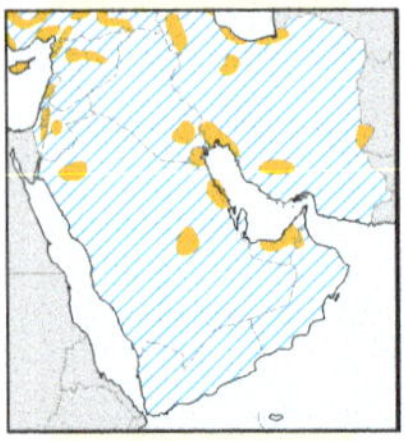

Caspian Reed Warbler *Acrocephalus (scirpaceus) fuscus* rb, mb, PM

L: 13. Very similar to Marsh Warbler, with similarly long wing-projection but less obviously pale edges to tertials, with ***longer bill and darker legs***; greyer, less rufous above and whiter below, thus even more similar to Blyth's Reed and Marsh Warblers. **Voice** Monotonous and fairly even-pitched mixture of scratchy, grating and churring notes; ***resembles song of Sedge Warbler but slower and lacks the changes in pitch and tempo***. Calls short *tchk* or hard, rolling *chrrrur*. **Habitat** Reedbeds and waterside vegetation. **Note** Passage hatched. [Alt: Eurasian Reed Warbler]

Marsh Warbler *Acrocephalus palustris* PM

L: 13. Difficult to identify; is brown above with ***slight olive tinge*** and slightly warmer rump; creamy-buff below with slightly buffier flanks; short buffish white supercilium and pale eye-ring. Like all *Acrocephalus* has a rounded tail, long undertail-coverts and rather sloping forehead. ***Marsh Warbler best told in spring by the pale-fringed dark tertials and long primary projection, eight primaries showing, each with pale-fringed tip***. Also buffier, less warm, than Caspian Reed (although juvenile Marsh rustier and thus extremely similar), with paler legs and shorter bill, though latter hard to judge. **Voice** ***Loud, musical and full of mimicry***; in fast tempo. Calls short *chek* and distinctive short buzzy *terrrr*. **Habitat** Any cover on passage. **Note** Passage hatched.

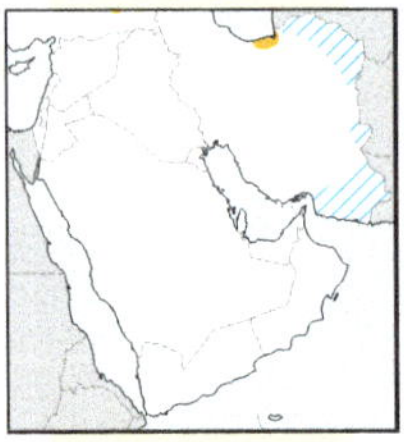

Paddyfield Warbler *Acrocephalus agricola* V

L: 12. Slightly smaller than Caspian Reed Warbler and differing in ***shorter bill, more prominent whitish-buff supercilium from bill to well behind eye*** (where most conspicuous), ***often with suffused darkish border above***, greyish nape shawl, dark-centred tertials with paler edges (unlike Blyth's Reed); ***shorter wings, and longer tail, which when landing or on the ground is often raised slightly and constantly flicked***. Underparts show warmer wash to flanks and undertail than in other unstreaked *Acrocephalus*. Bill brown with flesh-coloured lower mandible but distal part often darker in adult (so can appear dark-tipped). Crown feathers often raised; iris pale in adults. **Voice** Calls simple *chik, chik*, and a rolling *churrr*. **Habitat** Wetlands, grass or scrub. **Note** Passage hatched; vagrant Bahrain, Oman, UAE.

Moustached Warbler
ad
mimicus
ad
Sedge Warbler
1st-winter
ad
Blyth's Reed Warbler
ad
Caspian Reed Warbler
1st-winter
ad
Marsh Warbler
1st-winter
ad
Paddyfield Warbler
ad

Booted Warbler *Iduna caligata* V

L: 11.5. Small with ***short, rather fine bill, rounded forehead***, recalling *Phylloscopus*, but crown slightly peaked and bill broad-based, with pale tips and edges to outer tail feathers. Tail sometimes raised quickly upwards. Lacks paler wing-panel shown by larger Eastern Olivaceous Warbler. Confusable with Sykes's Warbler but ***bill shorter, distinctly finer and usually shows dark-tipped lower mandible***; upperparts grey-brown (lacking olive); longer, ***better-defined supercilium, sometimes with dark upper border, extends beyond eye. Legs flesh-brown*** with darker feet. **Voice** Song has rhythm of Eastern Olivaceous, but faster, more melodious without quick repetition of phrases; lacks mimicry. Call harder than Eastern Olivaceous, recalling weak Lesser Whitethroat's *tek-tek*. **Habitat** Scrub, tamarisk and woodland, often near water; often low cover, even weeds, on passage. **Note** Known passage hatched but poorly known and rare; vagrant Near East, Gulf states.

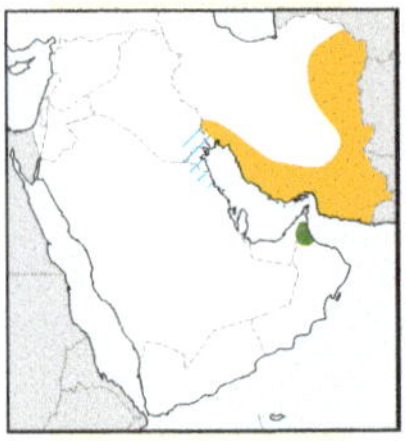

Sykes's Warbler *Iduna rama* rb, ?pm

L: 11.5. Only subtly different from Booted Warbler, having plainer face, ***longer bill, with pale lower mandible, longer tail, usually shorter primary projection (accentuating tail-length), greyish-pink legs*** and marginally paler underparts lacking any buff. Also has more horizontal stance and *Acrocephalus*-like appearance and habit. **Voice** Song bubbly with scratchy *Sylvia* subsong-like sequences, starts with **tiju-tiju-tiju** (***of tit-like character***). Call hard, clicking *tak*. **Habitat** Breeds in tamarisk, damp scrub, grazed mangroves; trees or scrub on passage. **Note** Known passage hatched, though rare; vagrant Qatar.

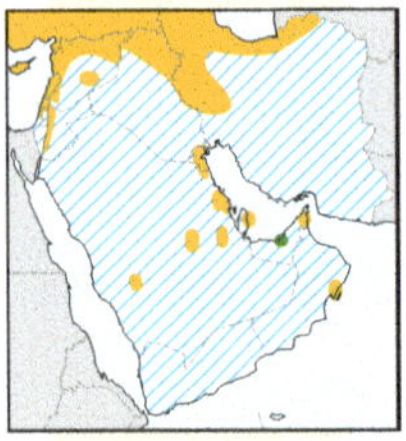

Eastern Olivaceous Warbler *Iduna pallida* rb, PM, wv

L: 12.5. Recalls *Acrocephalus* warbler but greyer with square-ended or slightly rounded tail and ***shorter undertail-coverts. Upperparts olive-brown, tinged greyish*** on head and mantle, underparts buffish-white (juvenile suffused yellowish at least on breast) with white throat; ***wings with marked pale panel in spring at least***, tail brown inconspicuously fringed and tipped white on outer feathers. Pale eye-ring and variably distinct ***supercilium from bill to rear of eye*** (longer, more contrasting supercilium in Booted and Sykes's Warblers); ***long, rather heavy, broad-based bill with yellow-pink lower mandible; legs greyish-brown***. Often deliberately flicks or ***pumps tail downwards*** (lacks circular tail-movements of Upcher's Warbler). Usually feeds in canopy. **Voice** Song like that of Caspian Reed Warbler in quality and rhythm, repetitive and unmusical but jaunty. Calls short *tch* or *tchek* or more drawn-out *che-ch-ch* or agitated *trrrrr*. **Habitat** Scrub, gardens, woodland, mangroves. **Note** Passage hatched.

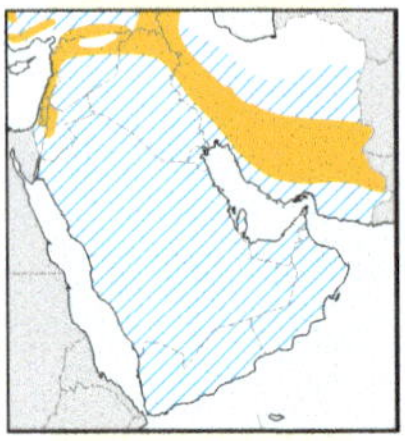

Upcher's Warbler *Hippolais languida* PM

L: 14. Similar to Eastern Olivaceous Warbler but larger with stronger bill, and proportionally ***longer, much darker sooty-brown tail with broader white tips and edges, especially visible from below. Tail uniquely waved up and down and sidewards in circular movements***. Upperparts greyer in fresh plumage (more like Eastern Olivaceous in abraded and juvenile plumage) with darker wings and prominent ***pale wing-panel on closed wings in fresh plumage***. Head more rounded than Eastern Olivaceous, though pale supercilium weaker and often extends behind eye (depending on lighting); legs cold grey. Larger Olive-tree Warbler has longer wing projection, powerful yellow-orange bill and short supercilium usually only visible in front of eye. Upcher's Warbler usually prefers mid- and lower levels of bushes. **Voice** Song repeated phrases like Eastern Olivaceous but louder more melodious, resembling Common Whitethroat. Call similar to Eastern Olivaceous; mostly silent on passage. **Habitat** Scrub, wooded areas. **Note** Passage hatched.

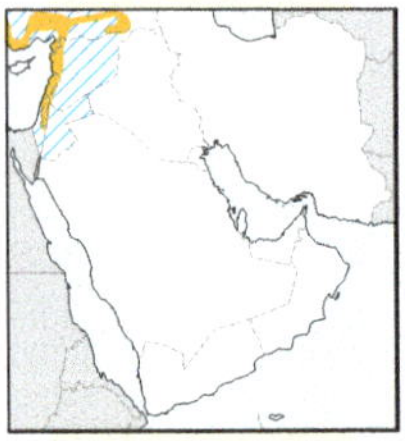

Olive-tree Warbler *Hippolais olivetorum* No records

L: 16. Large, brownish-grey ***with prominent wing-panel and long, dagger-like, yellow-orange bill***; forehead flat, sometimes raises crown feathers, recalling Icterine Warbler. Head pattern weak, faint supercilium ***usually only in front of eye***. Underparts whitish-buff; off-white throat contrasting with rather dark head and ear-coverts. Undertail-coverts with darker 'V's, hard to see. Tail dark, edged and tipped whitish on outermost feathers; legs bluish-grey. Resembles Upcher's Warbler but larger, with stronger and paler bill, longer wing projection and more prominent wing-panel. Often dips tail, sometimes waved and spread. **Voice** Song loud, ***harsh and raucous***, rhythm similar to Eastern Olivaceous but slower and lower-pitched. Call often repeated *tuc-tuc*. **Habitat** Scrub, trees. **Note** Passage hatched; vagrant Kuwait, Oman, Saudi Arabia, Yemen.

Booted Warbler
ad
Sykes's Warbler
ad
Eastern Olivaceous Warbler
ad fresh
ad worn
Upcher's Warbler
ad worn
ad fresh
Olive-tree Warbler
ad worn
ad fresh

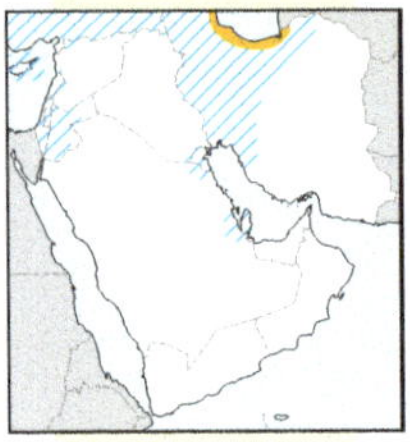

Icterine Warbler *Hippolais icterina* V

L: 13.5. Fairly large yellow and olive warbler, showing peaked crown, heavy yellow-orange bill (red gape when singing), ***very long wings with pale wing-panel in fresh plumage (less so in autumn adult) and bluish-grey legs***. Bare-faced appearance with conspicuous yellowish eye-ring and pale lores making eye rather prominent; yellowish supercilium short, rarely to rear of ear-coverts. Differs from ***Phylloscopus*** warblers in larger size, peaked crown, stronger bill, prominent wing-panel, and lack of contrasting supercilium and eye-stripe. Juvenile has browner upperparts and paler, more yellowish-white underparts with wing-panel more conspicuous than worn adult. Occasional individuals occur with greyish-olive upperparts and whitish underparts, recalling Eastern Olivaceous or Upcher's Warblers, but note longer wing projection and voice. **Voice** Varied fast song loud, highly mimetic, with harsh chatter and creaky 'violin' sounds, *de-de-dwiie* or *djehk-hyyii*; also includes characteristic trisyllabic *de-te-roy*; alarm call a hoarse, sparrow-like *tettettettett*. **Habitat** Trees, thickets; gardens, parks. **Note** Known passage hatched, but rare; vagrant Iraq, Oman, Qatar, UAE, Yemen.

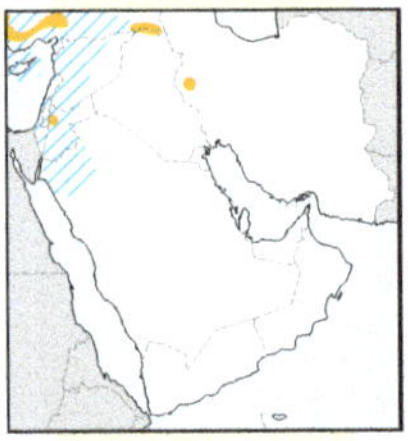

Eastern Bonelli's Warbler *Phylloscopus orientalis* V

L: 12. Size similar to Willow Warbler but ***mantle and more rounded head distinctly grey***, contrasting dark tertials (fringed yellowish) with ***obvious green edges to wing and tail feathers***; underparts silky-white; ***face 'washed-out' with large dark eye*** accentuated by narrow white eye-ring, usually ***pale lores, pale ear-coverts and indistinct supercilium***. If seen well, ***yellowish-green rump distinctive***. Confusable with pale Chiffchaff, but has plain head and longer primary projection; told from Booted Warbler by face pattern and well-marked tertials. **Voice** Call a short, flat, ***Crossbill-like chip***. **Habitat** Woodland, scrub. **Note** Passage hatched; vagrant Iraq, Kuwait, Oman, Saudi Arabia (Gulf), UAE (sole record under review).

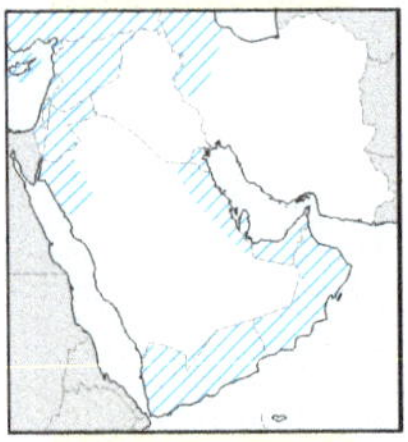

Wood Warbler *Phylloscopus sibilatrix* pm

L: 12.5. Large, ***long-winged Phylloscopus***, brightly coloured, showing ***vivid green upperparts with yellow supercilium, ear-coverts, throat and upper breast contrasting with gleaming white underparts; yellow-fringed secondaries form pale panel contrasting with blackish, whitish-fringed tertials***; black alula, yellow bill and pale legs. From Green Warbler by different wing pattern including lack of wing-bar; from other *Phylloscopus* also by very long wings, striking plumage, particularly head pattern and brightly fringed secondaries. Rare greyer individuals, recalling Eastern Bonelli's Warbler, told by strength of supercilium, long wings (still contrastingly fringed) and larger size. Rather active, often high in trees. **Voice** Call a clear melancholy *deeu-deeu-deeu* but usually silent on migration. **Habitat** Woodland and tall trees, also scrub. **Note** Passage hatched; scarce Arabia.

Dusky Warbler *Phylloscopus fuscatus* V

L: 11. Small, ***dark brown Phylloscopus*** with distinct head pattern resembling brownish Common Chiffchaff. ***Upperparts plain brown with rather distinct long, pale rufous-brown or whitish supercilium***, narrow and often palest in front of eye (reverse in Radde's Warbler). Underparts off-white, ***tinged pale rufous-buff on breast, flanks and undertail-coverts. Fine bill with largely yellowish lower mandible; thin legs flesh-brown***. From *tristis* Common Chiffchaff and Caucasian Mountain Chiffchaff by call, paler legs, paler bill (blackish in *tristis*), often rufous-tinged vent and undertail-coverts (not pale grey/off-white as in Common Chiffchaff/Caucasian Mountain Chiffchaff), also long striking supercilium (dull and shorter in *tristis*, but rather distinctive in Caucasian Mountain Chiffchaff). Often on ground. **Voice** Loud, hard, abrupt *chek* or *tack*, recalling Lesser Whitethroat, often persistent; harder and higher-pitched than Radde's Warbler. **Habitat** Low cover, grass or scrub, often in moist or marshy localities. **Note** Vagrant Kuwait, Oman, Saudi Arabia, UAE, Yemen.

Radde's Warbler *Phylloscopus schwarzi* V

L: 13. Robust *Phylloscopus* with large head, stout, deep-based rather pale bill and strong, straw-coloured legs. Upperparts olivaceous-brown with striking, wide, whitish supercilium (buffish and more diffuse in front of eye) accentuated by broad, dark eye-stripe, dark-mottled ear-coverts and dark margin above. Breast-sides, flanks and undertail-coverts yellowish-buff with centre of underparts dirty white; in worn plumage underparts become off-white with yellowish-buff undertail-coverts. First-winter often shows yellow-buff underparts (rarely or only faint in some Dusky Warblers). Calls frequently. **Voice** Call like Dusky Warbler, but weaker and lower-pitched *tuk* or more diagnostic double *tuk-tch*. **Habitat** Low vegetation. **Note** Vagrant UAE.

Icterine Warbler
1st-winter
ad
Eastern Bonelli's Warbler
ad
1st-winter
Wood Warbler
ad
Dusky Warbler
ad
ad worn
ad fresh
Radde's Warbler

Arctic Warbler *Phylloscopus borealis* — No records

L: 12. Rather large, ***dark greyish-green Phylloscopus*** with strong pale bill, striking head pattern with distinct ***long narrow supercilium from just in front of eye to well behind dark-mottled ear-coverts*** (sometimes upward kinked at end), accentuated by ***distinct dark eye-stripe from base of bill; short wing-bar*** (sometimes two visible in fresh plumage), plain tertials, dusky flanks and pale brown legs. Confusable with smaller Greenish and Green Warblers in worn, greyer plumage; distinguished from both by size, heavier build, head pattern, plumage tone and pale legs. **Voice** Characteristic short *dzik* (or *dzi-zik*), but usually silent on passage. **Habitat** Bushes, trees. **Note** Vagrant Oman, Saudi Arabia.

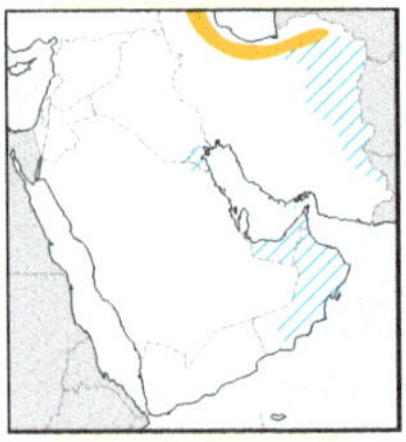

Green Warbler *Phylloscopus nitidus* — V

L: 11. Similar to Greenish Warbler with round head (like Common Chiffchaff), ***broad supercilium from base of pale bill to behind ear-coverts and short narrow wing-bar on greater coverts*** (chiffchaffs may show inconspicuous but rather long wing-bar). From Greenish Warbler by greener upperparts, and ***sulphur-yellow throat, upper breast, supercilium and ear-coverts*** contrasting with rather dark eye-stripe. In worn plumage more greyish, almost lacking yellow tone, and wing-bar may be lost; then very difficult to tell from Greenish Warbler, but from most other *Phylloscopus* by striking head pattern, pale bill and voice. **Voice** Song loud and variable, ending abruptly, *che-wee che-wee chui chi-di chi-dit*. Call White Wagtail-like *che-wee*, similar to Greenish Warbler. **Habitat** Bushes, trees. **Note** Passage hatched, but rare; vagrant Bahrain, Saudi Arabia, UAE.

Greenish Warbler *Phylloscopus viridanus* — No records

L: 11. ***Resembles a cold, greyish Willow Warbler***, with yellowish mostly restricted to ***long, broad supercilia, which usually meet over bill base*** (unlike Arctic Warbler) and may create capped, 'kind-faced' appearance. Also note ***dull greyish-white underparts and ear-coverts***, contrasting with distinct dark eye and eye-stripe; plain tertials, ***short wing-bar*** (sometimes two in fresh plumage, or worn off in summer), pale bill and dark brownish legs. From Willow Warbler and Common Chiffchaff by greyish-white underparts and cheeks, stronger head markings, short pale wing-bar (if present) and yellow lower mandible (especially from Chiffchaff). **Voice** Call similar to Green Warbler, *chi-wee*, or shorter *chi it*. **Habitat** Bushes, trees. **Note** Vagrant Iran, Oman.

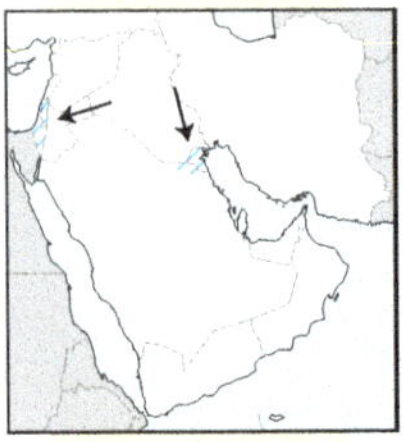

Yellow-browed Warbler *Phylloscopus inornatus* — V

L: 10. Very similar to Hume's Leaf Warbler, generally brighter ***with greener upperparts, whiter underparts, lacking buff suffusion; supercilium very long, pronounced and yellow-white***, rather evenly broad, ***usually two distinct yellow-white wing-bars contrasting with dark-centred greater coverts*** (pale-edged with ***less contrasting centres in Hume's***), also ***stronger contrast in tertial pattern; base of bill obviously paler*** (appears all dark in Hume's). Occasionally shows faint paler crown-stripe. In worn plumage greyer with paler wing-bars, and tertial-fringes narrower and whiter, when may best be separable by voice. **Voice** Thin song (may be heard on migration/winter) high-pitched, *tsee tseoo-tseee*, lacks buzzing quality of Hume's; call high-pitched lisping *tsweest* or *weest* with ***distinct upwards inflection***, frequently almost disyllabic *wii-ist*. **Habitat** Trees, bushes. **Note** Rare passage/winter as hatched; vagrant Iran, Oman, Qatar, UAE.

Hume's Leaf Warbler *Phylloscopus humei* — WV, pm

L: 10. Confusable with Yellow-browed Warbler; ***generally greyer, duller, less yellow-green*** with different call and song. In fresh plumage, upperparts and crown greyish-olive with very long, pronounced, ***buffish-white supercilium***, often broadest over and in front of eye; two wing-bars, lower one distinct, ***upper one short and ill-defined or absent*** (both distinct in Yellow-browed); ***underparts dusky-white, tinged buffish on ear-coverts, lower throat and sides of neck; bill dark*** (but pale base from directly below). Occasionally shows faint paler crown-stripe. In winter/spring worn plumage becomes greyer, wing-bars and tertial edges narrower, with upperwing-bar often lacking, then confusable with Greenish Warbler, but note smaller size, tertial pattern, bill colour and longer, broader wing-bar. **Voice** Calls sweet *wesoo* or shorter *dweed*, both ***lacking upwards inflection***; in winter more disyllabic, clean *sooit*, recalling Spotted Redshank but shortened. Song thin, high-pitched, drawn-out, ***descending buzzing* tzeeeee**. **Habitat** Trees, bushes. **Note** Winter hatched, but rare; vagrant Bahrain, Iraq, Kuwait, Oman, Qatar, Saudi Arabia. [Alt: Hume's Yellow-browed Warbler]

Arctic Warbler
1st-winter
Green Warbler
ad fresh
1st-winter
Greenish Warbler
ad fresh
1st-winter
Yellow-browed Warbler
ad fresh
1st-winter
Hume's Leaf Warbler
ad fresh
1st-winter

PLATE 76: *PHYLLOSCOPUS* WARBLERS III

Willow Warbler *Phylloscopus trochilus* PM

L: 11. Close to Common Chiffchaff, with ***longer primary projection; darker, pale-edged flight feathers and tertials; yellow wash on throat and breast; more distinct supercilium and pale-centred ear-coverts*** (*acredula* greyish above, whitish below; *yakutensis* similar, with grey-streaked breast). ***Legs pale pinkish-brown;*** tail-dipping much less pronounced. ***First-winter shows distinctly yellow underparts.*** **Voice** Song pleasant ***descending whistled verse dying away at end.*** Call rising, disyllabic *hoo-eet*. **Habitat** Woodland, scrub. **Note** Passage hatched.

Common Chiffchaff *Phylloscopus collybita* WV, PM

L: 11. Three subspecies, often inseparable, breed in Middle East: *collybita* (includes '*brevirostris*'); poorly known Iranian *menzbieri* and probably *abietinus*; at least two others visit. All show ***short supercilium***, broken eye-ring (conspicuous in autumn) and ***dark legs; tail frequently dipped.*** **Habitat** Woodland, scrub. **Note** Passage and winter hatched.

P. c. collybita (winters south to Near East/Iraq at least). Brownish-olive above, more olive on rump; short, weak supercilium. Underparts sullied white, streaked yellow on breast, with buff suffusion on flanks; undertail-coverts yellowish-white; underwing-coverts and bend of wing yellow. Wings and tail have indistinct yellowish-olive fringes. Weak bill dark horn, yellowish-brown at base of lower mandible. **Voice** Song repeated *chiff-chaff-chiff-chaff…*, varying geographically in structure and clarity. Calls plaintive *hoo-ee* or *huit*, with rising inflection, more monosyllabic than Willow Warbler; also downward-inflected disyllabic *sweeoo*.

P. c. 'brevirostris' (breeding form in Turkey; partial migrant), ***brown-tinged above*** (especially crown) ***and on breast sides***, with more distinct supercilium (especially over and in front of eye), ***off-white undertail-coverts*** and shorter primary projection. **Voice** Plaintive call, *biii* or *swee*, similar to Caucasian Mountain Chiffchaff. Song weak, thin *chiff-chaff*.

P. c. abietinus (winters abundantly south to Arabia). Paler and greyer above than *collybita*; buff and yellow on breast reduced, undertail-coverts whiter. **Voice** Wintering birds also call *peep*.

P. c. caucasicus (breeds Caucasus mountains, presumed visitor to Middle East). Darker green above; clean white belly and undertail-coverts. Legs blackish. **Voice** Song as *collybita/abietinus*; call without upward inflection.

P. c. tristis, Siberian Chiffchaff (a few reach UAE in winter). ***Greyish-brown above, lacking yellow except for bend of wing and underwing, olive-green tinge on rump, wing-coverts, scapulars and edges of flight and tail feathers. Thin supercilium often distinct, off-white to buff, rarely a yellow trace***, ear-coverts with rusty hue; entire underparts off-white, suffused buff on breast-sides and flanks; ***legs and bill almost black***. In fresh plumage ***may show indistinct wing-bar*** on greater coverts. **Voice** Song distinctive; fast, melodious, multi-syllabic rising and falling stream; almost unrecognisable as a chiffchaff. Call near-monosyllabic, plaintive *eep*.

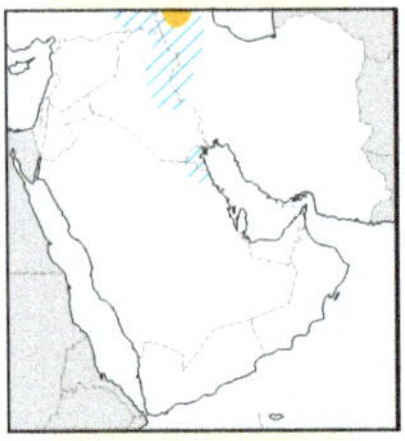

Caucasian Mountain Chiffchaff *Phylloscopus (sindianus) lorenzii* No records

L: 11. Recalls Common Chiffchaff, ***lacking obvious yellow or green*** (flight feathers may have subdued olive fringes, also faint tinge on scapulars; axillaries and underwing-coverts pale yellow). ***Upperparts earth-brown***, evenly brown crown slightly darker with ***broad white supercilium, bulging in front of eye*** (can 'bridge' forehead), buff behind eye; white lower eyelid, ***dark lores*** and buff ear-coverts. ***Entire underparts off-white***; tendency to show white bib surrounded by pinkish-buff, ***breast-sides and flanks tinged rusty***. Basal half of lower mandible yellow; legs and strong feet black. Appears rather short-winged. Tail-dipping and behaviour as Common Chiffchaff. **Voice** Fast, repetitive song more rambling than Common Chiffchaff. Call a melancholic *beee*. **Habitat** Mountain forest, subalpine and riverine scrub in summer. **Note** Winter reports hatched.

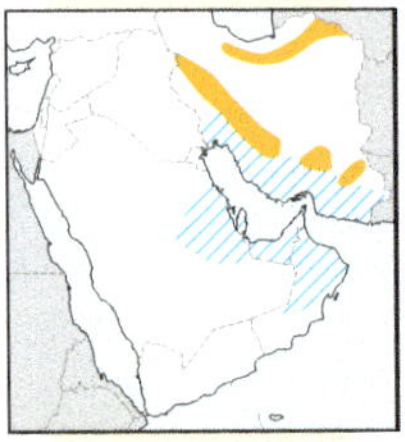

Plain Leaf Warbler *Phylloscopus neglectus* WV, PM

L: 9. Smallest *Phylloscopus*, with short tail and lacking obvious markings or any yellow; constantly flicks wings. Plumage like greyer chiffchaffs, with brownish-grey upperparts tinged olive, shawl greyish; indistinct, short, pale buff to cream supercilium, narrow white eye-ring, dusky lores and eye-stripe and buff-flecked ear-coverts. Underparts off-white, washed creamy on flanks. Legs and fine bill dark brown. Forages actively in cover; often hovers outside or below canopy. **Voice** Short, quiet tinkling song often heard in winter, *pt toodla toodla*. Call excitable, ***hard*** **t-jick *or* tdd**, commonly repeated. **Habitat** Scrub and trees in winter. **Note** Winter hatched.

Willow Warbler
acredula
yakutensis
nominate
1st-winter
Common Chiffchaff
tristis
'brevirostris'
abietinus
Caucasian Mountain Chiffchaff
Plain Leaf Warbler

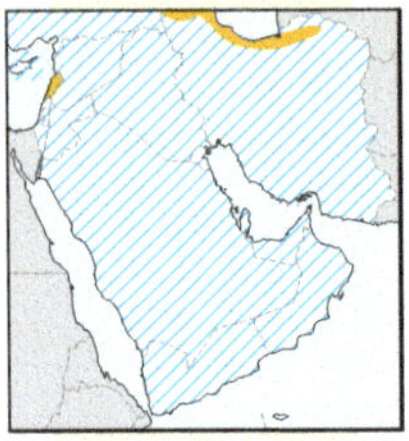

Eurasian Blackcap *Sylvia atricapilla* PM

L: 14. Similar in size to Common Whitethroat; plain grey-brown above with distinctive solidly ***black cap*** in adult male and ***warm brown cap*** in female and juveniles, lacks any white in outer tail feathers; these features preventing confusion with any other warbler. Sings from cover. **Voice** Joyful, liquid song similar to that of Garden Warbler, rich and varied, ending with a few clear melodious fluted notes. Call a hard insistent *tack*, or repetition, similar to Lesser Whitethroat; also short *churr churr*. **Habitat** Woodland, scrub. **Note** Passage hatched, occasional in winter.

Garden Warbler *Sylvia borin* pm

L: 14. Uniform grey-brown warbler with square-ended tail, ***short, relatively heavy bill***, practically no supercilium, often greyish sides of neck, ***rounded head and no white in tail or wings***. Most *Acrocephalus*, *Hippolais* and *Iduna* warblers have thinner, longer bills, *Acrocephalus* also have longer undertail-coverts and rounded tails. See juvenile Barred Warbler for separation. Usually keeps well hidden. **Voice** Song melodious, fast, with considerable jumps in pitch in irregular sequence, lacking Blackcap's ascending fluty finish. **Habitat** Woodland, scrub. **Note** Passage hatched, but scarce Arabia.

Asian Desert Warbler *Sylvia nana* WV, PM

L: 11.5. Small, relatively long-tailed, sandy grey-brown warbler ***with rufous-brown tertials, rump and closed uppertail; the spread tail tricoloured (dark brown with white at sides and centre rufous)*** often flicked half-cocked; ***iris and legs yellowish***, fine bill largely yellowish, sometimes having a pale area around eye. Terrestrial; hops and scuttles or flies low into cover, where remains hidden. Often accompanies Desert Wheatear in winter. **Voice** Call a dry, weak purring *drrrrrr*, descending and fading out; also *chrr-rrr* and high-pitched *che-che-che-che*. Song, sometimes uttered in flight, starts with purring call, followed by short, clear, melodious trill. **Habitat** Desert, semi-desert, hill-sides with low scattered bushes, sparsely scrubby saltflats and plains, rarely wooded areas. **Note** Passage and winter hatched.

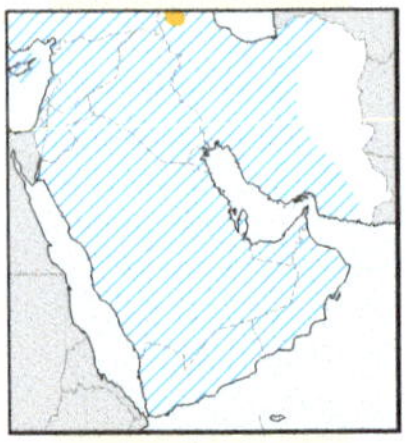

Barred Warbler *Sylvia nisoria* PM, wv

L: 15.5. Large warbler, ***adult with crescentic barring on underparts*** (reduced in female) ***and yellowish iris***; fairly long tail ***has conspicuous white corners*** when landing; tips of greater coverts and tertials whitish. First-autumn birds dark-eyed, recalling Garden Warbler, ***but larger, with white in tail, pale edges to wing-coverts and tertials and some dark barring on undertail-coverts***; this plumage often in birds through the following spring. Young Eastern Orphean Warbler lacks pale markings on wing-coverts and tertials, and has contrasting dark ear-coverts, but may show dark marks on undertail. Usually furtive, keeping hidden; movements rather heavy. **Voice** Song resembles that of Garden Warbler but phrases shorter; only sub-song usually heard in Arabia. **Habitat** Thickets, scrub, trees. **Note** Passage hatched, very rare Arabia in winter.

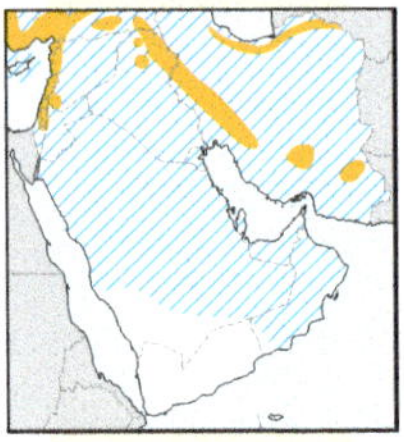

Eastern Orphean Warbler *Sylvia crassirostris* WV, PM

L: 15. ***Distinctly larger than Sardinian Warbler but with grey crown contrasting poorly with blackish ear-coverts, and larger bill; adult has yellowish iris*** (orange in adult Sardinian, which also has reddish eye-ring), but many singing males have dark, mud-coloured iris. Tail-sides white. First-autumn birds and immature females also ***lack pale iris***; former, as with some first-summer males, recall large, sluggish Lesser Whitethroat with contrasting dark lores and ear-coverts and dark markings on undertail-coverts. Often remains concealed. **Voice** Loud, varied, musical song recalls Common Nightingale's, given from dense cover. Calls include a loud Lesser Whitethroat-like *tak* and a *trrr*. **Habitat** Bushy hillsides, deciduous thickets, parkland. **Note** Passage hatched, winters east Arabia/south Iran.

Eurasian Blackcap
♀
♂
Garden Warbler
ad
Asian Desert Warbler
1st-winter
♂ spring
Barred Warbler
♂
1st-winter
♀
Eastern Orphean Warbler
♂
♀ 1st-winter
♀
1st-winter

Lesser Whitethroat *Sylvia curruca* PM, wv

L: 13.5. A small *Sylvia*, grey-brown above with browner wings, medium-grey crown and darker ear-coverts, which often vary in prominence; underparts show contrast between white throat and dusky-washed breast. Differs from Common Whitethroat in having ***dark ear-coverts and dark legs*** and in the ***absence*** of rusty fringes to wing feathers (though see *icterops* subspecies of Common Whitethroat). First-winter Lesser Whitethroat has slightly paler upperparts and often shows an indistinct whitish supercilium. Siberian subspecies, *blythi*, winters Iran and parts of Arabia, has paler, warmer brown upperparts and clean, whiter underparts, seemingly with little contrast between throat and breast. Does not songflight. **Voice** Song a fast, loud rattle *tell-tell-tell-tell-tell*. Call a rather short and hard *tek* often repeated. **Habitat** Dense undergrowth, trees. **Note** Passage hatched; winters south Arabia.

Desert Whitethroat *Sylvia minula* WV, PM

L: 13. Slightly smaller than Lesser Whitethroat with smaller bill, ***and paler, more washed-out plumage***. Upperparts grey-buff or sandy-brown with pale buffish edgings to wing feathers; buffish-grey crown and darker ear-coverts (but paler than in Lesser or Hume's Whitethroats); underparts clean sandy-white with white throat; more extensive white in outer tail feathers. More active than other whitethroats, frequently flicking tail. The subspecies *halimodendri* ('Central Asian Lesser Whitethroat'), which occurs only as a migrant, is considered closest to *minula* (genetically, vocally and morphologically). It is the same size as, but paler and sandier than Lesser Whitethroat (both *curruca* and *blythi*), having pale sandy or buff-brown upperparts; pale grey crown with clearly contrasting dark lores and ear-coverts; like *minula*, has more white in outer tail than *curruca*. **Voice** Characteristic call, ***buzzy tit-like* che-che-che-che-che**; hard *tek* only infrequently used; fast *sree-sree-sree* when agitated. **Habitat** Trees, especially acacia and *Prosopis*, in semi-desert, savanna and hills. **Note** Passage and winter hatched. [Alt: Desert Lesser Whitethroat]

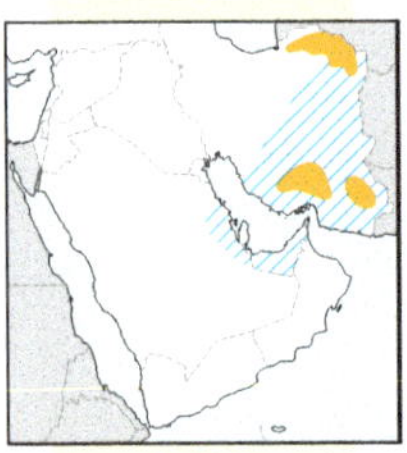

Hume's Whitethroat *Sylvia althaea* wv, pm

L: 14. ***Slightly larger than Lesser Whitethroat***, showing rather large head with ***stouter bill and darker plumage***. Dark ashy-grey crown (darker than in Lesser Whitethroat), merges into dull grey-brown back; ear-coverts also dark ashy, ***not contrasting with crown***, but contrasting strongly with white throat. First-winter birds can look particularly slaty above with paler feather-edgings on coverts and tertials. **Voice** Song a pleasant Blackcap-like warble. Call a hard *tek, tek*, a single *churrr*. **Habitat** Upland woody scrub in summer; trees and bushes on passage and in winter. **Note** Rare in hatched area in winter; vagrant Oman, Qatar.

Common Whitethroat *Sylvia communis* PM, wv

L: 14. Medium-sized warbler, similar to Lesser Whitethroat but differs in having ***chestnut fringes to coverts and secondaries***, slightly longer tail, ***orangey-coloured legs*** and ***white eye-ring***. Females and first-winter birds lack the grey wash to head of male and have dark (not orange) iris. Subspecies occurring, *icterops*, has brownish (barely chestnut) fringes to the coverts and secondaries, thus resembling Lesser Whitethroat more, but can always be told by larger size and orange legs. **Voice** Song, fairly short, scratchy warbling outburst from perch on bushtop or in display flight. Call a harsh *whet-whet-whet*; loud scolding alarm repeated drawn-out *jaairh*. **Habitat** Trees and scrub; patchy dense low vegetation. **Note** Passage hatched; rare south Arabia in winter.

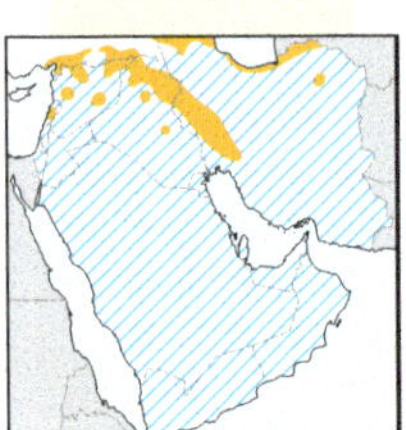

Ménétriés's Warbler *Sylvia mystacea* PM, wv

L: 12.5. Small, ***characteristically waves tail sideways or up and down. Male has black forehead and ear-coverts*** merging into grey crown; wings with sandy fringes and black alula. In fresh plumage shows white moustachial stripe contrasting with salmon-pink to brick throat, paler on breast (subspecies *mystacea*; breeding E. Turkey, N. Iran); pink much reduced or whitish in *rubescens* (SE Turkey, W. Iran, Iraq), with *turcmenica* intermediate (E. Iran), but any may be whitish below when abraded. Tail dark with much white in outer feathers; ***eye-ring varies from salmon-pink to red, bill bicoloured***. Legs usually pinkish-straw. ***Female and young paler, of plain appearance***; with pale sandy-brown upperparts, ***uniform buffish-white underparts*** and base of lower mandible straw or pinkish; eye-ring from brown to yellow in adult female, dull brown to yellow-brown in first-autumn/winter. Mostly active in low scrub or on ground. **Voice** Song rather quiet but fast, a mixture of melodious and soft grating notes. Calls include *chak* or *tret* or distinctive ***hard, quick-fire staccato rattle***; sometimes also a softer *tshshshshsh*. **Habitat** Scrub and thickets, riparian cover, often in broken country. **Note** Passage hatched; winters S. Iran and Arabia.

ad
1st-winter
ad
blythi
Lesser Whitethroat
spring
Desert Whitethroat
1st-winter
spring
Hume's Whitethroat
1st-winter
♀
communis
♂
icterops
Common Whitethroat
♀
♂ 1st-winter
♂
♀
♂ 1st-winter
Ménétriés's Warbler
♀ 1st-winter

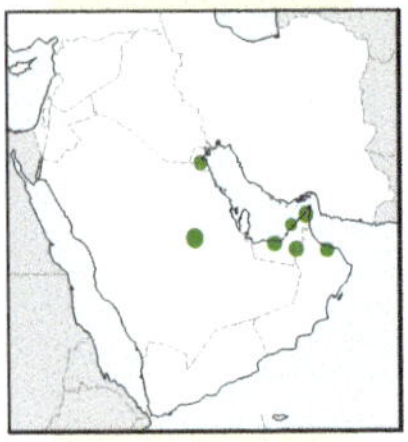

Bank Myna *Acridotheres ginginianus* RB (E/I)

L: 21. Resembles Common Myna but smaller, with ***body slate-grey*** (deep vinous-brown in Common Myna); the blackish head has short crest on forecrown; ***bare orange-red patch around eye and pale reddish bill diagnostic*** (Common Myna has both bill and eye-patch bright yellow). ***In flight, large rusty-buff patches across bases of primaries and on tail-corners***. Juveniles have body tinged brown, not slate; wing- and tail-patches more buffish-white, approaching Common Myna, but colour of bill and around eye always separates. Often confiding, rather noisy, in pairs or in flocks; feeds mainly on ground. **Voice** Garbled chattering notes from flock, sometimes a rather musical song, both distinguishable from Common Myna with experience. **Habitat** Towns, villages, fields, grassy areas; flocks roost in trees or reeds. Nests colonially in hole in bank, mud-well or masonry. **Note** Not native; breeding populations originate from escapes.

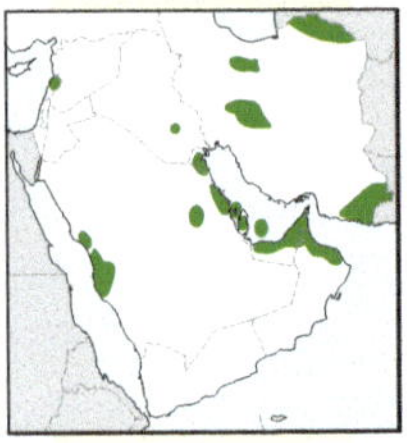

Common Myna *Acridotheres tristis* RB (E/I)

L: 23. Bold, noisy, gregarious myna. ***Body deep vinous-brown*** (slate in adult Bank Myna); ***bill and small bare patch below eye bright yellow*** (reddish in Bank Myna). ***In flight, has conspicuous large white patch across primary bases, on outer underwing-coverts and on tail-corners*** (rusty-buff in Bank Myna). **Voice** Song imitative and repetitive with strident, rough and liquid notes mixed, *piu-piu-piu, che-che-che, tliy-tliy-tliy, tuu-tuu-tuu, tititi, pryv-pryv*. Alarm a grating *traaah*. **Habitat** Urban settings, parks, gardens, fields; nests in hole in tree, palm crown, under eaves or other manmade setting. **Note** Resident; breeding populations outside E. Iran originate from escapes.

Pied Myna *Sturnus contra* RB (E/I)

L: 23. ***Black-and-white plumage diagnostic; black upperparts, conspicuous white rump and narrow white shoulder braces, with white ear-covert patch and eye red-bordered; pointed bill orange-red***. In flight has no white patch in wing or around tail. In young birds the black above is replaced by brown; underparts are buffish-white with breast streaked or spotted vinous-brown; ear-coverts whitish. **Voice** Calls wheezing; almost mechanical trilling song of long, rather monotonous but musical sequences. **Habitat** Open, cultivated moist ground near human habitation, dumps, grassy fields. **Note** Not native; UAE breeding populations localised.

Brahminy Starling *Sturnia pagodarum* rb (E/I)

L: 21. Small and undemonstrative, unlike Common Myna, may recall waxwing; seldom vocal, solitary or in pairs. Adult with ***black crest***, usually flattened, ***rufous-buff underparts finely white-streaked and blue skin-patch behind eye; bill yellow with blue base***. Vent, tail sides and tip white. Juvenile similar but much browner and lacking crest. **Voice** Often silent; short song (includes mimicry) a drawn out gurgle followed by bubbling yodel; alarm a Jay-like *churr*. **Habitat** Parks, gardens, woodland edge, scrub. **Note** Not native; breeding populations originate from escapes.

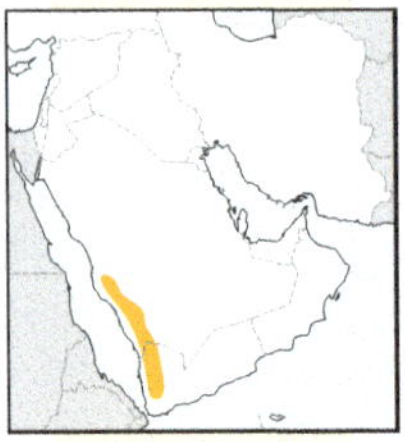

Violet-backed Starling *Cinnyricinclus leucogaster* V

L: 19. ***Male unmistakable with iridescent violet-purple upperparts, head and breast*** (but can look black or red), ***rest of underparts white***, eye yellow. ***Dark brown female has dark streaks on white breast below brownish throat***, belly whiter; ***eye yellow or chestnut***, but dark in otherwise similar juvenile. May flick wings singly when perched. Mostly in small groups, fairly shy. Flight direct. **Voice** Song a loud, metallic gurgling warble; call a ringing, grating musical squeal with rising inflection ending in quiet chuckle, latter also heard when flushed. **Habitat** Plains, hills and wadis with fruiting trees. **Note** Vagrant Oman, UAE. [Alt: Amethyst Starling]

Bank Myna
ad
Common Myna
ad
Pied Myna
ad
Brahminy Starling
ad
♀
Violet-backed Starling
♂

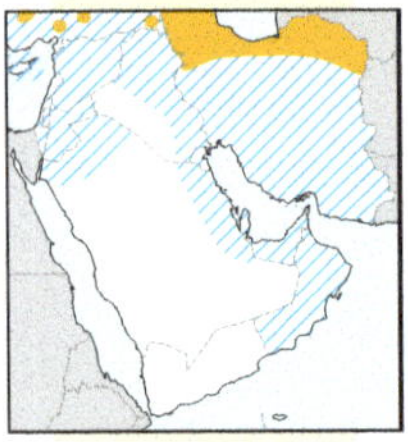

Rose-coloured Starling *Pastor roseus* PM, WV, sv

L: 21. ***Black-and-pink male unmistakable; long crest raised when singing***; female similar but more sombre. Juvenile lacks crest; from young Common Starling by ***shorter yellow bill with culmen curved*** (longer, darker, straight in Common Starling), ***pale lores, paler grey-brown upperparts with whitish-grey rump and dirty whitish underparts*** (Common Starling drab brown, with dark loral streak). Gregarious. **Voice** Calls resemble Common Starling's; rapid, high-pitched, musical chatter from feeding flocks. Song a lively chattering jumble mixed with melodious warbling. **Habitat** Open country, near agriculture, livestock, farmsteads. **Note** Passage hatched; some winter Arabia.

Common Starling *Sturnus vulgaris* rb, WV

L: 21. ***Short tail and short, pointed, triangular-shaped wings with fast, straight flight*** with frequent glides or brief wing-closures characteristic. ***Breeding adult has glossy-green, blackish plumage with numerous minute white spots***, though spots almost absent in summer in some subspecies breeding in region, notably E. Iran and Arabia, in which mantle and breast purple, and E. Turkey to W. Iran, which has bluish-green on mantle and bronzy sheen on head. ***Juvenile drab brown, moulting to blackish with prominent white spots in first-winter***, resembling winter adult. Gregarious, roosting flocks sometimes huge; feeds aerially or on ground; walks quickly. **Voice** Fast song, varied, whistles, strained whines and descending *seeeoo*, incorporating fine imitations. Flight call a short buzzing *tcheer*, alarm a hard *kjet*; at nest, grating *stahh*. **Habitat** Towns, villages, farmland, woods, parks, lawns. Post-breeding often roosts in reedbeds. Nests in hole in tree, building or nest-box. **Note** Passage and winter hatched.

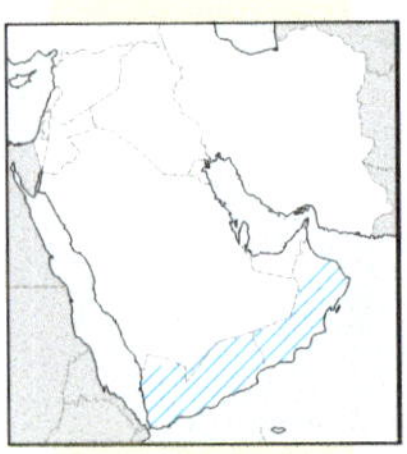

Wattled Starling *Creatophora cinerea* V

L: 22. Resembles Rose-coloured Starling but bill stronger and wing-tip slightly blunter. Juvenile, most often seen in region, cold grey-brown above, paler below, with pale fringes to coverts and tertials and ***whitish rump***; from young Rose-coloured Starling by pale upperwing-coverts, naked malar region and lack of pale fringes to flight feathers. ***Adult male has bare yellow head with black wattles***; outside breeding season resembles female, ***both sexes then showing creamy-white rump, fleshy-yellow bill, buffish greater coverts, white spot on leading primary coverts, yellowish area around eye***, diffuse loral spot and moustachial streak. **Voice** Soft squeaky whistle. **Habitat** Open bush and savanna. **Note** Rare and irregular visitor to hatched area; vagrant UAE.

Superb Starling *Lamprotornis superbus* E/I

L: 19. Unmistakable. Adults have ***glossy green upperparts and breast, whitish eyes, a narrow cream breast-band, rusty belly*** and white vent and underwing. Duller-plumaged juveniles lack cream breast-band and have dark eyes. Often swaggering and confiding. **Voice** A rising and falling chattering jumble of notes. **Habitat** Parks and gardens, open scrub, golf courses, near villages or other settlement. **Note** Not native; breeding populations originate from escapes (from Africa). In UAE breeds on Sharjah University Campus, only occasionally reported elsewhere and not yet considered naturalised.

juv
Rose-coloured Starling
♂ summer
juv
ad winter
Common Starling
ad summer
juv
♂ winter
♂ summer
Wattled Starling
juv
ad
Superb Starling

PLATE 81: THRUSHES I

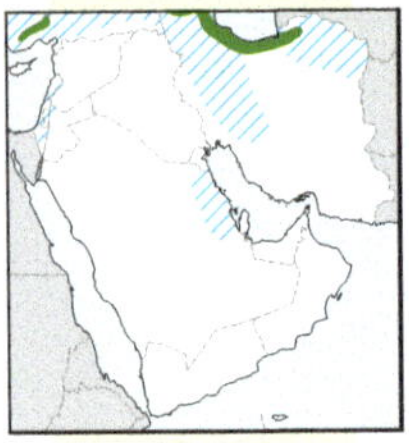

Ring Ouzel *Turdus torquatus* V

L: 24. Resembles Eurasian Blackbird; males told instantly by ***white breast-crescent and pale fringes to wing feathers***, which give 'frosty-winged' appearance in flight. Subspecies *amicorum* recorded (breeds E. Turkey, Iran, Caucasus) striking, with much more extensive white in wings. Female dark brown with ***obscure buffish breast-crescent*** (virtually absent in first-winter females), ***pale fringes to wing-feathers and to feathers of underparts, giving scaly appearance***. Often shy but will also sit prominently in open. **Voice** Song highly variable, normally 3–4 fluty notes, often followed by twittering. Most common call is a loud, hard scolding *tek-tek-tek*, with a soft shrill chatter in flight. **Habitat** Mountains with alpine meadows, upper limit of open forest. In winter/passage any area with trees or bushes; rocky hills, wadis. **Note** Partial migrant; winter hatched; vagrant Gulf States, except Saudi Arabia where rare as hatched.

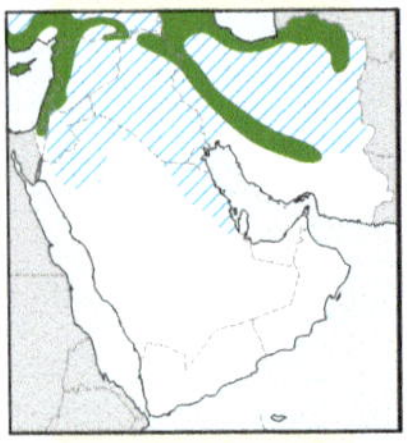

Eurasian Blackbird *Turdus merula* V

L: 25. Adult male entirely black with yellow bill; first-summer males show brown flight feathers and dark bill. Female dark brown with all-dark bill and slightly paler throat, breast with dark mottling (when seen closely). Juvenile similar to female but tends to show more mottling on underparts. Superficially resembles Common Starling but has longer tail, which often cocked, and usually hops (rather than walks). Feeds extensively on fruit in autumn and winter. **Voice** Fine songster with rich meandering flute-like notes. Various calls include *chak, chak, chak,* often accelerating into a fast outburst; also a thin *tsee* especially in flight. **Habitat** Woodlands, gardens, orchards with undergrowth; often near cultivation. **Note** Partial migrant; passage and winter hatched; vagrant Bahrain, Qatar, UAE.

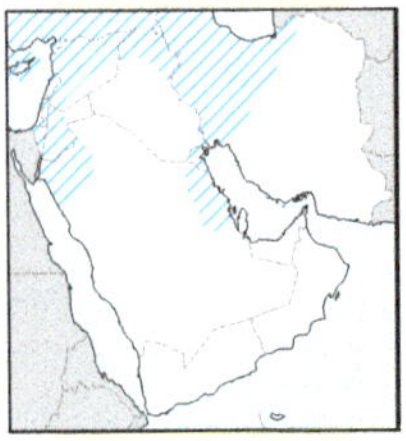

Fieldfare *Turdus pilaris* V

L: 25. Heavily built, size of Eurasian Blackbird, with ***rufous-brown mantle and coverts, former contrasting with pale ash-grey head and rump; black tail, rusty-yellow breast boldly streaked/spotted blackish, flanks with dense 'V'-shaped marks and underwing-coverts white***; young birds duller. Flight slightly undulating; in loose flocks, uttering characteristic call which separates it easily from larger Mistle Thrush. **Voice** Characteristic loud *dscach-dscach-dscach*; also thin *veid* uttered as contact note from flocks. **Habitat** Open country, fields with clumps of trees, hedges. **Note** Winter hatched, but rare Arabia; vagrant Bahrain, UAE.

Song Thrush *Turdus philomelos* WV, PM

L: 22. Small thrush, readily told from Redwing ***by blackish spots (not streaks) on whitish underparts, lack of supercilium and by voice***; underwing-coverts buff (rusty-red in Redwing). From Mistle Thrush ***by smaller size and different flight and voice***. Flight fast and straight. In winter seen singly or in small scattered groups, often shy. **Voice** Powerful song alternates between fluty and shrill sharp notes, usually repeated 2–4 times, *di-du-weet, di-du-weet, di-du-weet; dwi-dwi-dwi; du-drid-du-drid; peeoo-peeoo-peeoo-peeoo*. Call a short sharp *zit* or *zip*; alarm at nest sharp and noisy *telk-telk-telk*. **Habitat** Woodland undergrowth, parks, plantations; in winter also in open country with scattered trees and bushes. **Note** Winter hatched, but rare south Arabia.

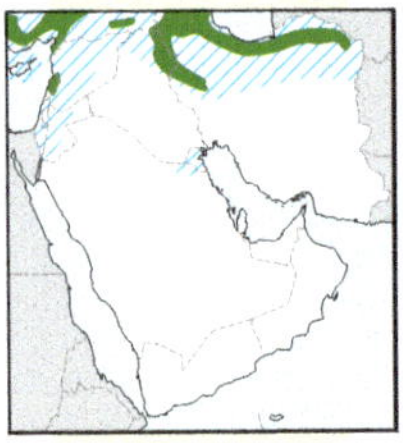

Mistle Thrush *Turdus viscivorus* V

L: 28. Large; resembling upright, outsize Song Thrush. White underwing-coverts conspicuous in flight; told from Fieldfare (similarly with white underwing-coverts) ***by grey-brown upperparts*** (without contrasts of rufous/pale ash-grey), ***white corners of uppertail*** (visible upon landing), ***whitish underparts evenly spotted blackish and voice***. Flight slightly more undulating than in Fieldfare (both often closing wings), and quite different from that in smaller Song Thrush. Appears pot-bellied, with stance more upright than other thrushes. Bold but often fairly shy; seen singly or in pairs, in winter also in small parties. **Voice** Eurasian Blackbird-like song, with shorter phrases, more monotonous and melancholic, *trueetroowu, trueetroowu, tih-ooh-woo- trooh*. Flight call a loud, dry churring *trrrrrr*, delivered repeatedly and more vehemently when alarmed. **Habitat** Wooded regions with clearings. In winter in open country, fields with trees, grassland. **Note** Winter hatched but rare Kuwait; vagrant Bahrain, Oman, Saudi Arabia, UAE.

♂
Eurasian Blackbird
♀
Ring Ouzel
♂
ad
♀
Fieldfare
ad
ad
Song Thrush
Mistle Thrush

Eyebrowed Thrush *Turdus obscurus* V

L: 19. Size of Redwing, ***lacking spots or streaks on breast and belly. Male has white supercilium, black loral streak, grey head, neck, throat and upper breast, pale rufous lower breast and flanks***; female and first-autumn birds have olive-brown head, supercilium and loral streak of male, some dark streaks on whitish throat, ochre-yellowish lower breast and flanks; underwing-coverts pale grey-brown; belly white. Small whitish spots on tip of outer tail feathers in adult hard to see; first-autumn birds have white-tipped greater coverts. Flight as Redwing, powerful and straight, often darting low over canopy. Generally shy. **Voice** Shares *zeeip* call with Redwing, sometimes shorter or with Eurasian Blackbird-like timbre; also *dackke-dsjak, psiiie*, first part recalling Fieldfare, latter (which sometimes uttered alone) Eurasian Blackbird. **Habitat** Wooded areas or open country with trees and bushes. **Note** Vagrant Oman, UAE.

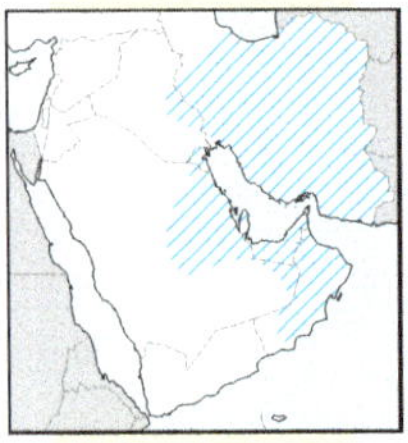

Black-throated Thrush *Turdus atrogularis* wv

L: 23. Slightly smaller than Eurasian Blackbird. Adult male with ***clear-cut black 'bib', sharply defined from whitish underparts***; in female white throat streaked blackish but ***lower border of black breast clear-cut***, upperparts browner. First-autumn male recalls adult but 'bib' slightly pale-mottled; first-autumn female has ***densely streaked upper breast with diffusely streaked flanks*** and sometimes parts of belly; short whitish eyebrow (unlike immature Dusky Thrush). Underwing-coverts rusty, tail grey-brown above. Fairly shy; behaviour similar to Fieldfare but flight faster and straighter. **Voice** Soft squawk *chork-chork* resembling quiet Fieldfare; also Eurasian Blackbird-like chatter, a quiet ***sip*** like Song Thrush and a squeaky *tscheeik*. **Habitat** Winters in open country with trees, grassy areas, parkland. **Note** Winter hatched, but irregular or rare; vagrant Qatar, Yemen. [Alt: Dark-throated Thrush]

Red-throated Thrush *Turdus ruficollis* No records

L: 23. Adult male as Black-throated Thrush but with ***bib red-brown***. Female and immatures separable from Black-throated by ***rusty-rufous tail***, at least on outer feathering. First-winter may also show slight rusty-buff on sides of streaked upper breast. Hybridisation with Black-throated Thrush can occur. **Voice** As Black-throated Thrush. **Habitat** Trees or scrub; cultivation, gardens. **Note** Vagrant Iran, Qatar.

Dusky Thrush *Turdus eunomus* V

L: 24. Size of Eurasian Blackbird. Upperparts dark olive-brown, the ***closed wings having a rufous panel*** (except first-autumn female); ***bold white supercilium, black lores and patch on ear-coverts, white throat and lower ear-coverts, and 1–2 narrow, boldly black-flecked breast-bands, spots or flecks extending boldly along flanks***; in female and first-autumn birds white throat sometimes streaked darker, breast-bands less bold and upperparts less dark; tail dark brown; underwing-coverts pale rufous-brown. Bill base bright yellow. **Voice** Flight calls include *srrii-i* recalling Redwing; also Fieldfare-like *kwae-waeg* or *tjshah-tjshah*. **Habitat** Open woodland or parkland. **Note** Vagrant Kuwait, Oman, UAE.

Redwing *Turdus iliacus* V

L: 20. Slightly smaller than Song Thrush but told easily by prominent buff-white supercilium and submoustachial streak below ear-coverts, streaked underparts, rusty-red flanks and underwing-coverts (latter often hard to see); also voice. Flight fast and straight. **Voice** Call a thin, metallic and drawn-out *steeef* or *zeeip*, often heard on migration at night. **Habitat** Woodland, cultivation, parkland. **Note** Winter hatched, rare Iraq, north Arabia; vagrant Bahrain, UAE.

Eyebrowed Thrush
♂
1st-winter
Black-throated Thrush
♀
♂
♀ 1st-winter
Red-throated Thrush
ad
Dusky Thrush
1st-winter
ad
Redwing
ad

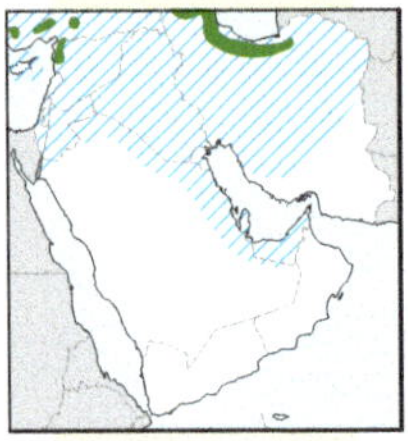

European Robin *Erithacus rubecula* WV

L: 14. Plump, upright, brown chat with ***diagnostic orange-red face and breast***, large dark eyes and short wings. Tail often cocked. Usually hops on or close to ground, often shy. **Voice** Song short but crystal clear and plaintive with abrupt changes in pitch and tempo. Usual call a sharp clicking *tic* or *tic-ik*, often repeated; also a thin drawn-out *tseer*. **Habitat** Shady gardens, copses, woodland, reedbeds. **Note** Winter hatched, but rare in south; vagrant Oman, Qatar.

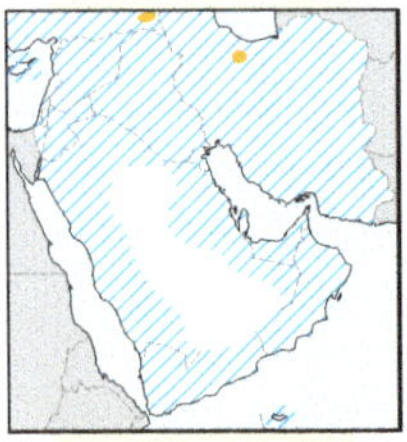

Bluethroat *Luscinia svecica* PM, WV

L: 14. Breeding male has ***blue chin and throat*** framed below by black and rusty bands on breast. Subspecies assignable only in adult males: throat-patch red in *pallidogularis* and *svecica*, white in *cyanecula*; but throat entirely blue in *magna*. Non-breeding male, female and first-winter have pale throat and ***black malar stripe joined to dark necklace. Rust-red sides to tail-base diagnostic***, most obvious in flight. Usually in cover on or close to ground; slinks about on foot, often darting back to cover. Tail often cocked. **Voice** Song melodic, many notes distinctly metallic or scratchy; mimetic. Calls a hard *tack*, soft *hweet*, an odd *dwzeer* and a repeated 'snipping' note. **Habitat** Mostly in reedbeds, tall grass, dense swampy cover. **Note** Passage and winter hatched; rare in north in winter. In UAE most are *pallidogularis*, with *magna* and *cyanecula* rare on passage/winter.

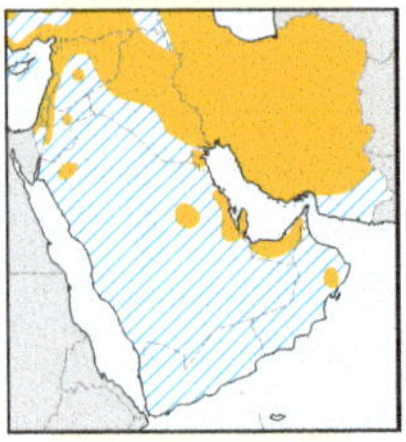

Rufous-tailed Scrub Robin *Cercotrichas galactotes* MB, PM

L: 15. Skulking; ***long tail, often cocked and spread***. Upperparts grey-brown in subspecies *syriacus* and *familiaris* (Iran, Arabia) contrasting with ***rufous rump and tail***, the latter ***showing prominent black subterminal-band and white tips***, obvious above and below; head distinctive with white supercilium contrasting with blackish eye-stripe. Juvenile has faintly mottled breast and flanks. Not shy but usually close to cover; often feeds on ground. **Voice** Song delivered from concealed or exposed perch or in butterfly-like songflight, is slow, clear, thrush-like, melancholic; often varied and musical, recalling lark or nightingale. Calls include a hard *teck*, a low rolling *schrrr*, and sibilant drawn-out *iiiip*; a distinctive penetrating *ssweep* when agitated. **Habitat** Semi-desert, cultivations, scrub, gardens, palm groves. **Note** Passage hatched, very rare into winter.

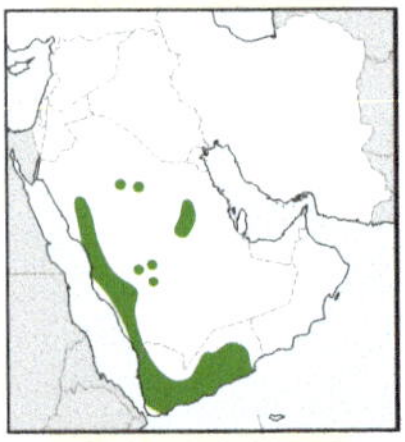

Black Scrub Robin *Cercotrichas podobe* V

L: 18. ***Entirely sooty-black***, with ***prominent white tips to undertail-coverts and outer tail feathers*** readily visible when ***lengthy tail is swept upwards over back*** and fanned. Skulking or close to cover, often on ground, but sings from exposed perch. **Voice** Song melodious with thrush-like whistles similar to Rufous-tailed Scrub Robin. Call a hoarse squeak, or liquid chatter. **Habitat** Desert fringe, dry scrub, cultivations, wind-breaks. **Note** Vagrant Bahrain, Kuwait, Oman, Qatar, UAE. Range expanding.

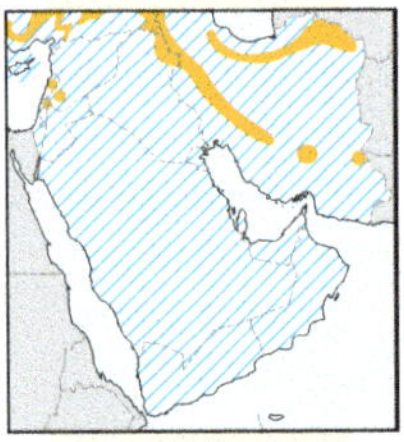

Rufous-tailed Rock Thrush *Monticola saxatilis* PM

L: 19. Rather small and short-tailed with longish pointed bill. Adult male easily told; in winter entire plumage edged dark and white producing scalloped effect. Female and first-winter birds browner, similar to winter male, resembling female Blue Rock Thrush but with ***pale spotting on upperparts, rusty tail and scaling on warmer buffier underparts***. Often shy and elusive. **Voice** Song similar to Blue Rock Thrush but less melancholic. Calls include a loud *chak*. **Habitat** Rocky, barren uplands; almost any habitat on passage. **Note** Passage hatched; very rare in winter Iran, south Arabia.

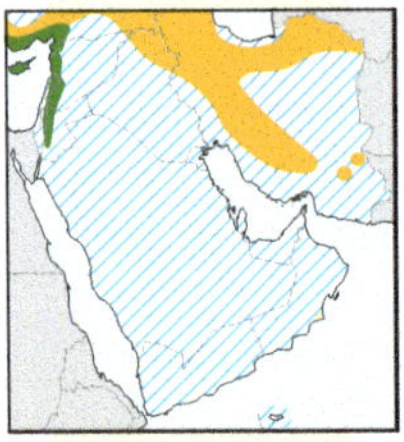

Blue Rock Thrush *Monticola solitarius* WV, PM

L: 21. Slightly larger and longer tailed than Rufous-tailed Rock Thrush and with longer bill. Male has ***all-dull inky-blue plumage*** (looks black at distance), in winter can show fine buffish fringes. Female resembles female Eurasian Blackbird but has longer bill, shorter tail and dull buff spotting and barring on underparts; some show bluish tinge to upperparts. Fairly shy but will sit in full view, remaining quite still. Sings in winter quarters. **Voice** Blackbird-like far-carrying song, melancholic, the short, fluty phrases interspersed with long pauses. Calls include a hard *chak* and high *tsee*. **Habitat** Rocky deserts, mountains and cliffs; often on buildings, even in cities on passage/winter. **Note** Passage and winter hatched.

European Robin
ad
Bluethroat
♀
♂
magna
♂
svecica
♂ 1st-winter
ad
syriacus
ad
Rufous-tailed Scrub Robin
Black Scrub Robin
ad
♂
♀
Rufous-tailed Rock Thrush
♂
♀
Blue Rock Thrush

PLATE 84: NIGHTINGALES & ALLIES

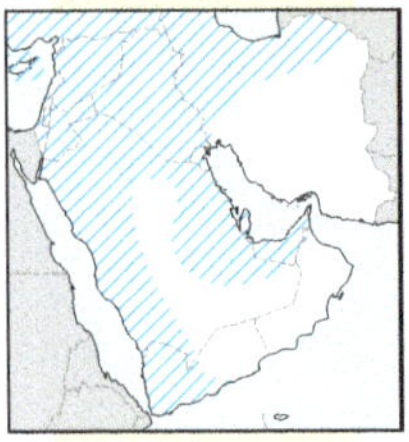

Thrush Nightingale *Luscinia luscinia* pm

L: 16. Dark olivaceous-brown, resembling small thrush, with ***rusty-red tail*** and pale underparts, whitish throat and eye-ring. Very similar to Common Nightingale but ***darker brown above, with duller rusty-red tail and darker brownish grey breast and flanks indistinctly (but variably) mottled; far-carrying song distinctive***. Usually skulking. Unlike Eastern Nightingale, does not pump or wave tail. **Voice** Sings from cover, often at night. Song recalls Common Nightingale's but even louder; includes ***characteristic hard chucks***, dry rattles and clear whistles; lower-pitched with delivery more mechanical, less variable *djüllock... djüllock... djüllock... drllrllrllrllrllrll-pst*, lacking distinctive crescendo of Common Nightingale. Calls include a high-pitched ***hiiid*** and a dry, rolling rattle. **Habitat** On passage in undergrowth, scrub, parks, gardens. **Note** Passage hatched; vagrant Oman.

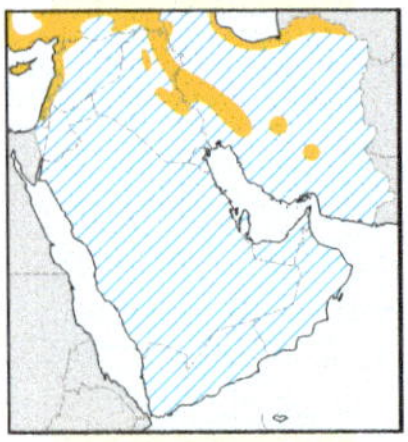

Common Nightingale *Luscinia megarhynchos*

All individuals recorded in UAE referable to subspecies *golzii*, which may represent a separate species, Eastern Nightingale, below. **Note** Map shows both taxa combined as geographical divide between them currently remains unclear.

Eastern Nightingale *Luscinia (megarhynchos) golzii* PM, wv

L: 16. Very similar to Thrush Nightingale but ***more russet-brown upperparts, paler, rusty-red tail***, and often more conspicuous whiter eye-ring; underparts 'cleaner', ***lacking mottled impression on breast and flanks of Thrush Nightingale***. Skulking, even when singing. The subspecies ***golzii*** breeding Iran, a common migrant through eastern part of the Middle East, ***shows pale fringes to the tertials and greater coverts, paler underparts and a pale supercilium***, with upperparts less russet, more greyish-brown. Characteristically droops wings, much like Rufous-tailed Scrub Robin; also ***pumping tail downwards, and waving it sideways part-spread***, much like Upcher's Warbler. **Voice** Beautiful song, by day and night, high-pitched comprising loud, rich, warbling whistles with ***distinctive crescendo*, lu-lu-lu-lu-lee-lee**, often a characteristic starting sequence, which not found in Thrush Nightingale; lacks strong chucks or frequent rattles typical of the latter. Calls similar to Thrush Nightingale include *wwheep* and a soft frog-like croak usually from dense cover. **Habitat** Deciduous woodland, scrub, gardens, wet and dry thickets. **Note** Map includes Common Nightingale. Passage hatched; occasional winter south Arabia; geographical divide between the two taxa, Common and Eastern Nightingale, currently remains unclear.

Red-flanked Bluetail *Tarsiger cyanurus* V

L: 14. Behaviour recalls European Robin, near ground and close to cover. Male blue above with brighter ***cobalt-blue shoulders and uppertail-coverts, dull bluish tail*** (looks dark in shade) ***and orange-buff flanks***; flight feathers fringed bluish. First-year birds and females have some bluish on uppertail-coverts and base of outer tail feathers (often hard to see), ***white eye-ring and narrow white throat-patch*** bordered by olive-grey sides of throat and breast, ***and orange flanks***, but these can be hidden. Legs dark. Jerks tail unlike shivering of Common Redstart. **Voice** Hard *tsak* or short series of quiet *tck, tck, tck* notes; also a frog-like note. **Habitat** Woodland, undergrowth. **Note** Vagrant UAE.

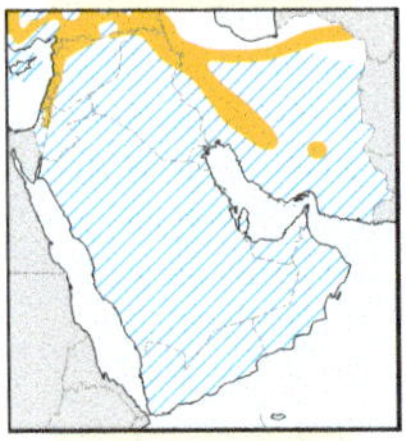

White-throated Robin *Irania gutturalis* PM

L: 16. Size and movements recall nightingales. Striking male has ***black sides of face and head framing pure white centre of throat, rusty-red underparts***, blue-grey upperparts, ***black tail*** and whitish supercilium. Scarcer variant with paler orange below and black line below white of chin. Grey-brown female has ***dark brown tail, ochre-buff sides of body***, whitish throat bordered grey-buff at sides of head and breast. Bill long. Generally skulking, often on ground, with wings held lowered. **Voice** Song, sometimes uttered in gliding flight, fast, consists of clear whistles and scratchy harsh rolling notes mixed together; calls a nightingale-like *kerr-r-rr-rr*; wagtail-like *tzi-lit* with alarm a hard tack. **Habitat** Stony hillsides and valleys with scrub in breeding season, usually 1,000–2,200m; on passage in scrub, woodland. **Note** Passage hatched.

Thrush Nightingale
ad
Common Nightingale
ad
Eastern Nightingale
ad
Red-flanked Bluetail
♀
♂
White-throated Robin
♀
♂

Eversmann's Redstart *Phoenicurus erythronotus* wv, pm

L: 16. Slightly larger than Common Redstart. Adult male has ***broad white patch on wing-coverts (including primary coverts), rusty-red mantle and most of underparts, including throat***. In winter both adult and first-winter male (which also has white in wing) have browner-grey crown and red parts fringed whitish. Grey-brown female told from similar Common and Black Redstarts by ***whitish wing-bars and edges to tertials. Does not shiver tail but jerks it up and down***. Often holds wings below level of slightly raised tail. **Voice** Alarm note a croaking *gre-er*; call a loud, whistling *few-eet* and a soft *trr*. **Habitat** Oases, scrub, woodland and gardens; in summer juniper woodland. **Note** Passage and winter hatched, but rare Arabia; vagrant Iraq, Kuwait, Saudi Arabia.

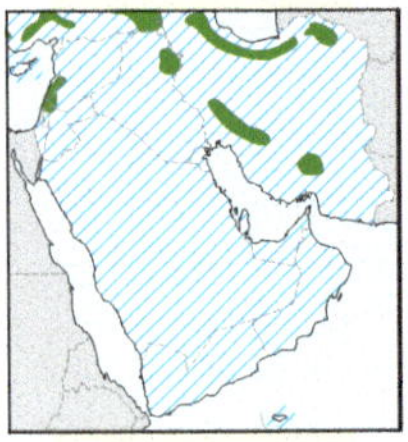

Black Redstart *Phoenicurus ochruros* WV, PM

L: 15. Several subspecies, all migrant, occur (subspecies *phoenicuroides* is that recorded in UAE): Near Eastern *semirufus* has ***black upperparts and throat to lower breast, sharply defined from deep red below***; NE Iranian *phoenicuroides* is less black above and black below extends only to mid-breast; male *ochruros* (Turkey to N. Iran) variable; some resemble *semirufus* or *phoenicuroides*, but many have ***black breast merging into reddish belly***. Males of all three breeding subspecies told from Common Redstart by darker upperparts, ***black extending below throat*** and lack of pure white forecrown. Female similar to Common Redstart but ***slightly darker and drabber, particularly below*** (Common Redstart more buffish below, with warmer flanks and sides of breast, and usually olive tone to mantle); some female *phoenicuroides* very similar to Common Redstart (but usually suffused rustier on lower breast and belly, with grey cast to upperparts, even if head and throat palish). ***Wintering male gibraltariensis (from Europe) grey above, all blackish below, some with white wing-panel; females more uniform, paler grey above and below, and lack wing-panel*** (recorded Cyprus, Turkey, Near East, N. Iraq at least). Often on ground; bobs body and shivers tail. **Voice** Alarm a dry *eet-tk-tk-tk*. Distinctive song short, fast and dry *jirr-te-te-te... chill-chill-chill-chill... kretsch... sree-we-we-we*, often uttered at night. **Habitat** 2,500–5,000m on stony slopes, rocks, cliffs. In winter to sea level in rocky areas, open woodland and villages. **Note** Passage and winter hatched.

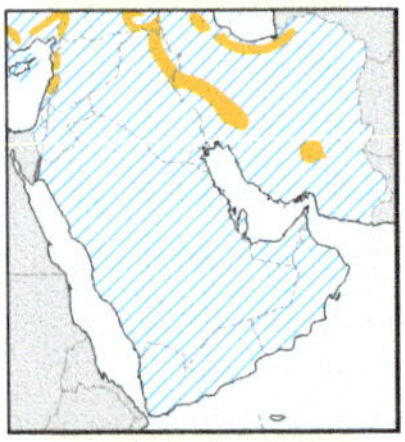

Common Redstart *Phoenicurus phoenicurus* PM

L: 14–15. Male with ***black cheeks and throat contrasting sharply with rusty-red breast*** and belly; crown grey with ***pronounced white forehead***. Male *samamisicus* (breeds Turkey eastwards; a few recorded on passage early spring in UAE) ***has white wing-panel*** and, often, darker upperparts; autumn male Common Redstart has black areas fringed pale. Female brownish-olive above, separated from similar Black Redstart by ***paler, warmer buffy-white underparts***. Hunts insects in flycatcher-like fashion; bobs body and shivers tail (as Black Redstart). **Voice** Call resembles Willow Warbler's soft *wheet*, often followed by *tuuk-tuuk*. Song short and melodious, *seeh-truee-truee-truee-see-see-seeweh*; frequently imitates other birds. **Habitat** Woodland, parks and scrub on passage. **Note** Passage hatched.

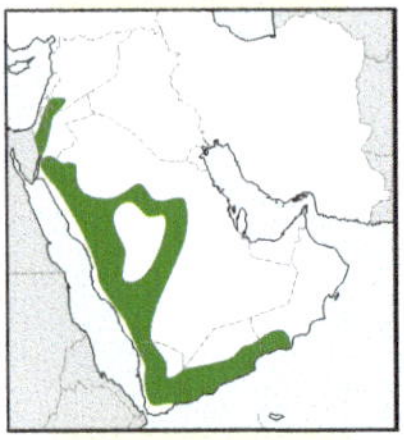

Blackstart *Cercomela melanura* V

L: 15. Slender, relatively long-legged chat with ***all-black tail which slowly lowered and spread, often coinciding with half-spreading of wings***. Nominate subspecies (N. Arabia, Near East) pale ash-grey above, whitish-grey below with whitish wing-panel; S. Arabian *erlangeri* almost uniform smoky-grey above, underparts little paler; wing-panel brownish. Perches freely on low branches or rocks, typically flirting wings and fanning tail; often inquisitive and approaching observer closely. **Voice** Short, mellow, subdued, simple song, sometimes uttered in flight, an often repeated *che-we-we* or *ch-lulu-we*. Alarm a short, deep *tjaet-aeteh*. **Habitat** Sparsely scrubby slopes, cliffs and bare rocky wadis. **Note** Vagrant UAE.

Eversmann's Redstart
♂ 1st-winter
♂ winter
♀
Black Redstart
♂
gibraltariensis
♂
ochruros
♀
♂
semirufus
Common Redstart
♂
♂ 1st-winter
♀
samamisicus
♂
Blackstart
ad

PLATE 86: STONECHATS

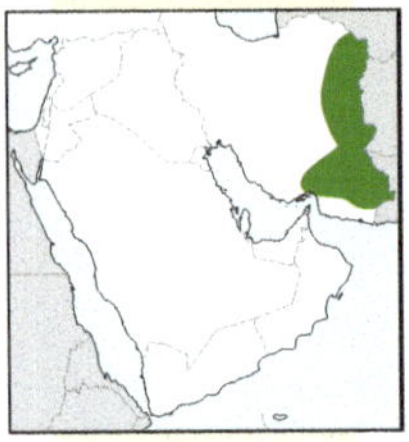

Pied Stonechat *Saxicola caprata* V

L: 13. Slimmer and slightly longer-tailed than other stonechats; tail flicked less. ***Jet black male has white belly, rump and narrow shoulder-patch*** easily seen in its low jerky flight. Female from other chats by ***unstreaked sooty earth-brown upperparts and breast***, creamy belly and rufous-orange rump, some being more rusty-brown on breast and having slight supercilium. **Voice** Song a short rich warble of whistling notes; alarm a curt *chuk* and a *chek-chek-trweet*. **Habitat** Cultivations, scrub, marshes. **Note** Some dispersal; vagrant Oman, Qatar, Saudi Arabia, UAE.

Whinchat *Saxicola rubetra* PM

L: 12.5. Short-tailed chat with slightly smaller head and longer primary projection than European Stonechat, reaching almost halfway down tail. From stonechats by ***combination of streaked brownish rump and white sides to base of tail. Male has bold, clear-cut white supercilium and white stripe between blackish sides of head and orangey throat***; female duller with paler throat and browner sides of head. ***White spot visible on primary coverts***, particularly in flight (almost absent in some first-autumn birds). Confusion possible in autumn with immature Siberian Stonechat. **Voice** Lilting song short, fast and abrupt, variable, usually a mixture of melodious and scratchy notes; often imitates other birds; alarm call *djü-tek-tek*. Silent on passage. **Habitat** Open country, marshes, scrub; on passage in fields and other open ground. **Note** Passage hatched.

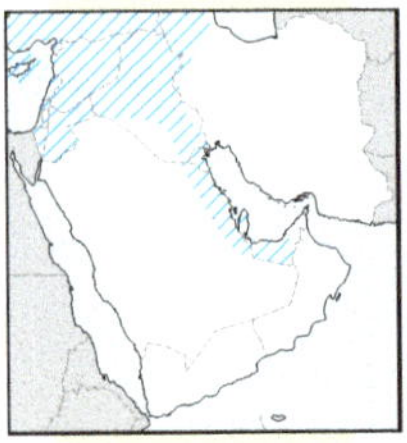

European Stonechat *Saxicola rubicola* WV

L: 12. Short-tailed, short-winged chat with large rounded head and upright stance, frequently flicks wings and tail. Adult male easily told from Whinchat ***by black head and throat, reddish breast, white neck-patch and black tail***. Female duller with dark brown head and usually throat, reddish-brown upper breast, with almost Whinchat-like supercilium variably present in autumn. Male European Stonechat has variable width grey-brown to white rump, if latter usually (but not always) narrow or with some dark streaks; ***grey underwing-coverts and axillaries*** (blackish in Siberian Stonechat) with white collar narrower and less extensive than in male *maurus* Siberian Stonechat. Female Europeans have dark earth-brown back, usually a ***streaked brownish rump***, and are dusky or whitish on throat (Siberian typically only whitish), although some perhaps indistinguishable from *maurus*. Female Siberian Stonechats have paler sandy to warm buff upperparts, often with a more obvious supercilium, and a pale unstreaked rump. Immature European Stonechat is darker, lacks pale rump of *maurus* and has less extensive white in coverts. **Voice** Short song has irregular, rapidly repeated series of double notes; alarm call *wheet-trak-trak* like pebbles hit together. **Habitat** Open terrain; sea level to over 3,000m, in cultivated areas or scrub-covered slopes. **Note** Breeding range apparently expanding eastwards; winter hatched, but rare Arabia.

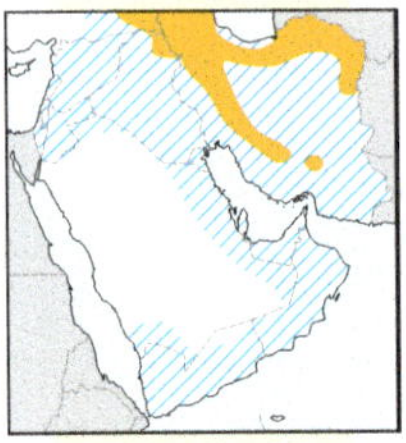

Siberian Stonechat *Saxicola maurus* WV, pm

L: 12. Male in breeding plumage has broader, more extensive white collar and ***conspicuous, broad, white to salmon-pink or orange-buff rump*** than European Stonechat. First-autumn Siberian Stonechat superficially recalls autumn Whinchat. Females lack the clear-cut, long, creamy-white supercilium of Whinchat, but often difficult to separate from European Stonechat (see above). Males of different subspecies described below.

S. m. maurus: (breeds NE Iran) has longer primary projection than short-winged European Stonechat (of W. and central Turkey); male *maurus* has ***unstreaked orange-buff to white rump*** (wider than in European males); ***black tail***; broader white half-collar; ***paler back and axillaries jet black*** (greyish-white in male European Stonechat).

S. m. armenicus: (breeds E. Turkey, NW and SW Iran) also long-winged but has less white at sides of tail (not always visible) than *variegata*; male has dark chestnut breast contrasting with pure white belly.

S. m. variegatus (breeds Caspian) has even longer primary projection (approaching Whinchat's); ***much white in sides of tail base*** (recalling Northern Wheatear); the palest subspecies, male with warm buff upperparts and large white patch on rump, sides of neck and shoulders.

Voice As European Stonechat. **Habitat** Open terrain; sea level to 3,000m, in cultivated areas or scrub-covered slopes; on passage/winter any open ground. **Note** Winter hatched. Distribution poorly documented by subspecies; *maurus* most frequently reported in region, *variegatus* scarce or vagrant Arabia; winter range of *armenicus* unclear.

♂
♀
1st-winter
Whinchat
Pied Stonechat
♀
♂
♂
European Stonechat
♀
Siberian Stonechat
variegatus
♀
♂ 1st-winter
maurus
♂
♂ 1st-winter

PLATE 87: WHEATEARS I

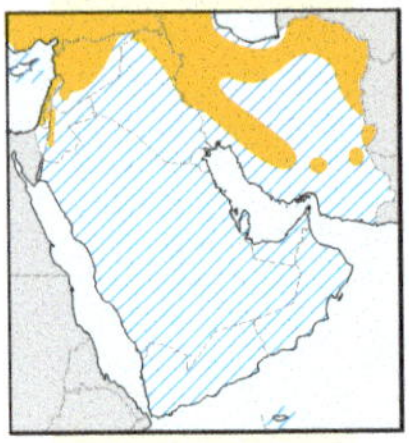

Isabelline Wheatear *Oenanthe isabellina* PM, WV

L: 16. ***Resembles large female Northern Wheatear, usually more robust in build***, with longer legs and apparently also bill, slightly shorter tail and more upright stance. ***Best separated by isolated black alula in paler, more uniform sandy wings, contrasting less with upperparts***, broader pale feather-edges to wings than Northern Wheatear (some first-autumn Northerns can show pronounced isolated dark alula); ***broader black terminal tail-band with shorter stalk to 'T'. Supercilium usually broadest and whiter in front of eye*** (in Northern narrower and buffier in front of eye). In low flight shows ***half-translucent, dark-tipped primaries; underwing-coverts and axillaries buffish-white*** (dusky-grey in female Desert Wheatear; dark grey, broadly tipped whitish in Northern Wheatear). Runs mostly; wags tail strongly and frequently. **Voice** Song, often in display flight, longer and more variable than Northern Wheatear with less scratchy notes and often ending with series of whistles; mimics well. Alarm call *tjack-tjack*, sometimes followed by slightly descending *hiu* or *diu*. **Habitat** Barren or grassy areas, steppe; on passage/winter also cultivation. **Note** Passage and winter hatched; absent from north in winter.

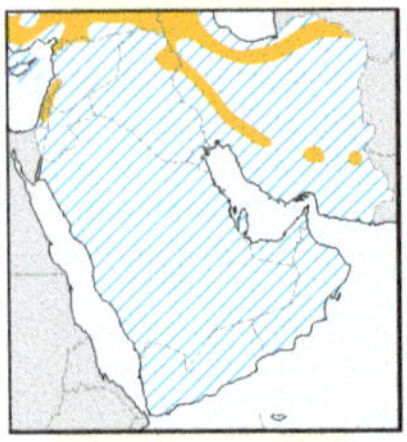

Northern Wheatear *Oenanthe oenanthe* PM

L: 15. Relatively short-tailed wheatear with ***blackish terminal tail-band of even width***. Male readily told by ash-grey crown and back, white supercilium, thin black eye-stripe and black ear-coverts. Female is greyer-brown than similar Eastern Black-eared and Pied Wheatears; first-autumn birds ***best told by shorter tail*** (wing-tips closer to tail-tip) ***and tail-band of even width***. In Eastern Black-eared and Pied tail longer, wings half or less of tail, which has more white and narrower black tail-band with black extending upwards on outer tail feathers. Restless; bobs body, wags tail and flicks wings. **Voice** Song short with fast chacking call-notes mixed with high-pitched whistles in irregular rhythm. Alarm a hard *tack-tack* or *hiid, tack-tack*. **Habitat** Uplands with rocky or stony slopes, often with bushes; on passage to sea level in any open area, including cultivations. **Note** Passage hatched.

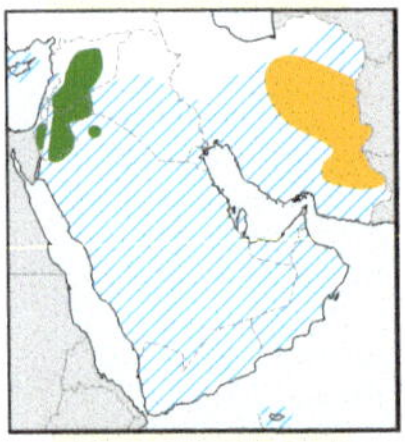

Desert Wheatear *Oenanthe deserti* WV, PM

L: 14.5. Easily told in flight from all other wheatears in region ***by almost wholly black tail*** (no white at sides). Male recalls black-throated form of Eastern Black-eared Wheatear, ***but black throat joins to narrower black of wings*** (often with whitish scapulars contrasting); in flight showing whitish wing-panel and white rump ***tinged buffy toward top***. Sandy-brown or grey-buff female often lacks black throat, may recall female Eastern Black-eared but always told by rump and tail pattern. Often confiding. **Voice** Short piping song plaintive, with downward inflection, occasionally includes rattling notes; calls include a soft whistle. **Habitat** Shrubby desert, barren stony areas. **Note** Passage and winter hatched.

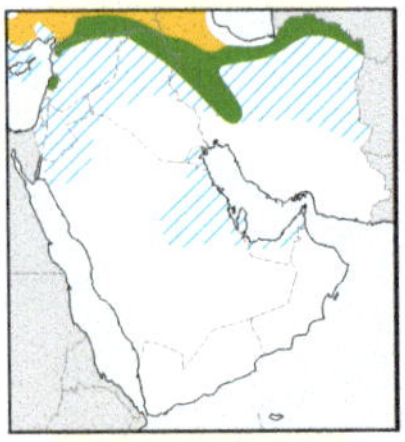

Finsch's Wheatear *Oenanthe finschii* V

L: 15. Heavier and stockier than Pied and Eastern Black-eared Wheatears (see plate 88); male told from former by ***narrow creamy-buff (milky tea) or, when worn in spring, silvery-white stripe down mantle and back to join white rump***; from latter by ***larger black 'bib' broadly connected with black wings*** (beware Black-eared with head sunk between shoulders); ***also from both by terminal tail-band of even width*** (no black extension up sides) ***and pale greyish flight feathers below, appearing translucent above***. Crown as mantle, but centre flecked dusky when worn. Some females have variable blackish on throat (sometimes lower throat only); pale-throated birds told from female Eastern Black-eared/Pied by ***sandy brown-grey upperparts***, contrasting with darker, browner wings, creamy breast lacking orange-buff tone of female Eastern Black-eared, paler flight feathers below and tail-pattern. ***Flight feathers and primary coverts often finely pale-tipped*** (in both sexes into second calendar year). Female often shyer than male. Ground-dwelling, perching infrequently in trees. Frequently ***bows low, cocking tail, repeatedly spreading and lowering it slowly***. Has descending zig-zag songflight. **Voice** Song short and rich with scratchy notes often mixed with clear whistles, phrases intermittent, including musical *ctsi-tsi-tseeoo*. Alarm call *tack*; also *che-che-che*. **Habitat** Dry rocky, stony uplands and foothills, sparsely vegetated semi-deserts. **Note** Passage and winter hatched; rare Arabia; vagrant Oman.

Isabelline Wheatear
1st-winter
ad
Northern Wheatear
♂ 1st-winter
♂
♀
♂
Desert Wheatear
♀
♂
♀
♀
♂
♂
Finsch's Wheatear

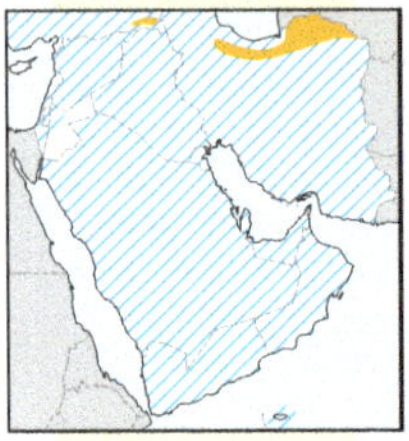

Pied Wheatear *Oenanthe pleschanka* PM

L: 14–16. Slender, often perches on bushes. ***From Eastern Mourning and male Arabian Wheatears by absence of white panel in open wing***. White-throated form, '*vittata*', closest to Eastern Black-eared Wheatear but mantle black. Autumn male has black back and throat fringed buffish, dark crown with buff-white supercilium, and buff underparts. Female like female Eastern Black-eared Wheatear, but ***upperparts usually duller, cooler brown-grey***; some show large dark greyish 'bib' in summer (absent in female Eastern Black-eared); autumn female also told by dark brown breast-sides merging with greyish 'bib'. First-autumn female told with difficulty from Eastern Black-eared by colder tone above; crown, mantle and shoulders usually ***scalloped with rows of pale fringes*** (absent or ordered erratically in Eastern Black-eared). First-winter female from Northern Wheatear by ***black extending up sides of tail;*** the black tail-band ***sometimes being of uneven width***. **Voice** Short musical song, often in flight; twittering phrases resembling lark or wagtail; often mimics. Calls hard *tack*, dry *trrrlt* or dry, sneezed *snerr*. **Habitat** On passage and winter rocky terrain, bare fields, wasteground. **Note** Passage hatched; occasional in winter south Arabia. White-throated form '*vittata*' regular in small numbers in UAE in spring.

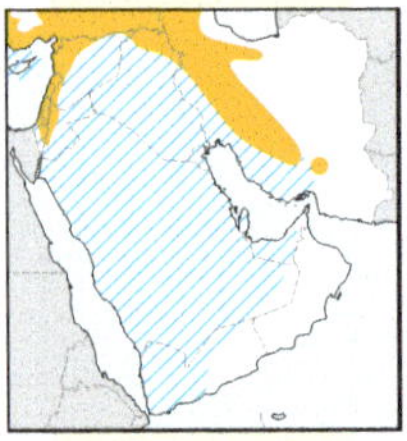

Eastern Black-eared Wheatear *Oenanthe (hispanica) melanoleuca* PM

L: 13.5–15. Small, build much as Pied Wheatear and sharing same (variation in) tail pattern. Male of black-throated form can be confused with male Finsch's Wheatear but ***black of throat/ear-coverts not joined with black of wings and shoulder***; mantle whitish (summer) or buffish-grey (autumn). White-throated male told from '*vittata*' form of Pied Wheatear by pale mantle. Female Eastern Black-eared difficult to separate from Pied Wheatear, but has whitish chin (dusky in Pied), usually with slight rustiness below and ***sandier mantle***. **Voice** Song resembles Pied's, rather variable, dry, scratchy. Calls include a hard *tack*, sneeze and a characteristic buzzing (like an angry fly!). **Habitat** Sparsely vegetated, stony slopes; any open area on passage. **Note** Passage hatched.

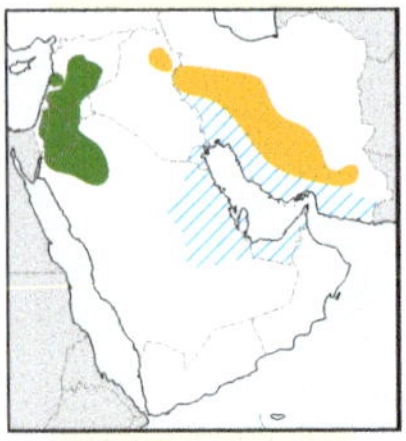

Eastern Mourning Wheatear *Oenanthe lugens* wv, pm

L: 13.5. Sexes similar; resembles stocky male Pied Wheatear when perched, ***but undertail-coverts apricot (all ages); in flight shows prominent whitish wing-panel***. Compared with Pied Wheatear the black bib is smaller, underparts whiter (tinged buff in Pied), primary coverts narrowly tipped white and black band on tail lacking the black extension up the outer feathers of Pied. In autumn lacks the pronounced pale feather fringes to black throat and mantle of male Pied. **Voice** Song a lively twitter; call *check-check*; alarm *peet-peet*. **Habitat** Coastal and inland bluffs in desert and semi-desert in winter, sometimes near remote habitation. **Note** Winter hatched, but scarce Arabia. Iranian *persica* occurs UAE; siimilar *lugens* occurs Near East/NW Saudi Arabia.

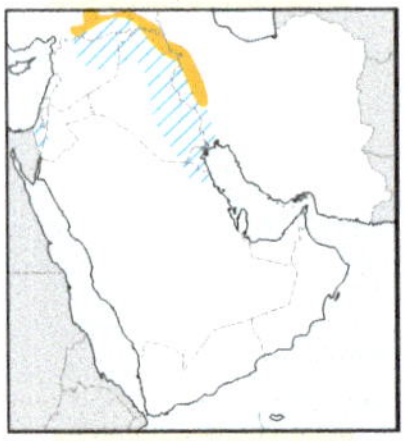

Kurdistan Wheatear *Oenanthe xanthoprymna* V

L: 15. Male distinctive, with whitish supercilium, black throat, sides of head and neck merging with blackish-brown wing-coverts, ***rufous rump, with white sides to tail-base in adults (rufous in immatures)***; black band at tip of tail narrow; ***little white in wings***. Female confusable with Red-tailed Wheatear (which see), though sometimes shows dark throat. Usually solitary; has bounding hops and slight downward tail-flicks; often flies with tail closed. Favours rocky outcrops. **Voice** Brief song a slow throaty warble. Calls include *steu-steu-steu*, alarm note a short dry *zuk* or *zvee-tuk*. **Habitat** In winter hillsides, cultivations, ruins. **Note** Winter hatched, but distribution poorly known; vagrant Oman, UAE. [Alt: Rufous-tailed Wheatear]

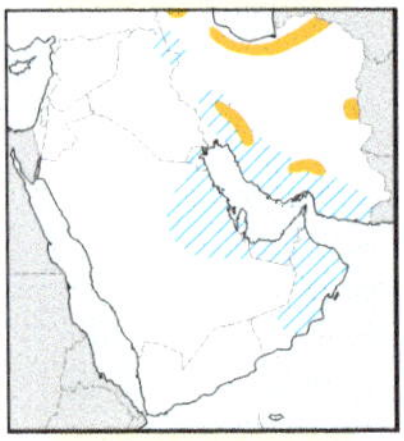

Red-tailed Wheatear *Oenanthe chrysopygia* WV, PM

L: 15. Sexes alike; ***upperparts drab greyish, especially head, face pattern rather bland (like autumn*** flava ***wagtail), rump and sides of tail orangey-rufous*** (rump sometimes paler); ***tail tip fringed rufous when fresh; underwing coverts pale/off-white*** (dusky-grey in Kurdistan Wheatear), ***vent and flanks rufous-orange*** (paler orange to buff or whitish in female Kurdistan Wheatear). Long slim bill; silhouette can recall Blue Rock Thrush. **Voice** Warbling song loud with adept mimicry. Undemonstrative and ordinarily silent in winter; occasionally giving grating alarm. **Habitat** Stony or barren hillsides, low scrubby vegetation. In winter cultivations, ruins, rubble and dumps in sand desert. **Note** Known winter distribution hatched.

♂
♂
♂ 1st-winter
'vittata'
Pied Wheatear
♂
♀
1st-winter
♀
♂
♂
♀
♂
Eastern Black-eared Wheatear
Kurdistan Wheatear
ad
ad
persica
dark-throated ♀
lugens
♀
Eastern Mourning Wheatear
♂
♂
Red-tailed Wheatear

PLATE 89: WHEATEARS III

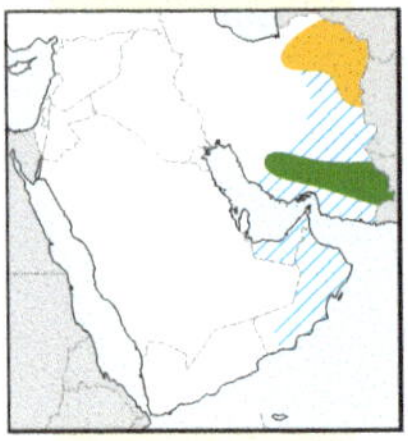

Variable Wheatear *Oenanthe picata* wv

L: 15. Three subspecies occur, but only *picata* likely to be seen in region, the other two (*capistrata* and *opistholeuca*) being rare or vagrant to Iran. Subspecies ***picata:*** (breeds Iran, winters south to UAE & Oman) resembles small Hume's Wheatear but ***more slender. Male has crown, upperparts and 'bib' dark charcoal (matt, without gloss, unlike blacker Hume's)***; rest of underparts white, undertail-coverts sometimes buff in autumn. Female usually matt sooty-brown above where male charcoal blackish, 'bib' often rufous-tinted but throat sometimes blackish; rest of underparts creamy-buff to white. ***Perches low, in winter often in trees (like redstart) rather than on rocks.*** **Voice** Song rather scratchy warble; mimetic. Ordinarily silent in winter. **Habitat** Barren, boulder-strewn country; hillocks with sparse woody vegetation, steep river banks; in winter arid stony plains with trees and outcrops, cultivation. **Note** Winter hatched (subspecies *picata*), but mostly scarce Arabia; vagrant Saudi Arabia. [Alt: Eastern Pied Wheatear]

Hume's Wheatear *Oenanthe albonigra* RB

L: 16.5. Closely resembles *picata* subspecies of Variable Wheatear, but ***larger, with 'bull-headed' appearance, velvety black plumage more glossy and white on back extending farther up between wings, where border to black mantle is rounded*** (square-cut in Variable Wheatear). Black throat has slight side-extension thus less 'bib-shaped' than Variable Wheatear. Underwing less contrasting, but tail pattern like Variable Wheatear. Juvenile like adult, but plumage matt blackish-brown (with yellow gape line). Stance more upright than in Variable Wheatear; prefers open rocky terrain, only rarely on plains or in more wooded areas. Often inquisitive, at other times shy. On territory year-round, but sometimes descending from higher altitudes. **Voice** Loud, melodious ringing song, a short lyrical jumble, recalling Whinchat. Ventriloquial. Call sharp, short and high-pitched; alarm harsh and grating. **Habitat** Mountain slopes, foothills down to sea-level and boulder-strewn barren hills with scant vegetation; often on buildings, overhead wires. Nests in hole in rock or scree. **Note** Winter dispersal hatched; vagrant Bahrain, Kuwait.

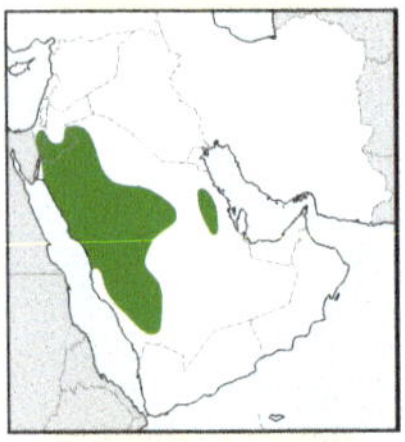

White-crowned Wheatear *Oenanthe leucopyga* V

L: 17. Large wheatear with sexes similar; glossy ***black with black underparts down to legs, and, in many adults, a white crown***; immature and some adults have black crown but size, long bill ***and white sides of tail with black corners diagnostic*** (no black terminal-band). Some black-crowned birds show a few white feather-tips, eventually developing a white crown. In male Hooded Wheatear, which has similar tail pattern, black below extends only to centre of breast. **Voice** Variable song has whistling and tuneful notes, sometimes scratchy, often with imitation of other locally occurring species; common phrase *viet-viet-dreeit-deit*, slightly descending but much variation. Call *peeh-peeh*. **Habitat** Rocky deserts, ravines in rocky mountains, usually without vegetation, often around human settlement. **Note** Some winter dispersal, rare Kuwait; vagrant Bahrain, Iran, Iraq, Oman, Qatar, UAE.

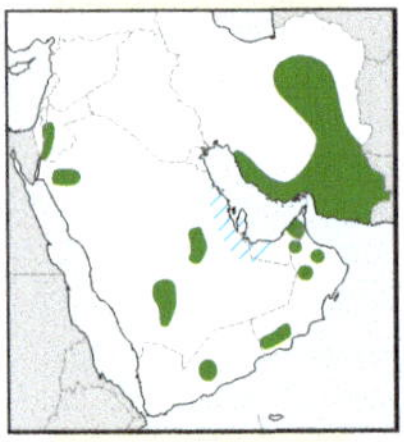

Hooded Wheatear *Oenanthe monacha* rb, pm

L: 17. Large, slender, long-tailed and long-winged wheatear with long bill and buoyant, almost butterfly-like flight, recalling Spotted Flycatcher when catching prey, sometimes in long sallies. Sexes differ. Adult male distinctive, with whitish crown, black below extending to centre of breast and, except for black central tail feathers, nearly all-white tail with just black corners. Autumn and juvenile male has creamy-buff crown, whitish fringes to black throat, wing-coverts and mantle, with lower underparts, rump and sides of tail tinged buffish. Female sandy brownish-grey above, merging into cream-buff rump, tail-coverts and sides of tail, in which central feathers and tail-corners are dark brown; whitish-grey underparts washed buff at sides of breast, flanks and undertail-coverts. In autumn female and juvenile the rump and underparts may appear reddish-buff with almost reddish-brown sides of tail, but absence of dark terminal tail-band separates from Red-tailed Wheatears. **Voice** Song has short melodious phrases, interspersed with some stone-clicking notes; brief throaty thrush-like warble heard infrequently, relatively simple, lilting and slightly sad. Female utters a *whit-whit* or repeated *jiirp* like a fledgling Eurasian Blackbird; also a *wit-awheet-wheet-wheet* or *whee-whee-whee-wheeoo*. **Habitat** Desolate, barren rocky ravines, gorges and deserts. Nests in hole in rock. **Note** Winter dispersal hatched.

♀
♂
opistholeuca
picata
♀
♂
♀
♂
capistrata
♂
Variable Wheatear
ad
ad
imm
juv
Hume's Wheatear
White-crowned Wheatear
♀
♂
♂ winter
Hooded Wheatear

PLATE 90: FLYCATCHERS

Blue-and-white Flycatcher *Cyanoptila cyanomelana* V

L: 17. Larger, longer-tailed and heavier-billed than Spotted Flycatcher. ***Male dark blue above, with shining blue crown, bluish-black sides of head, throat and breast, sharply demarcated from white belly***; ***white sides to tail-base*** usually visible in flight only. ***Dark olive-brown female is paler below with distinct creamy throat-patch***, white belly and undertail-coverts; indistinct pale eye-ring; first-winter male has ***bright blue rump and tail, latter tipped black, but white sides at base***; wings bright blue with black-tipped primaries. Upright stance with frequent slow tail movements and wing-flicks; swoops to ground to feed, then returns to perch. **Voice** Grating *tchach* or *tek-tek*. **Habitat** Any area with trees. **Note** Vagrant Oman, UAE.

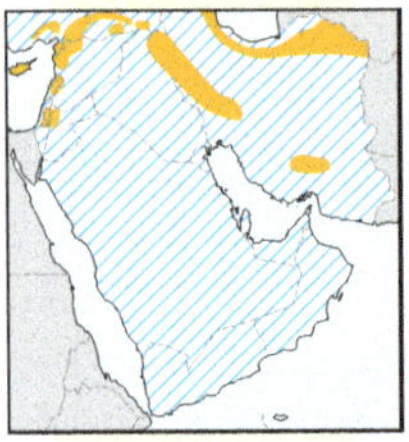

Spotted Flycatcher *Muscicapa striata* PM

L: 14. Brownish-grey, long-winged flycatcher ***with streaked forehead, crown and breast*** (though may not be obvious), ***variable pale edges to greater coverts and tertials*** (depending on subspecies and wear), ***and blackish bill*** and legs. Faint eye-ring whitish. Sexes similar. Makes short aerobatic flights (may hover) from exposed branch to catch prey, often returning to same perch. Perches upright, often flicking wings. **Voice** Call a sharp *tzeet*; alarm *isst-tek*. **Habitat** Gardens, parks, woodland. **Note** Passage hatched.

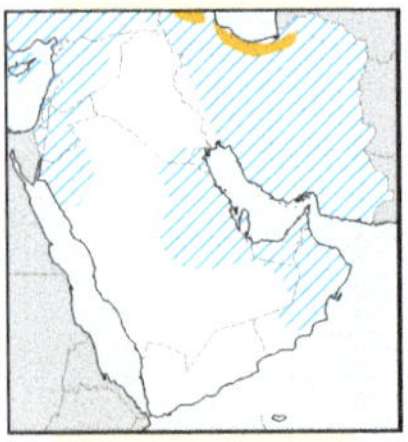

Red-breasted Flycatcher *Ficedula parva* wv, pm

L: 12. Small flycatcher with conspicuous white patches at sides of tail-base, whitish eye-ring, and straw or yellowish base to bill (when seen well); adult male has ***reddish-orange throat and sometimes upper breast*** and lead-grey sides of head and neck; female and second-year male have buffish-white throat. Often cocks tail. See similar Taiga Flycatcher below. **Voice** Call a dry, rolling *terrrr*; also a thin *tsri*, a *tek* or, in alarm, *tee-lu*. **Habitat** On passage/winter anywhere with trees or scrub. **Note** Passage hatched; winters SE Arabia.

Taiga Flycatcher *Ficedula albicilla* V

L: 12. Very similar to Red-breasted Flycatcher, differs in having ***bill all dark or barely paler on base of lower mandible*** (straw or yellowish-based in Red-breasted), ***coal-black uppertail-coverts and differing call***. Females, as with winter males and juveniles, colder less buff below than Red-breasted, again distinguished by dark bill and rump, and call. Although not diagnostic, first-winter Taiga also typically shows broad, cold white fringes on the outer web of the tertials, broadening and expanding to a round blob at the shaft (as first-winter Pied Flycatcher), whereas in Red-breasted both webs have warmer yellowy-buff fringes, with a thorn shape at the shaft. **Voice** Call is ***an insect-like buzz*** rather than dry roll of Red-breasted. **Habitat** Wooded areas. **Note** Vagrant Oman, UAE.

Eurasian Pied Flycatcher *Ficedula hypoleuca* V

L: 13. Male black-and-white ***with black hindneck***, small white forehead-spot (often divided in centre) and ***narrow white streak at base of primaries*** (sometimes absent). Some males (rare variant) are grey-brown above but still with white spot on forehead. First-winter has dark tertials, white-bordered on outer webs, with rounded blob at end and only ***small white streak at base of primaries***. For separation from Semi-collared Flycatcher see under species. Often flicks wings. **Voice** Calls on passage *tuk*, and a ***short metallic*** **twink**, different from call of Semi-collared. **Habitat** Trees and woods. **Note** Passage hatched; vagrant Oman, UAE.

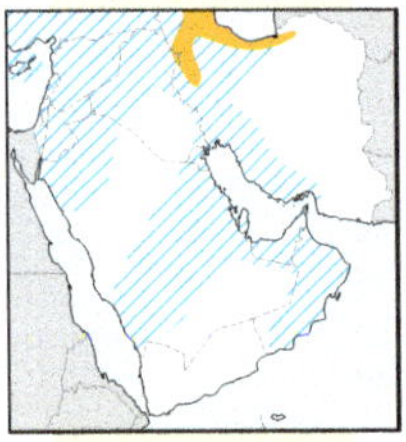

Semi-collared Flycatcher *Ficedula semitorquata* pm

L: 13. Resembles Eurasian Pied Flycatcher, also flicks wings. Male told by ***white half-collar onto sides of neck, larger white spot at base of primaries*** (absent, or just a narrow streak in Pied), more white at sides of tail and white-tipped median coverts (rarely so in Pied); extent of white in rest of wing and on forehead greater than in Pied. Female greyer above than Pied; ***often*** (but not always) ***more white at primary bases and on tips of median coverts (as well as greater coverts)***; however, this second wing-bar on median coverts is not reliable for first-autumn birds. **Voice** Call single dry *thuk* (close to Red-breasted Flycatcher), also hard *tack*, alarm call *eeet*. **Habitat** Woods, parks, large gardens. **Note** Passage hatched.

Blue-and-white Flycatcher
♀
♂
Spotted Flycatcher
1st-winter
neumanni
ad
inexpectata
♂ 1st-winter
Red-breasted Flycatcher
♂
1st-winter
Taiga Flycatcher
♂
♀ / 1st-winter
Eurasian Pied Flycatcher
♂ variant
1st-winter
♀ spring
♂ spring
Semi-collared Flycatcher
1st-winter
♀ spring
♂ spring

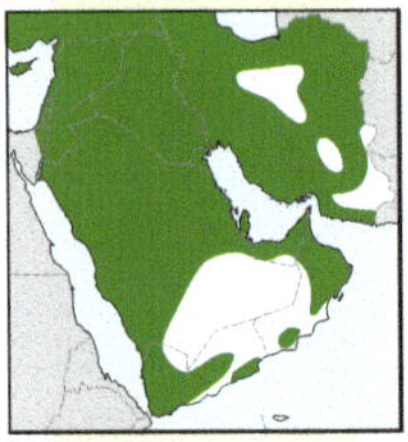

Indian House Sparrow *Passer (domesticus) indicus* RB

L: 14.5. Perhaps a separate species from House Sparrow, this is the taxon that occurs in Arabia. ***Male has grey crown, chestnut-brown sides of head, variably large black 'bib', very pale, clean underparts, strikingly white cheeks***, with prominently dark streaked back. Female and juvenile buffish-brown above, ***greyish on crown, boldly streaked darker on mantle*** and pale greyish below; supercilium creamy. Feeds and roosts in flocks. **Voice** Simple song monotonous, cheeping or chirping notes of varying pitch. **Habitat** Towns, villages, farmland. Colonial; untidy nest built in rock crevices, buildings, tree or bush. **Note** Resident much of Arabia east to SE Iran. Race *hufufae*, sometimes included in *indicus* 'group', occurs E. Arabia. Map includes other House Sparrow races.

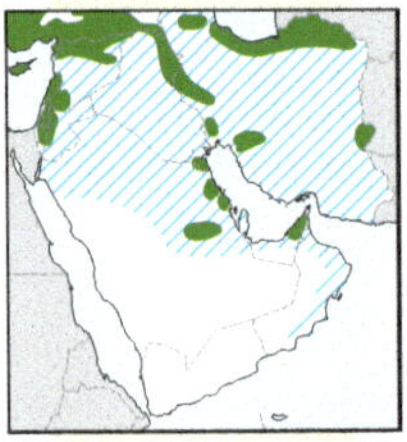

Spanish Sparrow *Passer hispaniolensis* rb

L: 15. Resembles House Sparrow (with which it hybridises); larger bill sometimes evident. ***Male has rufous-brown crown (pale fringed duller in winter) and large black 'bib' extending to bold black streaks on breast and flanks (all pale fringed in winter); back boldly streaked black, merging at sides with black of breast.*** Female not always safely separable, though supercilium and underparts are whiter, with breast and flanks grey-streaked. Often gathers in large, compact flocks on passage/winter. **Voice** Song as House Sparrow, but faster and audible at long range; calls slightly higher than in House Sparrow. **Habitat** Rural settings, open cultivation in winter. **Note** Has bred Bahrain, Qatar; passage and winter hatched, rare SE Arabia.

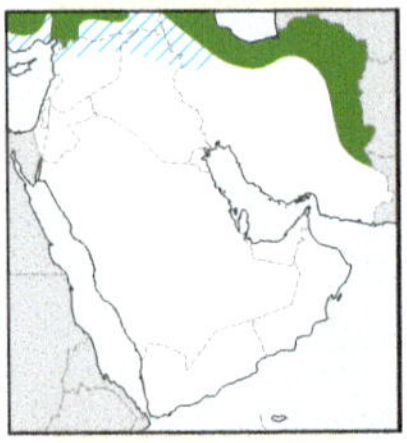

Eurasian Tree Sparrow *Passer montanus* V

L: 14. Smaller, more delicate than House Sparrow with chestnut crown, black spot on whiter cheeks and a smaller black 'bib'; two white wing-bars and rump grey-brown. **Voice** Flight call a hard *tek-tek-tek*; also *tchu-wit*, *pilp* and a hard *chik*. **Habitat** Parks, gardens or woodland, often near habitation. **Note** Winter hatched; vagrant UAE.

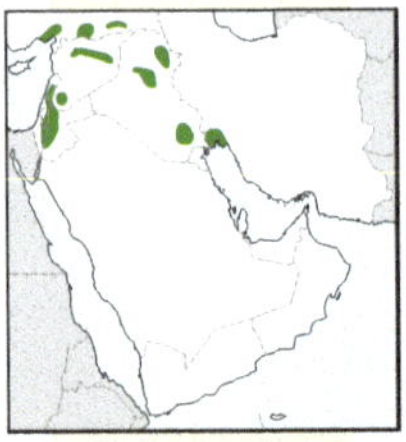

Dead Sea Sparrow *Passer moabiticus* V

L: 12. Small; distinctive head pattern with dark grey cheeks and crown contrasting with long white and rusty supercilium, and pale yellow moustache bordering small black bib. Female resembles small House Sparrow with slightly shorter wing-tip, but undertail-coverts distinctly dark-spotted. **Voice** Song rolling *tri-rirp*, *tri-rirp* or *tlir-tlir-tlir*, more tuneful and rhythmic than House Sparrow; calls include a short *chilp* like House Sparrow, harder, metallic, high-pitched *tlip* or *trrirp* and occasional buzzing *tzzeeer*. **Habitat** Trees, scrub, reeds, usually near water. **Note** Some winter dispersal; vagrant Bahrain, Kuwait, UAE.

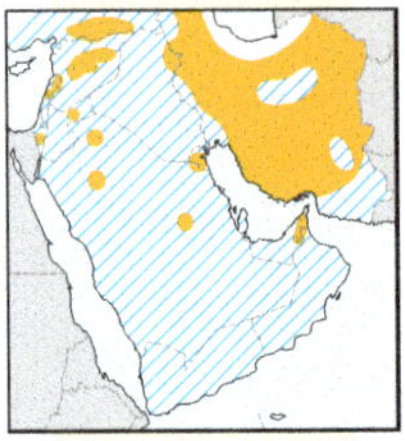

Pale Rockfinch *Carpospiza brachydactyla* mb, PM

L: 15. ***Unstreaked, grey-brown, lark-like sparrow with whitish wing-panel and wing-bars, long primary projection, short dark tail with white tip*** (obvious in flight), strong pale bill (with curved culmen) and prominent dark eye in pale face. Sexes similar. In flight has a lark-like appearance. Often gregarious outside breeding season. Feeds in manner of lark, otherwise hops almost upright. **Voice** Distinctive insect-like buzzing song monotonous and persistent *tss tss tss tseeeeeeeeei*. Flight call a soft trill, recalling distant European Bee-eater; also *piyee* or *twee-ou*. **Habitat** Rocky and scrubby areas at low to moderate altitudes; cultivations on passage. Untidy, domed twig and grass nest built in low bush. **Note** Passage hatched; some winter SW Arabia.

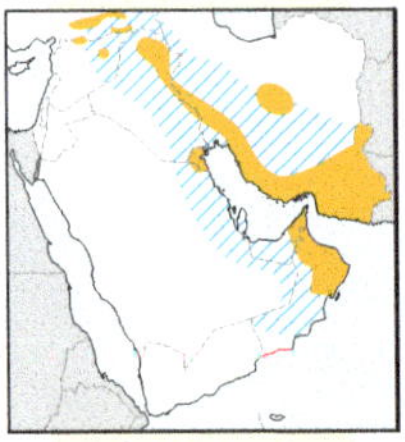

Yellow-throated Sparrow *Gymnoris xanthocollis* MB, PM

L: 13.5. Olive grey-brown unstreaked sparrow, with long, stout-based, pointed bill – black in breeding male, pinkish-brown in female and non-breeding male. Male has yellow spot on lower; chestnut lesser coverts and broad white wing-bar. Female and juvenile lack chestnut lesser coverts and yellow throat-patch but have distinctive bill shape and prominent wing-bar; grey legs, pointed bill and all-dark tail separates from Pale Rockfinch. Perches in trees; finch-like movements on ground; dipping flight rather pipit-like. May migrate by day in flocks. **Voice** Quiet chirruping song softer, more melodious and rhythmic than House Sparrow. Call a sparrow-like *cheep*, *chilp* or *chirrup*. **Habitat** Open dry woodland, date groves, cultivated areas. Nests in hole or crevice in tree. **Note** Passage hatched; rarely in winter Oman, UAE.

Indian House Sparrow
♀
♂
♂ transitional
Spanish Sparrow
♂ hybrid
Spanish × House
♀
♂
♂ winter
♂ winter
Eurasian Tree Sparrow
ad
♀
Dead Sea Sparrow
♂
Pale Rockfinch
ad
♀
♂
Yellow-throated Sparrow

Forest Wagtail *Dendronanthus indicus* pm, wv

L: 17. Unique in having ***two broad blackish breast-bands, one across upper breast, the other below, broken***; underparts otherwise white, washed very pale yellow or creamy. Upperparts olive-brown, wings dark with ***broad buff-white wing-bars***; also a thin white supercilium; tail blackish with white outer feathers. Pipit-like movements on the ground, ***usually in woodland; does not wag tail but sways body from side to side***. Perches freely in under-canopy, where remains when disturbed from floor. **Voice** Call a loud, abrupt Chaffinch-like *pink* or *pink-pink*. **Habitat** On ground in clearings in damp forest or plantation, often near water. **Note** Very rare passage and winter visitor; vagrant Kuwait, Oman.

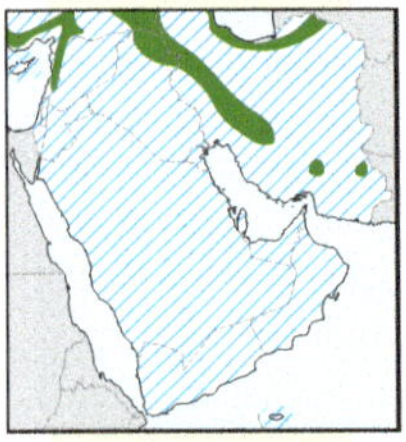

Grey Wagtail *Motacilla cinerea* WV, PM

L: 18. In all plumages has ***yellow vent and undertail-coverts, greyish back, bold white translucent bar at base of flight feathers (in flight), very long tail and extremely undulating flight***. Male in summer ***has black throat and white sub-moustachial stripe***; female has less black or even whitish throat; immature has white throat and buff, not yellow breast. ***Legs pinkish-brown*** (black in other *Motacilla* wagtails). Tail-wagging more pronounced than in other *Motacilla* wagtails. **Voice** Call distinctive, resembles White Wagtail's but harder, more metallic and high-pitched, ***a piercing* tzi-lit *or* tsiziss, *or loud* chink,** usually given in bouncing flight. Song a distinctive series of sharp mechanical notes. **Habitat** Wooded streams, wadis, pools, trickles. **Note** Passage and winter hatched.

White Wagtail *Motacilla alba* WV, PM

L: 18. Easily told from other wagtails by ***grey, black-and-white plumage***, with grey back contrasting with black nape (male) or crown (female), white face and ear-coverts contrasting sharply with black throat and breast. Juvenile, and especially winter birds, have whitish underparts ***with prominent black sagging half-moon on breast***. Flanks show some greyness, concolorous with or paler than mantle. Flight undulating, when often calls. **Voice** Disyllabic call slightly metallic, a hard *tse-lit* with stress on second syllable. Song composed of call-notes, a lively garbled twittering. **Habitat** Open areas, often near habitation, cultivation, livestock. **Note** Passage and winter hatched.

Masked Wagtail *Motacilla (alba) personata* WV, pm

L: 18. NE Iranian breeding subspecies, *personata*, has ***black of head and breast merging at sides of neck and ear-coverts, leaving smaller white area around face*** (same pattern in both sexes, but darker in male); white edges on wing-coverts, tertials and secondaries also much broader than in White Wagtail. **Voice** Call more metallic than White Wagtail, approaching Grey Wagtail. **Habitat** Open areas, often near habitation, cultivation and especially near cattle. **Note** Winter range hatched.

Forest Wagtail
ad
Grey Wagtail
♀
♂ summer
White Wagtail
♀
imm
♂ summer
♂ winter
Masked Wagtail
juv
ad

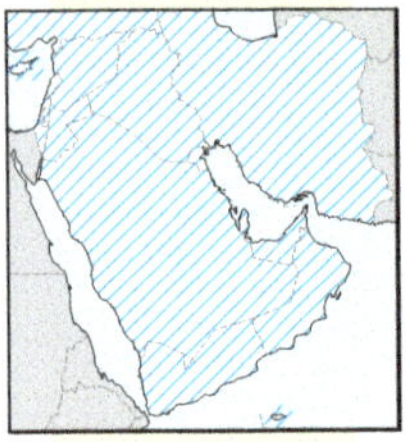

Yellow Wagtail *Motacilla flava* PM, WV

L: 16.5. This species complex has many subspecies. In spring and summer the head colour of males varies as shown below; hybrids occur, making some identifications uncertain. Females all similar: ***yellowish below with whitish throat*** and mostly unidentifiable to subspecies. **Voice** Calls: *flava* high *see-u*; *thunbergi* slightly sharper; *beema* more buzzy; *cinereocapilla* harsher, ringing *psreee-u*. **Habitat** Wet or dry grassland, open fields. **Note** Passage hatched for all subspecies; generally scarce in region in winter.

Sykes's Wagtail *M. f.* ***beema*** Pale greyish crown and ear-coverts, long, broad white supercilium and whitish chin. Some first-autumn lack yellow below (rear of underparts normally tinged yellow in immatures of other subspecies), are greyish above (normally tinged olive-brown in other subspecies) and show fairly prominent whitish wing-bars, thus resembling first-autumn Citrine Wagtail (call also rather similar). **Note** Passage throughout, some winter south Arabia (incl. UAE).

Grey-headed Wagtail *M. f.* ***thunbergi*** Dark grey crown, with no sharp demarcation to upperparts but contrasting with darker ear-coverts, usually no supercilium or white on chin and throat. **Note** Widespread passage, abundant UAE; some winter Arabia.

Yellow-headed Wagtail *M. f.* ***lutea*** Whole head yellow, some with olive crown and ear-coverts, upperparts yellow-olive. **Note** Passage Near East and Arabia, scarce UAE.

White-headed Wagtail *M. f.* ***leucocephala*** White head, tinged blue or grey on rear-crown and ear-coverts. Broad pale yellow wing-bars, moss-green mantle, tertials broadly edged pale yellow, 'thighs' white-feathered. **Note** Vagrant NE Iran, Kuwait, UAE.

Hybrids: '***superciliaris***' (*feldegg* x *flava* or *beema*) Dark slaty-blue head with clean white supercilium and sub-moustachial, broad yellowish wing-bars. **Note** Passage Near East and Arabia (incl. UAE).

'***dombrowskii***' (*feldegg* x *flava* or *beema*) Similar to *flava*, but with blackish ear-coverts. **Note** Scarce on passage Near East and Arabia (incl. UAE).

'***xanthophrys***' (*feldegg* x *lutea*) Supercilium yellow, crown, lores and ear-coverts contrasting solidly dark, underparts yellow. **Note** Rare on passage or winter Near East and Gulf States (incl. UAE).

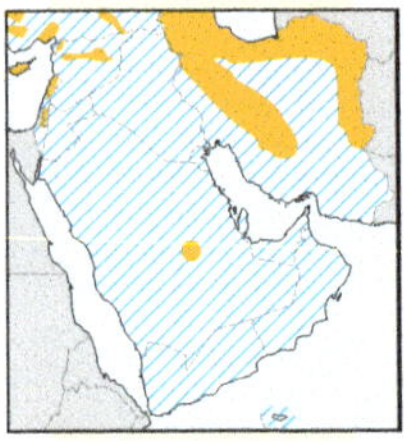

Black-headed Wagtail *Motacilla (flava) feldegg* PM, WV

L: 16.5. Male has head and nape glossy black, sharp demarcation to dark mossy-green mantle, entire underparts bright yellow; duller in winter. Female superficially like male, often almost monochrome, being dark-headed with greyish upperparts and only yellow-sullied underparts. **Voice** Song loud, simple, *sree-srriep*, stress on second syllable. Call strong ***ringing* psreee-u**. **Habitat** As Yellow Wagtail. **Note** Passage hatched; regular winter south Arabia only. Has attempted to breed UAE.

Eastern Black-headed Wagtail *M. f.* ***melanogrisea*** Resembles *feldegg* (with which often grouped) but back paler olive, underparts paler yellow, chin and sides of throat white. **Note** Breeds E. Iran (see *feldegg* map); passage Arabia (Incl. UAE), Near East (rare).

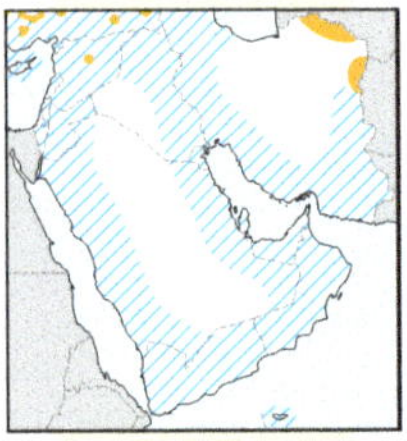

Citrine Wagtail *Motacilla citreola* PM, WV

L: 18. Male unmistakable; subspecies *citreola*, has ***bright yellow head*** and underparts, ***black neck-band and grey upperparts***; *werae* (Turkish breeders) has no (or reduced) neck-band and less-rich yellow underparts; ***broad pure white double wing-bars characteristic at all ages***. Female from Western Yellow Wagtails by ***greyish upperparts*** (tinged olive-brown in Yellow), ***yellow supercilium surrounding grey-brown cheeks merging on sides of neck and yellowish throat***. First-autumn birds have grey upperparts, ***pale surround to ear-coverts, often pale forehead and lores, whitish underparts without yellow***, prominent wing-bars and broadly fringed dark tertials. **Voice** Call pronounced ***buzzy* tsreep**; sometimes a double *zielip*; some eastern '*flavas*' call rather similarly. Song recalls Western Yellow Wagtail's, generally shriller. **Habitat** Breeds 1,500–2,500m in swampy meadows or near streams. Outside breeding season near fresh water, lagoons or sewage ponds. **Note** Map includes Black-backed Citrine Wagtail; passage and winter hatched, but very rare in north in winter.

Black-backed Citrine Wagtail *Motacilla c. calcarata* No records

Male in summer has ***pitch black mantle*** and coverts largely white. Female doubtfully distinguishable from subspecies of Citrine Wagtail (above). **Note** Migrant breeder NE Iran (mapped under Citrine Wagtail); scarce passage Oman.

Yellow Wagtail
'superciliaris'
♂
melanogrisea
♂
'dombrowski'
♂
thunbergi
♂
feldegg
♂
lutea
♂
Black-headed Wagtail
feldegg
♀
leucocephala
♂
juv/1st-winter
Yellow Wagtail
beema
beema
♀
beema
♂
calcarata
♂ summer
citreola
1st-winter
♂ summer
winter
Citrine Wagtail

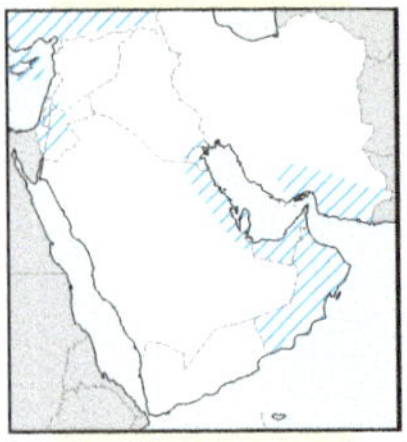

Richard's Pipit *Anthus (novaeseelandiae) richardi* — WV, PM

L: 18. Large, robust pipit; adult from smaller, sandier Tawny Pipit ***by longer tail and legs, stouter bill, typically more upright stance, prominently dark-streaked brown upperparts, streaked breast and characteristic call.*** From first-autumn Tawny Pipit, some still with well-streaked breast, ***by pale lores*** (dark in Tawny, bleaching whitish). More undulating flight and longer tail give more wagtail-like flight than Tawny; ***frequently hovers just before landing*** (Tawny rarely does). **Voice** Flight call loud, harsh throaty *schreeip*; also less distinctive Tawny-like *tjiirrup*. **Habitat** Grassland, fields, marshes. **Note** Passage hatched but often rare; winters SE Arabia.

Blyth's Pipit *Anthus godlewskii* — WV, PM

L: 17. ***Very like Richard's Pipit but slightly smaller, shorter tailed (noticeable in flight) and shorter legged***, with slightly shorter, deep-based but pointed bill (Richard's culmen downcurved at tip); supercilium often shorter. Outer tail feathers also differ (see Plate). Crown neatly streaked, lores pale. Mantle and breast streaked, on latter often a neat fine gorget; flanks sometimes buffy. ***Hindclaw of medium length (long in Richard's). Median wing-coverts of adult (if not worn), and first-autumn birds with some renewed to adult type (but not otherwise), have dark centres sharper, more squarely cut-off against pale tips than in Richard's, producing a prominent pale bar*** (in Richard's, dark centres protrude centrally, with pale edging more diffuse). Juveniles more streaked above and on breast than adult, with whiter fringes and tips to wing-coverts more clear-cut. First-autumn birds from young Tawny Pipit by pale lores, more streaked upperparts, warmer, browner plumage, distinctly streaked breast, and narrower, more clear-cut pale tertial-fringes. Drops into grass without hovering. **Voice** ***High-pitched*** **shreuu** ***or*** **spzeeoo*****, unlike Richard's flight call, closer to slightly hoarse*** **sweeuu** ***call of*** **flava** ***wagtails; also a short*** **chep, chip, chup** ***or*** **pip** ***reminiscent of Tawny Pipit; sometimes joined into unique*** **schreuu-chup chup.** **Habitat** Fodder fields, other grassy and open ground. **Note** Known passage and winter hatched; vagrant Bahrain, Oman.

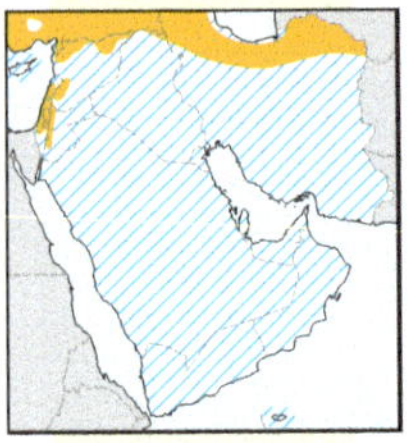

Tawny Pipit *Anthus campestris* — PM, WV

L: 16.5. Slim, largish, upright pipit with ***relatively long tail, legs and bill. Adult has poorly streaked sandy upperparts, nearly unstreaked breast***, plain sandy wings with conspicuous dark-centred median coverts, and bold whitish supercilium. First-autumn birds, still with streaked breast, ***lack bold flank-streaks of most smaller pipits.*** Legs pinky-salmon. Flight undulating; runs quickly, stopping suddenly. **Voice** Typical flight call ***sparrow-like*** **chilp** ***or*** **chirrup.** Song, usually in undulating songflight, simple, thin and metallic, *zriiliu, zseer-lee* or *ziu-ziirliu*. **Habitat** Sparsely vegetated ground, cultivation, plains; any open country on passage. **Note** Passage hatched; winters S. Iran, most Arabia.

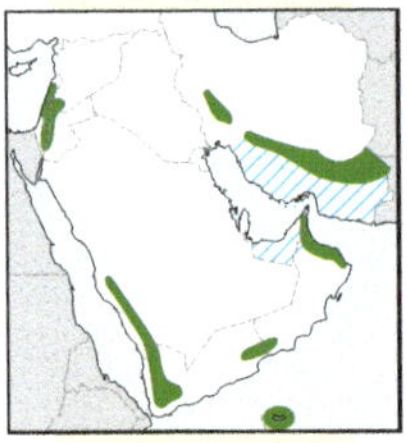

Long-billed Pipit *Anthus similis* — RB, wv

L: 17. Large; from Tawny Pipit by ***creamier-buff flanks and vent***, less pronounced malar streak and ***creamy-buff outer tail feathers*** (which bleach whitish to resemble Tawny). ***Legs pinky-yellow.*** Bill longer with more drooping tip than Tawny, with at least culmen dark. ***Upperparts grey-brown, tinged olive in some***, pale edges and dark centres to wing-coverts and tertials generally less pronounced than in Tawny, supercilium often narrower and ear-coverts plainer brown. ***Tail looks long, broad and dark*** in flight; ***when perched tail often flicked upwards and fanned outwards.*** Seldom well-streaked on breast and upperparts. From Richard's by plainer head pattern, less upright stance, ***less streaked plumage and buff outer tail feathers*** (always white in Richard's). **Voice** Flight call loud, clipped *tjuip* or *che-vee*; also rich *tchup* and quiet, soft *tchut*. Rising and falling song given in undulating songflight, *duiit-diuuu, peet-trueet* or *shreep chew-ee*. **Habitat** Rocky hills, mountain slopes with scattered vegetation, to 3,000m. **Note** Partial and altitudinal migrant; winter hatched.

Richard's Pipit
ad
1st-winter
Blyth's Pipit
1st-winter
Tawny Pipit
juv
ad
Long-billed Pipit
ad
1st-winter

PLATE 95: SMALLER PIPITS

Buff-bellied Pipit *Anthus (rubescens) japonicus* — WV

L: 15.5. Resembles dark winter Water Pipit; ***upperparts dark olive or grey-brown, faintly streaked***; ***breast and flanks usually rather boldly streaked*** (in Water Pipit, sparser, finer, particularly on flanks); ***legs pale brown or reddish-brown*** (normally darker in Water Pipit); ***white eye-ring more obvious, malar streak often broader, ending in black patch at sides of neck***. In spring underparts pinkish-buff, breast lightly spotted, ***but flanks still strongly streaked***; upperparts almost unstreaked olive-grey. Larger and darker than young Red-throated with faintly streaked upperparts. **Voice** High-pitched, short *tripp*, lacking shrill quality of Water Pipit but similar to Meadow Pipit. **Habitat** Grassy fields, near water. **Note** Passage and winter hatched, but rare; vagrant Iran, Kuwait, Oman, Qatar.

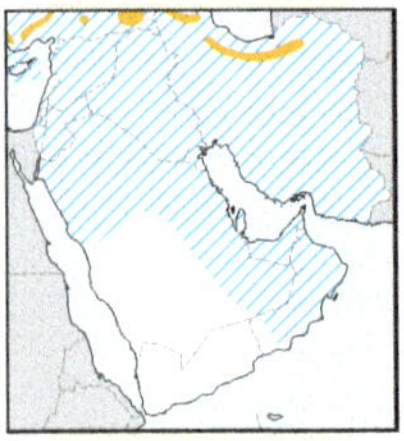

Water Pipit *Anthus spinoletta* — PM, WV

L: 16. Largish with ***dark- or red-brown legs*** and greyish indistinctly streaked mantle. In winter browner above, with ***whitish underparts sparsely streaked, almost unstreaked in some***. In summer ***breast variably tinged rosy-pink and practically unstreaked***; creamy-white supercilium generally pronounced. **Voice** Call sharpish ringing *tsrieh* or *bzisp*. **Habitat** On passage and in winter lowland grassland, wetlands, lakesides. **Note** Passage and winter hatched, often abundant.

Meadow Pipit *Anthus pratensis* — WV, pm

L: 14.5. Similar to Tree Pipit but more slightly built, with ***long hindclaw***; from first-autumn Red-throated Pipit by ***almost unstreaked rump, more broken breast- and flank-streaks, call*** and less contrasting mantle pattern. Flies with irregular undulations and changes of direction; on ground flicks tail nervously. **Voice** Flight call ***thin, nervous,* sit–sii–sit *or* tsis–sip**. **Habitat** Open county, marshes, coasts. **Note** Passage and winter hatched; vagrant Oman.

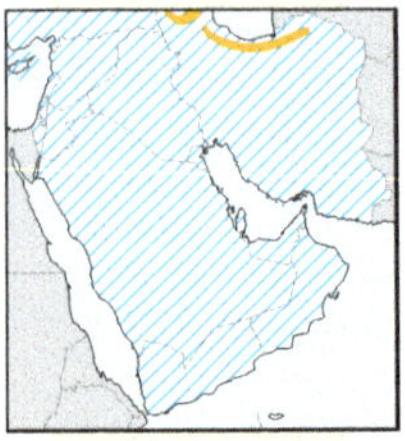

Tree Pipit *Anthus trivialis* — PM, wv

L: 15. Slightly stockier than Meadow Pipit with slightly shorter tail and deeper-based bill. Warmer, less olive, above, warmer buff on breast and whiter belly; breast often boldly striped, but flanks more finely streaked than Meadow; dark malar and creamy sub-moustachial streak generally bolder in Tree Pipit but moustachial streak fainter. ***Pale hindclaw short, legs pinky-salmon*** (dirty-flesh to orange-brown in Meadow). In autumn, from Red-throated Pipit ***by voice, unstreaked rump, fine flank-streaks***, less variegated streaking above and smaller blackish spot at end of malar. Bounding flight more direct than Meadow Pipit; on ground pumps tail, unlike Meadow's nervous flicking. **Voice** Flight call distinctive, short *bzeez* or *speez*; alarm a repeated *stit*. Contact note distinctive quiet *tip*. **Habitat** Open woodland, parkland and grasslands. **Note** Passage hatched; rare in south in winter.

Olive-backed Pipit *Anthus hodgsoni* — wv, pm

L: 14.5. Resembles Tree Pipit, ***with broad supercilium, white behind eye, buff in front, edged black above, pronounced small white and black spots on rear of ear-coverts*** (only sometimes weakly present in Tree); ***upperparts greener, in subspecies yunnanensis vaguely or diffusely streaked*** unless worn (Tree Pipit clearly streaked); ***breast more boldly streaked black***; some also show ***bold blackish flank-streaks***. Subspecies *hodgsoni* similar but mantle obviously streaked (extreme vagrant to UAE). Behaviour as Tree Pipit, though tail-wagging usually more pronounced. **Voice** Flight call a ***thin* tzeez**, usually drawn-out, sometimes slightly stronger at start. **Habitat** As Tree Pipit. **Note** Winter hatched, rare; vagrant Bahrain, Iran, Saudi Arabia.

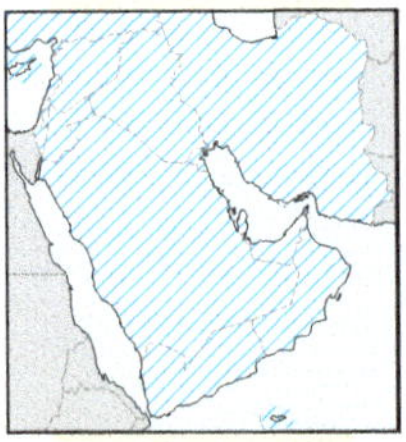

Red-throated Pipit *Anthus cervinus* — WV, PM

L: 15. Adult has variable amount of ***pinkish- or reddish-buff on face, supercilium, throat and upper breast***; retained in autumn though rarely in winter; some females have only beige throat. From Meadow Pipit in autumn/winter ***by call***; ***boldly streaked rump***, breast and flanks creamy or white (usually pale buff in Meadow), ***flanks usually with 2–3 bold, almost unbroken, stripes***; ***malar streak ends in large, dark triangle on side of throat***; upperparts with ***bolder blackish and creamy white stripes***. Flight similar to Meadow Pipit. **Voice** Distinctive, thin, high-pitched, ***drawn-out* pseeee**, slowly dying away. **Habitat** Marshes, grassland, cultivation; usually near water. **Note** Passage and winter hatched; rare in north in winter.

ad summer
Buff-bellied Pipit
ad winter
Water Pipit
ad summer
ad winter
Meadow Pipit
ad
Red-throated Pipit
ad summer
Tree Pipit
ad
ad summer
Olive-backed Pipit
ad
1st-winter

PLATE 96: FINCHES & SILVERBILL

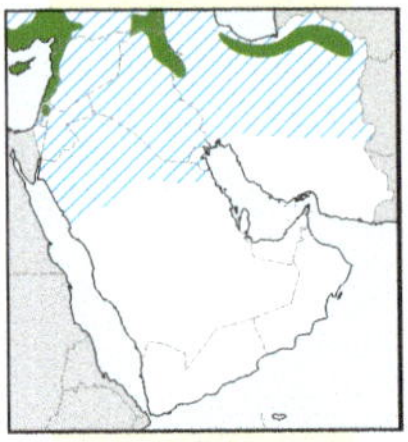

Common Chaffinch *Fringilla coelebs* V

L: 15. Male readily identified by ***blue-grey head, chestnut back, pinkish breast and double white wing-bars***. Superficially resembles Brambling but note absence of orange in plumage, prominent white wing-bars ***and lack of white rump***. More nondescript female and first-winter birds told from all other finches by white wing-bars, white outer tail feathers and unstreaked brownish-olive upperparts (including rump). Undulating flight with wings closed every few beats. Largely feeds on ground with rather jerky shuffling movements. Forms flocks on passage and in winter. **Voice** Song a short, accelerating trill, which ends in a flourish. Calls vary considerably but has typical loud metallic *fink*, repeated loud *huit* and, in flight, a soft *jup jup*. **Habitat** Woodland, orchards; in winter, hedges, open fields. **Note** Winter hatched; vagrant Bahrain, UAE.

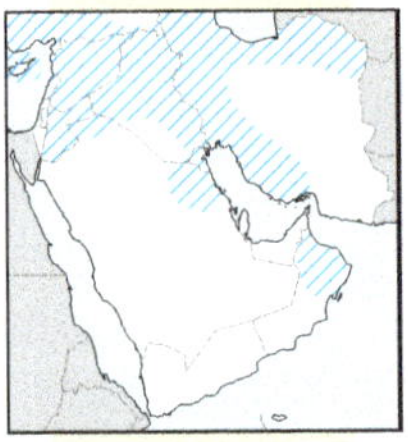

Brambling *Fringilla montifringilla* V

L: 15. Resembles Common Chaffinch but readily told in any plumage by ***rusty-orange breast, shoulder and wing-bar, white belly, mottled upperparts and, in flight, white rump***. In winter (when only likely to be seen in the region) ***shows pale patch on nape and yellowish bill with dark tip***. Often in flocks; mixes freely with Chaffinches and buntings in winter feeding areas. Flight much as Chaffinch but more erratic. **Voice** Distinctive, in flight or perched, a loud, unmusical nasal *jehp*, or softer repeated *tip* or *pip*. **Habitat** Woodland, fields or other cultivation, often amongst or under trees. **Note** Winter hatched; vagrant Bahrain, UAE.

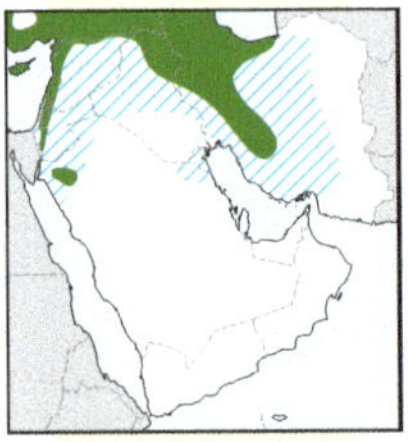

European Goldfinch *Carduelis carduelis* V

L: 14. Colourful and readily identified by bold red, black-and-white head pattern, striking broad yellow bar on black wings, particularly obvious in flight, when white rump also conspicuous. Juvenile has pale greyish body-plumage and unmarked head; identified by yellow wing-bar, white rump and call. **Voice** Call, often uttered in flight, a characteristic, repeated, liquid *tick-le-lit* with stress on last note; rasping notes sometimes heard from flocks. Song a liquid twitter, with calls mixed into song. **Habitat** Scrub, woodland, orchards, cultivated areas; open country, weedy wasteground. **Note** Passage and winter hatched; vagrant Bahrain, Oman, UAE. A popular cagebird, with escapes frequent.

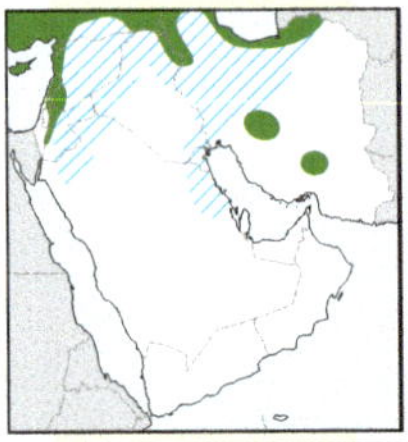

Common Linnet *Carduelis cannabina* V

L: 13. ***In male, greyish head, chestnut back and crimson-red forehead and breast characteristic***. Female and juvenile lack red and chestnut, being warm brown above with dark streaks, visible also on breast and flanks (particularly in juvenile, which also lacks grey on head). ***In all ages has white fringes to primaries and outer tail feathers, dark grey-brown bill and streaked centre of throat***. Flight irregularly undulating. **Voice** In flight, a nasal, sharp *tett* or *tetteret*, often followed by short trill or whistle, *trreu* or *tru-kee-wou*. **Habitat** In winter in wastelands, cultivations and coastal areas. **Note** Winter hatched; vagrant UAE.

Eurasian Siskin *Carduelis spinus* WV

L: 12. Small, finely built finch; ***male yellowish-green with black forehead and chin, broad yellow wing-bar, sides of tail and rump, particularly obvious in flight***. Female (which lacks black on head) is olive-brown above, white below but has tinge of yellow-green, which, together with yellowish in wings, tail and on rump, aid identification. Back and flanks streaked dark in all ages. Often feeds hanging upside-down in cone-bearing trees. Appears small and short-tailed in undulating flight, often in tight flocks. **Voice** Call a drawn-out, ***high-pitched tsee-u***; from feeding flocks a fast, dry *kettkett* and twittering notes. **Habitat** Parks, plantations, Casuarinas. **Note** Passage and winter hatched, sporadic and rare in south; vagrant Bahrain, Oman, Qatar.

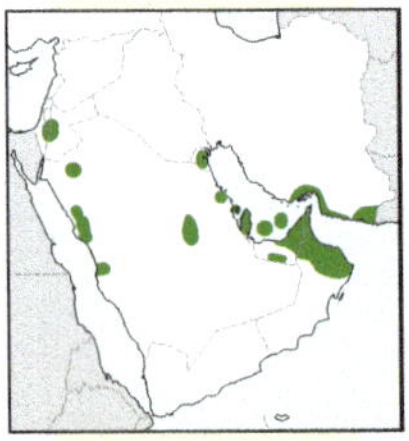

Indian Silverbill *Lonchura malabarica* RB

L: 11. Small with ***large conical silver-grey bill, prominent eye, pointed black tail, whitish rump and uppertail-coverts***. Fairly tame and frequently in small groups, sitting close when perched; often waves and flicks tail. Flight undulating. **Voice** Rapid, tinkling *cheet cheet cheer* flight call; short high-pitched, trilling *zip-zip*; harsh *tchwit* and conversational *seesip-seesip*; song a short trill. **Habitat** Hills and wadis, grassland, scrub, cultivation, palm groves, gardens. Nests in bush, crevice or old nest of sparrow; builds suspended nest. **Note** Native to UAE, but breeding populations further west originate from escapes.

Common Chaffinch
♂
♀
♂
Brambling
♂
♀ winter
♂ winter
European Goldfinch
juv
ad
Common Linnet
1st-winter
♀
♀
♂
Eurasian Siskin
♂
♀
♂
Indian Silverbill
ad

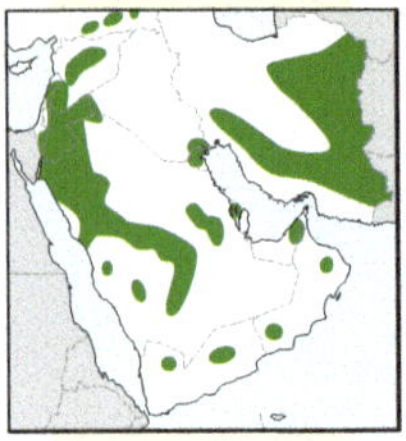

Trumpeter Finch *Bucanetes githagineus* wv, ?rb

L: 14. Small, ground-dwelling finch with large head, stout bill and rather short tail. Male distinctive with ***grey head, pinkish wash on forehead, underparts, rump and wings and orange-red bill***. Non-breeding male, female and juvenile/first-winter plain sandy grey-buff with slightly paler rump, blackish-grey wings and tail with paler feather-edgings and pale yellowish-brown bill; some breeding females develop faint pink wash in plumage; ***legs in all plumages orangey-flesh***. **Voice** Song distinctive, drawn-out nasal, wheezing buzz *cheeeee*; call short *chee* or *chit*; in flight a soft *weechp*, most calls with buzzing quality. **Habitat** Bare rocky and stony hillsides and wadis, stony desert; visits pools and waterholes. **Note** Nomadic. May breed UAE.

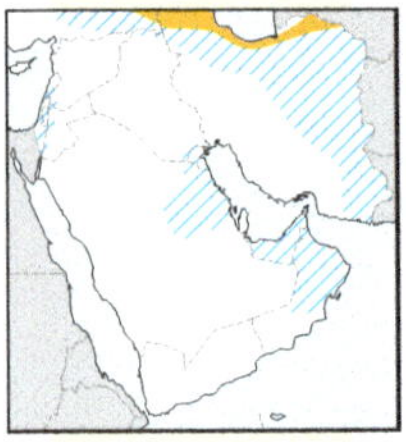

Common Rosefinch *Carpodacus erythrinus* pm

L: 14. Compact finch with ***stout bullfinch-like bill*** and round head. Adult male easily told by ***red head, breast and rump contrasting with brown upperparts***. Females, first-summer males (which often sing) and juveniles are dull olive-brown, lightly streaked above and ***more heavily streaked below, with two whitish or buffish wing-bars***, without wing-panel in flight feathers; uniform head shows conspicuous dark beady eye. Often sings from prominent position, otherwise rather inconspicuous. **Voice** Song diagnostic, a clear, lively loud whistle *vii-dji vii-di-djiv-viuuu* (rendered as 'pleased to meet youuu'). **Habitat** On passage anywhere with trees or scrub. **Note** Passage hatched; rare winter Arabia.

Yellowhammer *Emberiza citrinella* V

L: 16.5. Medium-sized bunting with grey bill ***and chestnut rump at all ages. Male's yellow crown and throat diagnostic; female's yellowish moustache, belly, sometimes supercilium and centre of crown separates from similar female Pine Bunting***. First-autumn birds told (sometimes with difficulty) from Pine Bunting by yellow on belly or vent and on primary edges; a minority may lack any yellow on underparts. Yellowhammer hybridises with Pine Bunting in W. Siberia; hybrid males often (but not always) show white parts of head sullied yellow. **Voice** Metallic, sharp *staeup*, and fast *twe-tic* in flight. **Habitat** Farmland, open country with bushes. **Note** Winter hatched; vagrant Bahrain, Kuwait, UAE.

Pine Bunting *Emberiza leucocephalos* V

L: 16.5. Identical to Yellowhammer in build and size. ***Male's black-bordered white patch on crown and cheeks in otherwise chestnut head and throat diagnostic*** (subdued in winter). Grey-brown female has streaked crown, mantle, breast and flanks; chestnut rump like Yellowhammer; ***some show a little whitish on crown and chestnut on whitish throat. Many females, and all first-autumn birds, resemble Yellowhammer, but yellow in plumage replaced by white, including belly and fringes to primaries***; lesser coverts more uniform grey-brown and bill more frequently bicoloured than Yellowhammer's (dark grey upper, pale grey lower mandible). Some males from hybrid-zone (W. Siberia) often have yellow tinge in plumage (head, fore-body and primaries), others appear mainly white-headed; female hybrids probably indistinguishable. See also Rock and White-capped Buntings. Often in flocks in winter, or mixing in with Yellowhammers. **Voice** Some calls identical with Yellowhammer's; also utters a nervous *trr-rrr-rrr-ick*. **Habitat** As Yellowhammer. **Note** Winter hatched, but rare; vagrant Oman, Saudi Arabia, UAE.

White-capped Bunting *Emberiza stewarti* V

L: 14.5. Small bunting with unstreaked reddish-chestnut lower back and rump, and relatively small, fine, dark bill. Male superficially recalls male Pine Bunting or even sparrow, but has whitish-grey fore-body and head with black bib and eye-stripe and characteristic chestnut band on lower breast (limited to sides in winter). Female resembles small first-autumn female Pine Bunting but has chestnut scapulars in grey-brown surroundings and the neater, sharper-streaked underparts have rufous-chestnut patch at sides of lower breast. Juvenile lacks rufous at sides of breast. **Voice** Calls include sonorous *tsik*, a pleasant soft rolling *turrit* or *ru-ti-ti*; also sharp *tit* and *tsip-ip*. **Habitat** Open hills with scattered bushes; descends to lower ground in winter. **Note** May breed NE Iran; vagrant UAE.

Trumpeter Finch
♂
♀
♂
Common Rosefinch
♀
♂
hybrid
Pine Bunting ×
Yellowhammer
♀
Yellowhammer
♀ 1st-winter
♂
Pine Bunting
♂ spring
♀
♂ winter
♀
White-capped Bunting
♂ winter
♂ spring

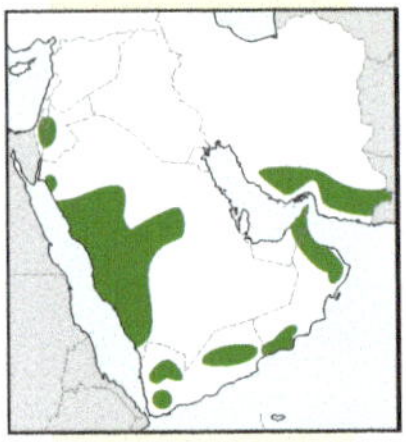

Striolated Bunting *Emberiza striolata* RB

L: 14. ***Small rufous-coloured bunting with orange-yellow lower mandible (adult) and rufous-edged outer tail feathers; streaks on upperparts and wing-coverts thin and vague. Typical male has head striped dark ash-grey and white,*** in some individuals poorly defined; ***throat and upper breast are steel-grey with black speckles; upperparts, wing-coverts, lower breast and belly bright rufous.*** Female duller with more diffuse head pattern. Often fairly shy. **Voice** Song *wi-di-dji-du-wi-di-dii* or *witch witch a wee*. Calls include squeaky *tzswee*, nasal *dwiib*; *dweek* and *sweee-doo*. **Habitat** Oases, desolate rocky wadis and hills with little vegetation. Nests in hole in building, wall or rock crevice. **Note** Vagrant Bahrain, Qatar.

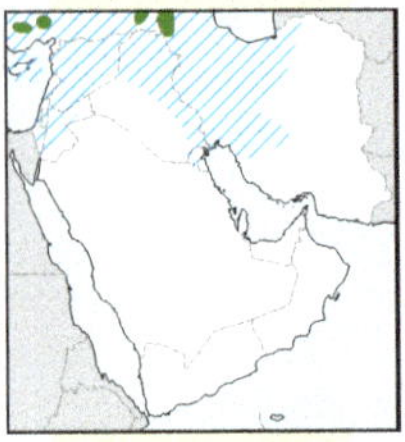

Common Reed Bunting *Emberiza schoeniclus* V

L: 15.5. Breeding male told by ***black hood and bib, white half-collar and white moustache***. Female has ***broad black malar streak and whitish supercilium*** on otherwise brown head. First-autumn birds resemble female but cheeks more mottled, breast and flanks more streaked. Separated from Little Bunting ***by brownish (not chestnut) ear-coverts, bill with convex culmen, inconspicuous eye-ring, chestnut lesser coverts and by call.*** When perched, jerks and spreads tail nervously. In flight, undulations short and irregular, compared with the more steady flight of most other buntings. **Voice** Calls include a fine, drawn-out *tsii-u* and a metallic, slightly voiced *bzü*; in flight a low nasal *bäh*. **Habitat** Reedbeds, swampy thickets. **Note** Passage and winter hatched; vagrant Bahrain, Oman, Saudi Arabia, UAE.

Little Bunting *Emberiza pusilla* V

L: 13. Small and secretive. ***Adult told by reddish-chestnut crown, supercilium and ear-coverts, framed by black lateral crown-stripe and ear-coverts surround narrow whitish eye-ring; white underparts with narrow black streaks on breast;*** whitish wing-bars usually obvious. First-autumn told from Common Reed Bunting by ***uniform pale rusty cheeks, bill shape (straight culmen), whitish eye-ring, black malar streak barely reaching bill, grey-brown lesser coverts, more defined blacker streaks on whiter underparts and call.*** Separated from Rustic Bunting by ***absence of rufous streaks below and grey-brown rump*** (rusty in Rustic). **Voice** Call a single, hard, metallic *tik*, like Rustic Bunting but slightly sharper. **Habitat** On passage, damp bushy and grassy areas. **Note** Vagrant to most countries of Middle East.

Rustic Bunting *Emberiza rustica* V

L: 14.5. Fairly robust bunting, streaked rufous below and with longish bill with straight culmen. Male unmistakable with black head broken by bold white supercilium and nape spot, bright rusty nape merging on sides of neck with rusty red breast-band; flanks with extensive rusty streaks; rump rusty-red. Female has less blackish head with paler median crown-streak and ear-coverts. First-autumn birds confusable with Common Reed Bunting but note rufous streaks on breast and flanks, rufous rump, bill colour and shape, bolder supercilium behind eye, whiter wing-bars and call. **Voice** Call a short, distinct *zib* or *zik*, less sharp than Little Bunting. **Habitat** On passage generally damp, bushy or wooded areas. **Note** Vagrant to most countries of Middle East.

Corn Bunting *Emberiza calandra* WV, pm, ?rb

L: 18. Bulky with fairly large head, ***large conical bill and no white in tail***. Underparts heavily streaked, often merging into irregular black spot on breast; often pale submoustachial stripe and variable ill-defined malar streak. ***Heavy, rather fluttering flight, legs often dangling***; lacks white on trailing edge of wings shown by Skylark, with which it can be confused in flight. Often perches on telegraph wires or bush-top; flocks outside breeding season. **Voice** Song, often from exposed perch, accelerates, ending with monotonous almost insect-like sound *tick-tick-tzek- zee-zri-zizizizi*. Calls single hard, almost clicking *twik* or *tritt*, sometimes quickly strung together in series. **Habitat** Open farmland, irrigated grasslands, bushy areas. **Note** Passage and winter hatched.

Striolated Bunting
♀
♂
♂
Common Reed Bunting
♀
Little Bunting
1st-winter
♂ winter
1st-winter
♂
♂ spring
Rustic Bunting
♂ 1st-winter
♀ 1st-winter
Corn Bunting
♂
ad

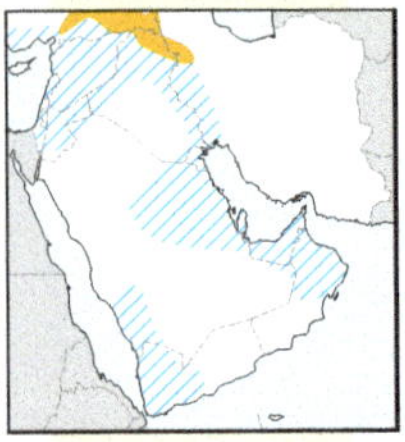

Eastern Cinereous Bunting *Emberiza (cineracea) semenowi* pm

L: 16.5. Male has yellowish underparts, including belly and undertail-coverts, upperparts olivey. Olive grey-brown female has ***throat yellowish and streaked crown yellowish olive-brown***; underparts dusky streaked on breast with belly off-white **Voice** Call short, metallic *kjip* or *djib*. **Habitat** Dry rocky slopes with sparse vegetation; also semi-deserts or bushy wadis on passage. **Note** Eastern Cinereous breeds SE Turkey, Iraq and Iran and is known on passage in Arabia, including UAE in spring.

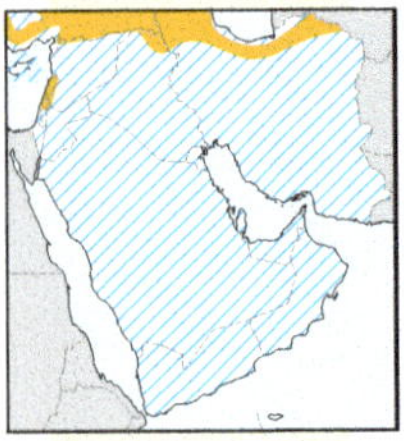

Ortolan Bunting *Emberiza hortulana* PM

L: 16.5. Male told from similar Cretzschmar's Bunting by ***olive-grey head and breast-band framing pale yellowish throat, pink bill, boldly streaked mantle and blackish-brown tertials with clear-cut light chestnut*** (adult) ***or creamy-buff notched edges***. Female duller and more streaked; separable from female Cretzschmar's by pale yellow-buff throat, brownish, not rufous-tinged, rump and pale buff vent and undertail-coverts. First-autumn birds more streaked above and on breast than female; sometimes separable from similar Cretzschmar's by colour of rump, vent and undertail-coverts. Migrates in flocks. **Voice** Call a hollow, soft *plet* or *büb* and a slightly falling *sliie*. **Habitat** Open country, bare cultivation with trees and scrub, semi-desert, oases. **Note** Passage hatched.

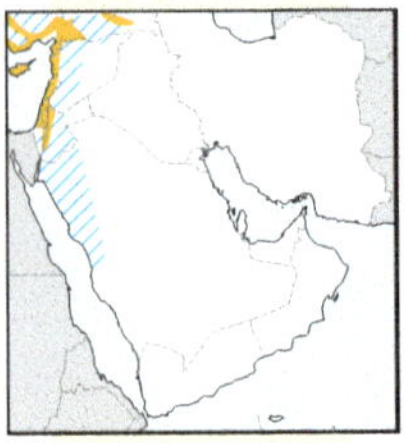

Cretzschmar's Bunting *Emberiza caesia* ?V

L: 16.5. Resembles Ortolan Bunting. ***Male has bright blue-grey head and breast-band framing rusty-orange throat; rump rufous, vent and undertail-coverts rufous-buff***; blackish-brown tertials have clear-cut ***rufous notched fringe***. Female duller, often with traces of grey on head and some grey on streaked breast; ***from female Ortolan by rusty throat*** (sometimes yellowish-buff) and warmer rump, vent and undertail-coverts. First-autumn birds resemble Ortolan ***but yellow-buff throat often tinged rufous (adult plumage developed quickly)***. **Voice** Calls Ortolan-like, but sharper, *blep*; metallic *tlik* or *tlev*. **Habitat** On passage cultivation, semi-desert. **Note** Passage hatched; vagrant Iran, Oman, Qatar, Yemen, with recent claim UAE.

Yellow-breasted Bunting *Emberiza aureola* V

L: 15. Medium-sized with pale, stout, conical bill and white in tail-sides. Non-breeding male, female and first-autumn birds told ***by prominent creamy supercilium, faint pale central forecrown-stripe with dark brown lateral crown-stripes*** (less defined in first-autumn), ***pale cheeks strongly dark-bordered, cream mantle braces and fairly distinct white wing-bars***. Rump grey-brown and heavily streaked. Underparts yellowish-white or at least sullied buffy (mainly first-autumn) with whitish undertail-coverts; flanks and sometimes breast faintly streaked. **Voice** Two main calls heard on passage, a short *tik*, not unlike that of Little Bunting, and a softer, metallic *tsiu* or *tsip*. **Habitat** Open country with trees or bushes; grassland, cultivations. **Note** Vagrant Bahrain, Iran, Oman, Saudi Arabia, UAE.

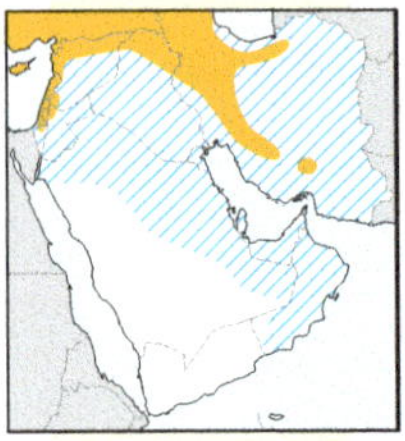

Black-headed Bunting *Emberiza melanocephala* PM

L: 16.5. Heavy-bodied, stout-billed bunting ***without white in tail. Male has black head, yellow underparts and unstreaked chestnut upperparts***; black of head subdued by pale fringes in autumn. Female and first-autumn birds lack black, bright chestnut and yellow, then appearing almost identical to Red-headed Bunting ***although upperparts usually warmer-toned*** (less olive grey-brown) and underparts generally tinged more yellow, especially undertail-coverts. Often in groups on passage, particularly in spring. **Voice** Song typically bunting-like, starting with short harsh notes, ending with more ringing *tsi-tia-tia-tia-terlu-terlu-terlu*. Call a sparrow- or Tawny Pipit-like *tjilp*; also a metallic *tlev*. **Habitat** Bushy and grassy country, open farmland. **Note** Passage hatched.

Red-headed Bunting *Emberiza bruniceps* V,E

L: 16.5. Size of Black-headed Bunting, also lacking ***white in outer tail feathers. Male with red-brown head and breast*** (subdued by yellow and white in autumn), ***with rest of underparts and rump yellow***. Female and first-autumn birds very like Black-headed and some not identifiable, but note ***Red-headed has grey-brown, often olive-tinged upperparts including rump and scapulars*** (tinged rufous in many Black-headed Buntings). Head of Red-headed Bunting slightly more uniform with less streaked forecrown and paler ear-coverts than in Black-headed Bunting; underparts, particularly undertail-coverts, tend to be tinged less yellowish than in Black-headed, some being uniform buffish-grey below. **Voice** Song like Black-headed Bunting. Call *bzisf*. **Habitat** Farmland and open country on passage. **Note** Vagrant Kuwait, Oman, Saudi Arabia, UAE. Frequent escape.

Eastern Cinereous Bunting
♀
♂
Ortolan Bunting
1st-winter
♀
♂
♂ 1st-summer
Cretzschmar's Bunting
♂ 1st-winter
♀
♂
Black-headed Bunting
1st-winter
♀
♂
♀
♂ winter
Yellow-breasted Bunting
juv
♂ 1st-summer
1st-winter
♀
♂
Red-headed Bunting

Yellow-billed Stork *Mycteria ibis* E/I

L: 100. W: 160. Easily told in adult plumage by ***slightly drooping, orange-yellow bill, bare red face and long, orange-red legs***. Plumage resembles White Stork's but ***tail black*** (visible in flight) and ***mantle and tips of wing-coverts tinged pink***. In subadult plumage sandy-buff with ***some pinkish on underwing-coverts***, much duller bill and legs and greyish to pale orange facial skin. In first-winter shows greyish wash, brownish underwing-coverts, yellowish-grey bill and greyish-brown legs. **Habitat** Wetlands. **Note** Escaped birds in Arabian Gulf (Qatar, UAE).

Marabou Stork *Leptoptilos crumenifer* E/I

L: 150. W: 240. Unmistakable very large stork with huge long stout bill and long legs. Adult has ***naked head, neck and chest-pouch, slate-grey upperparts and whitish underparts***. Juvenile has downy cover to bare head and neck. In flight shows ***blackish underwing with conspicuous white armpits***. Soars like a vulture on broad, flattish wings with curved trailing edge and deep fingers, looking, as at rest, thick-necked with head sunk between shoulders. **Habitat** Wetlands, grassland. Escapes Al Ain Zoo area.

Goliath Heron *Ardea goliath* E

L: 145. W: 220. Very large with long legs and heavy bill. Adult has chestnut head and hindneck, bluish-grey upperparts and ***rich chestnut underparts and underwing-coverts. Immature birds are paler rufous-orange on head*** and have rufous edgings to feathers on upperparts; white underparts are dark-streaked. ***Distinguished from Purple Heron by much larger size, stouter head and neck, lack of black on crown, absence of black line across cheeks, greyish bill*** (yellowish in Purple), ***dark legs*** (yellowish-green in Purple) and upperwing more uniform grey. In flight, wingbeats slow and heavy with legs protruding well beyond tail. **Habitat** Extensive reedbeds; also mudflats, mangroves and coastal islands. **Note** Feral breeding previously in UAE (Al Ain zoo); vagrant Oman (breeds Iraq and Red Sea).

Yellow-necked Spurfowl *Pternistis leucoscepus* E/I

L: 40. A largish gamebird with diagnostic ***bare yellow skin on neck and reddish skin around eye***. Bill dark; underparts heavily streaked. Runs from danger, but ***in flight shows large creamy patch in primaries***. Males have spurs on rear of leg. **Voice** A loud grating descending series, *k-wrrrrrrrk, k-wrrrrrrrk*, notes upslurred and fading away. **Habitat** Dry bushy country. **Note** Feral birds sometimes encountered in UAE, but perhaps none breeding successfully.

Helmeted Guineafowl *Numida meleagris* E/I

L: 63. W: 97. Size of domestic chicken. ***Note horny, brown protuberance on crown***. Juvenile brownish with buff and white spotting and less pronounced head pattern. Gregarious, runs fast with head held high or flies rapidly with occasional glides. **Voice** Far-carrying, raucous *kek, kek, kek, kek, kaaaaaa, ka, ka, ka, ka, kaaaaa, ka ka* uttered by flocks or a roost. **Habitat** Scrub, thick cover, gardens. **Note** Non-native escape.

Grey Crowned Crane *Balearica regulorum* E/I

L: 100. Unmistakable. Very tall with characteristic ***golden head plumes, pale cheeks and relatively short bill. Neck is lowered in hunched, laboured flight, with extensive pale panel on both upper- and underwing coverts***. **Voice** Evocative bugling, and single *wonk* or *ka-wonk*. **Habitat** Grassland, marshy ground. **Note** Non-native escape; successful feral breeding recorded UAE.

Speckled Pigeon *Columba guinea* E/I

L: 35. Adult distinctive, with reddish chestnut-brown upperparts, prominently ***white-dotted wing-coverts***, pale grey rump and conspicuous ***bare red eye-patch***. Underparts grey, iris pale yellow, legs yellowish. Juvenile duller with less pronounced spotting, upperparts more sombre brown, as in worn adults. **Voice** Distinctive, often loud, in fast-rising series increasing in pace and volume, *ooroo-coo*. **Habitat** Open country, also towns, where ground-feeding in manner of Feral Pigeon. **Note** Vagrant Yemen. Non-native escape in UAE.

PLATE NOT TO SCALE
ad summer
ad
ad winter
Yellow-billed Stork
ad
Marabou Stork
Yellow-necked Spurfowl
ad
Goliath Heron
ad
Helmeted Guineafowl
ad
imm
ad
ad
Speckled Pigeon
Grey Crowned Crane

Lesser Masked Weaver *Ploceus intermedius* E/I

L: 12.5. Male with ***black forehead and throat to upper breast***, underparts otherwise yellow. Female olive above, with yellow breast and white belly; ***both sexes with grey legs***, and eyes pale yellow (dark in juvenile). **Voice** Rapid metallic or squeaky babble. **Habitat** Trees, scrub, gardens. Nest spherical with short entrance tube. **Note** Non-native escape; breeds UAE.

Village Weaver *Ploceus cucullatus* E/I

L: 15. ***Large***; breeding male dark-faced with yellow-and-black spotted back and wings. Female with yellow throat and white belly. ***Note large bill and red eye*** in both sexes. Gregarious. **Voice** Wheezy chatter, often given in unison. **Habitat** Parkland. **Note** Non-native escape; breeds UAE.

Golden-backed Weaver *Ploceus jacksoni* E/I

L: 13. Male has red eye, ***golden-yellow back, ebony head and deep chestnut underparts***; brown-eyed female and non-breeding male olive above with dusky streaks, unmarked underparts and supercilium yellow; colonial. **Voice** Song a distinctive drawn-out wheezing chatter. Call an abrupt note, somewhat like cicada. **Habitat** Parkland, scrub; nest often built on branch tip over water. **Note** Non-native escape; breeds UAE, occasional Oman.

Streaked Weaver *Ploceus manyar* E/I

L: 14. Short-tailed and thickset with heavy bill. Male with ***yellow crown and prominent streaks on breast***. Female and non-breeding male brown, ***streaked on breast***, and upperparts, including crown; ***with dark submoustachial streak*** and yellowish supercilium. **Voice** Song unmusical strained jumble; calls loud *chirt*. **Habitat** Wetland scrub and reeds. **Note** Non-native escape; breeds Kuwait, Qatar, Saudi Arabia, UAE.

Bengal Weaver *Ploceus benghalensis* **(Not illustrated)** E/I

L: 14. Similar to Streaked Weaver but with whitish chin and ***black breast; unstreaked below***. **Habitat** Reeds, scrubby cover. **Note** Non-native escape; breeding attempted UAE.

Baya Weaver *Ploceus philippinus* E/I

L: 15. Stout, long-billed weaver. Male with yellow crown and nape, black mask and unmarked underparts. Female unmarked buffy below or only indistinctly streaked. **Voice** Dry wheezy notes or chatter. **Habitat** Scrub, cultivation, riparian cover. **Note** Non-native escape; breeds Saudi Arabia, formerly noted also UAE but no recent reports.

Vitelline Masked Weaver *Ploceus vitellinus* E/I

L 13: Similar to Lesser Masked; male has ***red eye, mask in only a narrow band above bill onto forehead and pinkish-red legs***. Female has ***red eyes, yellow breast and pinky legs***. **Voice** Song buzzy and scratchy, unmusical. **Habitat** Trees, parks. **Note** Non-native escape; occasionally reported UAE, but not confirmed as nesting.

Chestnut Weaver *Ploceus rubiginosus* E/I

L: 15. Unlike other weavers, male being largely ***rich chestnut, with clear-cut black head***, white edged wing feathers and red eye. Female/non-breeding male has brown upperparts and ***lightly streaked tawny-buff breast-band***. **Voice** Song unusual, with sudden dry fizzing given during bout of chattering. **Habitat** Trees, tall scrub. **Note** Recorded UAE, but not yet confirmed as nesting.

♀
♂
Lesser Masked Weaver
♀
♂
Village Weaver
♀
♂
Golden-backed Weaver
♀
♂
Streaked Weaver
♀
♂
Baya Weaver
♀
Vitelline Masked Weaver
♂ summer
♀
♂ summer
Chestnut Weaver

Yellow-crowned Bishop *Euplectes afer* E/I

L: 10. Small and compact; breeding ***male black below, with black hindneck collar and sulphur-yellow crown, back and rump***. Non-breeding male, female and juvenile streaked on sides below with yellowy supercilium. **Voice** Buzzing and chipping notes. **Habitat** Grassland, marshes, scrub. **Note** Non-native escape; breeds Oman, UAE.

Southern Red Bishop *Euplectes orix* E/I

L: 11–12. Sparrow-sized but short-tailed. Breeding male fiery red above with ***black on forehead and throat. breast-band red, belly black***. Female, juvenile and non-breeding male flat-crowned; buffy and finely streaked below, bill small and pointed. (Similar ***Northern Red Bishop*** *E. franciscanus* (not illustrated) occasionally reported; ***male has throat red and black on head reaching crown***. Females doubtfully separable.) **Voice** Calls thin, high, squeaky *cheet*; song buzzy chirping. **Habitat** Fields, grassy scrub, waterside canes. **Note** Non-native escape; breeds UAE.

Scaly-breasted Munia *Lonchura punctulata* E/I

L: 12. Adult chestnut or dull tan above with white ***underparts brown-scaled***. Juvenile plain brown above and below. Adult bill gun metal, duller in juvenile. Flight bouncing; often in rapidly moving flocks. Feeds on grass seed-heads. **Voice** Calls high piping *bee-bee*, recalling Siskin; song soft, thin whistles and slurred notes. **Habitat** Waste and grassy areas. **Note** Non-native escape; breeds Oman, Saudi Arabia, UAE.

Red Avadavat *Amandava amandava* E/I

L: 9–10. ***Male crimson with numerous white spots on underparts and wing-coverts. Female brown above, buffish below with fulvous-yellow belly; rump red, tail blackish***. Non-breeding male as female but greyer below. Coral-red bill (both sexes) with black culmen. Juvenile as female but with buff wing-bars and dark bill; lacks red rump. Often in small, low-flying flocks. **Voice** Song high-pitched, continuous twittering; call thin, high *teee* or *tsi*. **Habitat** Damp scrub, reeds. **Note** Non-native escape; breeds Bahrain, Iran, Kuwait, Saudi Arabia, formerly bred UAE, now rarely reported.

Red-billed Quelea *Quelea quelea* E/I

L: 12. Stocky and short-tailed. Male in breeding dress distinctive; non-breeding male and female ***streaked above, with prominent supercilium and all-red bill***. **Voice** Chipping chatter. **Habitat** Grasslands, reeds. **Note** Non-native escape; not yet confirmed breeding UAE.

Chestnut Munia *Lonchura atricapilla* E/I

L: 12. All ***chestnut with black head, throat and neck, and blue bill***. Immature similar but paler. **Voice** A weak *pee, pee*. **Habitat** Weedy wasteground, cultivation. **Note** Non-native escape; breeding not yet confirmed UAE.

Pin-tailed Whydah *Vidua macroura* E/I

L: 11. Male in breeding dress ***black and white with 20cm tail; bill always red***. Female has ***prominent head pattern*** and, when breeding, black bill, otherwise red. Non-breeding male loses tail streamers and resembles female. **Voice** A jumble of squeaks and chirps. **Habitat** Scrub, cultivation, gardens. **Note** Non-native escape; breeding not confirmed UAE. (A brood parasite; usual host species absent UAE, though Indian Silverbill may be prove suitable.)

Java Sparrow *Padda oryzivora* E/I

L: 16. Distinctive; ***all-grey with large red bill, black head and throat, white cheeks***; wine-coloured belly, white undertail coverts and black tail. Immature browner, with less contrasting cheeks and duller bill. **Voice** Short *clip* in flight. **Habitat** Parks, gardens. **Note** Non-native escape; not yet confirmed breeding UAE.

Yellow-crowned Bishop
♂
♀
Southern Red Bishop
♂
♀
imm
Scaly-breasted Munia
ad
Red Avadavat
♀
♂
Red-billed Quelea
♂ winter
♀
♂ summer
♂ transitional
♂ summer
♂ winter
Chestnut Munia
ad
♀ summer
juv
Pin-tailed Whydah
Java Sparrow
ad

REFERENCES AND FURTHER READING

UNITED ARAB EMIRATES

Aspinall, S. J. 1996. *Status and Conservation of the Breeding Birds of the United Arab Emirates.* Hobby Publications, Warrington, UK and Dubai, UAE.

Aspinall, S. 2010. *The Breeding Birds of the United Arab Emirates (3rd edition).* EAD, Abu Dhabi, UAE.

Hellyer, P. & Aspinall, S. J. (eds.) 2005. *The Emirates – A Natural History.* Trident Press, UK.

Pedersen, T. & Aspinall, S. (comp.). 2010. EBRC Annotated Checklist of the birds of the United Arab Emirates. Sandgrouse Supplement 3. OSME, Sandy, UK.

Richardson, C. 1990. *The Birds of the United Arab Emirates.* Hobby Publications, Warrington, UK and Dubai, UAE.

UAE Checklist. 2010: http://www.uaebirding.com (site contents continuously updated)

MIDDLE EAST

The Ornithological Society of the Middle East, OSME (www.osme.org) publishes its own journal *Sandgrouse* twice yearly and the topical articles are recommended reading. The newsletter of the Atlas of the Breeding Birds of Arabia project, *Phoenix,* is also a useful source of up-to-date findings (contact: arabianbirds@dsl.pipex.com).

Berthold, P. 1999. *Bird Migration: a general survey. 2nd Edition.* OUP, Oxford.

Christie, D. A., Shirihai, H. and Harris, A. 1996. *The Macmillan Birder's Guide to European and Middle Eastern Birds.* Macmillan, London.

Cramp, S., Simmons K. E. L. and Perrins, C. M. 1977–1994. *The Birds of the Western Palearctic* (BWP). Vols 1–9. OUP, Oxford.

Cramp, S., Simmons, K. E. L. and Perrins, C. M. 2004. *The Birds of the Western Palearctic: interactive* (BWPi). BirdGuides, Sheffield.

Delany, S., Scott, D., Dodman, T. and Stroud, D. 2009. (Eds). *An Atlas of Wader Populations in Africa and Western Eurasia.* Wetlands International, Wageningen, The Netherlands.

Dickinson, E. C. 2003. (Ed). *The Howard and Moore Complete Checklist of the Birds of the World. 3rd edition.* Christopher Helm, London.

Evans, M. I. 1994. *Important Bird Areas in the Middle East.* BirdLife Conservation Series No 2. BirdLife International, Cambridge.

Gill, F., Wright, M. and Donsker, D. 2010. *IOC World Bird Names (version 2.3).* Available at http://www.worldbirdnames.org

Grimmett, R., Inskipp, C. and Inskipp, T. 1998. *Birds of the Indian Subcontinent.* Helm, London.

Hüé, F. and Etchécopar, R. D. 1970. *Les Oiseaux du Proche et du Moyen Orient.* Éditions N Boubée et Cie. Paris, France.

IOC. 2010. (International Ornithological Congress). Update v2.3 (see Gill et *al.* 2010)

Jennings, M. C. 2010. Atlas of the Breeding Birds of Arabia. *Fauna of Arabia 25.*

Jonsson, L. 1992. *Birds of Europe with North Africa and the Middle East.* Christopher Helm, London.

Porter, R. F., Aspinall, S. J., Preddy, S. & Blair, M. J. 2010. OSME Region List of bird taxa. v2.3. www.osme.org/orl

Porter, R. F., Christensen, S. and Schiermacker-Hansen, P. 2006. [*Birds of the Middle East*] [In Arabic]. Translated by Saeed Mohamed. SPNL. Beirut, Lebanon.

Rasmussen, P. and Anderton, J.C. 2005. *Birds of South Asia: The Ripley Guide.* 2 vols. Lynx Edicions, Smithsonian Institution, Barcelona, Spain and Washington, USA.

Redman, N., Stevenson, T. and Fanshawe, J. 2009. *Birds of the Horn of Africa.* Christopher Helm, London.

Svensson, L. 1992. *Identification Guide to European Passerines.* 4th Edition. Privately published.

Svensson, L., Mullarney, K. and Zetterström, D. 2009. *Collins Bird Guide, 2nd Edition.* HarperCollins, London.

van Duivendijk, N. 2010. *Advanced Bird Id Guide.* New Holland, London.

Williamson, K. 1976. *Identification for Ringers 1. The genera* Cettia, Locustella, Acrocephalus *and* Hippolais. Revised edition. BTO, Tring.

Williamson, K. 1976. *Identification for Ringers 2. The genus* Phylloscopus. Revised edition. BTO, Tring.

Williamson, K. 1976. *Identification for Ringers 3. The genus* Sylvia. Revised edition. BTO, Tring.

BAHRAIN

Hirschfeld, E. 1995. *Birds in Bahrain: A Study of their Migration Patterns 1990–92.* Hobby Publications, Dubai, UAE.

Mohamed, S. A. 1993. *Birds of Bahrain and the Arabian Gulf.* Bahrain Centre for Research Studies, Bahrain.

Nightingale, T. and Hill, M. 1993. *The Birds of Bahrain.* Immel, London.

CYPRUS

Flint, P. R. and Stewart, P. F. 1992. *The Birds of Cyprus, 2nd Edition.* BOU, Tring, Herts.

Richardson, C. 2009. *Birds of Cyprus Checklist 2003–2008 (2009).* BirdLife Cyprus. (Earlier versions by other editors). NB Annual bird reports also published by BirdLife Cyprus.

IRAN

Mansoori, J. 2008. [*Field Guide to the Birds of Iran*] [In Farsi]). Fahzanek Books, Tehran, Iran.

Scott, D. A. and Adhami, A. 2006. An Updated Checklist of the Birds of Iran. *Podoces* 1 (1/2): 1–16.

IRAQ

Salim, M. A., Porter, R. F., Schiermacker-Hansen, P., Christensen, S., and al-Jbour, S. 2006. [*Field Guide to the Birds of Iraq.*] [In Arabic]. Nature Iraq & BirdLife International, Amman, Jordan.

ISRAEL

Mild, K. 1996. *Bird Songs of Israel and the Middle East.* Sound recordings of 114 spp; two cassettes. Privately published.

Perlman, Y. and Meyrav, J. 2009. *Checklist of the Birds of Israel.* Soc. Prot. Nat. Israel/Israel Orn. Center, Tel-Aviv, Israel.

Sandgrouse 21(1). 1999. Special Levant Publication, including: Shirihai, H., Andrews, I. J., Kirwan, G. M. and Davidson, P. A checklist of the birds of Israel and Jordan.

Shirihai, H. 1996. *The Birds of Israel.* Academic Press, London.

JORDAN

Andrews, I. J. 1995. *The Birds of the Hashemite Kingdom of Jordan.* Privately published.

Andrews, I. J., Khoury, F. and Shrihai, H. 1999. Jordan Bird Report 1995–97. *Sandgrouse* 21(1): 10–35.

Sandgrouse 21(1). 1999. Special Levant Publication, including: Shirihai, H., Andrews, I. J., Kirwan, G. M. and Davidson, P. A checklist of the birds of Israel and Jordan.

KUWAIT

Gregory, G. 2005. *The Birds of the State of Kuwait.* Privately published.

LEBANON

Ramadan-Jaradi, G., Bara, T. and Ramadan-Jaradi, M. 2008. Revised checklist of the birds of Lebanon 1999–2007. *Sandgrouse* 30(1): 22–69.

OMAN

Gallagher, M. and Woodcock, M. W. 1980. *The Birds of Oman.* Quartet Books, London.

Oman Bird Records Committee. 2003. *Oman Bird List (Edition 6)*. The official (and regularly updated) list of the birds of the Sultanate of Oman.

Sargeant, D. A., Eriksen, H. and Eriksen, J. 2008. *Birdwatching Guide to Oman (2nd edition)*. Al Roya Publishing, Muscat, Oman.

PALESTINE

al-Safadi, M. M. 2006. Observations on the breeding birds of the Gaza Strip, Palestine. *Sandgrouse* 28(1): 22–33.

QATAR

Oldfield, C. and Oldfield, J. 1994. *A Birdwatcher's Guide to Qatar*. Privately published.

SAUDI ARABIA

Bundy, G., Connor, R. J. and Harrison, C. J. O. 1989. *Birds of the Eastern Province of Saudi Arabia*. HF&G Witherby, London.

Jennings, M. C. 1981. *The Birds of Saudi Arabia: A Check-list*. Privately published.

SYRIA

Baumgart, W. 1995. *Die Vögel Syriens: Eine Übersicht*. Max Kasparek Verlag, Heidelberg, West Germany.

Murdoch, D. A and Betton, K. 2008. A checklist of the birds of Syria. OSME *Sandgrouse* Supplement 2. OSME, Sandy.

Syrian Society for Conservation of Wildlife & BirdLife International. 2009. [*Field Guide to the Birds of Syria*]. [In Arabic]. SSCW & BirdLife International, Damascus, Syria.

TURKEY

Green, I. and Moorhouse, N. 1995. *A Birdwatchers' Guide to Turkey*. Prion, Perry, UK.

Kirwan, G. M., Martins, R. P., Eken, G. and Davidson, P. 1999. A checklist of the Birds of Turkey. *Sandgrouse* Supplement 1. OSME, Sandy.

Kirwan, G. M., Boyla, K. A., Castell, P., Demirci, B., Özen, M., Welch, H. and Marlow, T. 2008. *The Birds of Turkey*. Christopher Helm, London.

Porter, R. F., Christensen, S. and Schiermacker-Hansen, P. 2009 [*Turkiyive Ortadogu'nun Kuslari*]. [In Turkish]. (Birds of Turkey and the Middle East). Translated by Kerem Ali Boyla & Kazım Çapacı. Doğa Derneği, Turkey.

Roselaar, C. S. 1995. *Songbirds of Turkey: An Atlas of Biodiversity of Turkish Passerine Birds*. Pica Press, Robertsbridge, UK.

YEMEN

Multiple authors. 1987. {16 papers on the birds of northern Yemen}. *Sandgrouse* 9.

Multiple authors. 1996. {23 papers on the birds of southern Yemen and Socotra}. *Sandgrouse* 17.

SOCOTRA

Porter, R. 2003. Socotra: Yemen's special island. *Sandgrouse* 25 (2): 93–102

www.socotraproject.org (for latest archipelago checklist).

CHECKLIST OF THE BIRDS OF THE UNITED ARAB EMIRATES

This checklist contains all the species recorded in the UAE, except non-naturalised escapes. It follows the order adopted by the Ornithological Society of the Middle East (see Introduction) but may differ from the sequence in the main text, as the latter has sometimes been changed for presentational reasons.

- ❑ Chukar Partridge *Alectoris chukar*
- ❑ Sand Partridge *Ammoperdix heyi*
- ❑ Black Francolin *Francolinus francolinus*
- ❑ Grey Francolin *Francolinus pondicerianus*
- ❑ Common Quail *Coturnix coturnix*
- ❑ Eastern Greylag Goose *Anser anser rubrirostris*
- ❑ Greater White-fronted Goose *Anser albifrons*
- ❑ Lesser White-fronted Goose *Anser erythropus*
- ❑ Mute Swan *Cygnus olor*
- ❑ Bewick's Swan *Cygnus (columbianus) bewickii*
- ❑ Whooper Swan *Cygnus cygnus*
- ❑ Egyptian Goose *Alopochen aegyptiaca*
- ❑ Common Shelduck *Tadorna tadorna*
- ❑ Ruddy Shelduck *Tadorna ferruginea*
- ❑ Cotton Pygmy Goose *Nettapus coromandelianus*
- ❑ Gadwall *Anas strepera*
- ❑ Eurasian Wigeon *Anas penelope*
- ❑ Mallard *Anas platyrhynchos*
- ❑ Northern Shoveler *Anas clypeata*
- ❑ Northern Pintail *Anas acuta*
- ❑ Garganey *Anas querquedula*
- ❑ Eurasian Teal *Anas crecca*
- ❑ Marbled Duck *Marmaronetta angustirostris*
- ❑ Red-crested Pochard *Netta rufina*
- ❑ Common Pochard *Aythya ferina*
- ❑ Ferruginous Duck *Aythya nyroca*
- ❑ Tufted Duck *Aythya fuligula*
- ❑ Red-breasted Merganser *Mergus serrator*
- ❑ Cory's Searwater *Calonetris (diomedia) borealis*
- ❑ Wedge-tailed Shearwater *Puffinus pacificus*
- ❑ Persian Shearwater *Puffinus persicus*
- ❑ Sooty Shearwater *Puffinus griseus*
- ❑ Flesh-footed Shearwater *Puffinus carneipes*
- ❑ Jouanin's Petrel *Bulweria fallax*
- ❑ Wilson's Storm Petrel *Oceanites oceanicus*
- ❑ Leach's Storm Petrel *Oceanodroma leucorhoa*
- ❑ Little Grebe *Tachybaptus ruficollis*
- ❑ Great Crested Grebe *Podiceps cristatus*
- ❑ Black-necked Grebe *Podiceps nigricollis*
- ❑ Greater Flamingo *Phoenicopterus roseus*
- ❑ Lesser Flamingo *Phoeniconaias minor*
- ❑ Red-billed Tropicbird *Phaethon aethereus*

- ❑ Black Stork *Ciconia nigra*
- ❑ Western White Stork *Ciconia ciconia*
- ❑ Glossy Ibis *Plegadis falcinellus*
- ❑ Eurasian Spoonbill *Platalea leucorodia*
- ❑ Eurasian Bittern *Botaurus stellaris*
- ❑ Little Bittern *Ixobrychus minutus*
- ❑ Cinnamon Bittern *Ixobrychus cinnamomeus*
- ❑ Black-crowned Night Heron *Nycticorax nycticorax*
- ❑ Striated Heron *Butorides striata*
- ❑ Squacco Heron *Ardeola ralloides*
- ❑ Indian Pond Heron *Ardeola grayii*
- ❑ Western Cattle Egret *Bubulcus ibis*
- ❑ Eastern Cattle Egret *Bubulcus (ibis) coromandus*
- ❑ Grey Heron *Ardea cinerea*
- ❑ Purple Heron *Ardea purpurea*
- ❑ Western Great Egret *Egretta alba*
- ❑ Intermediate Egret *Egretta intermedia*
- ❑ Little Egret *Egretta garzetta*
- ❑ Western Reef Heron *Egretta gularis*
- ❑ Great White Pelican *Pelecanus onocrotalus*
- ❑ Dalmatian Pelican *Pelecanus crispus*
- ❑ Masked Booby *Sula dactylatra*
- ❑ Red-footed Booby *Sula sula*
- ❑ Brown Booby *Sula leucogaster*
- ❑ Great Cormorant *Phalacrocorax carbo*
- ❑ Socotra Cormorant *Phalacrocorax nigrogularis*
- ❑ (Western) Osprey *Pandion haliaetus*
- ❑ European Honey Buzzard *Pernis apivorus*
- ❑ Crested Honey Buzzard *Pernis ptilorhynchus*
- ❑ Black-winged Kite *Elanus caeruleus*
- ❑ Black-eared Kite *Milvus (migrans) lineatus*
- ❑ Pallas's Fish Eagle *Haliaeetus leucoryphus*
- ❑ Egyptian Vulture *Neophron percnopterus*
- ❑ Eurasian Griffon Vulture *Gyps fulvus*
- ❑ Lappet-faced Vulture *Torgos trachielotus*
- ❑ Short-toed Snake Eagle *Circaetus gallicus*
- ❑ Western Marsh Harrier *Circus aeruginosus*
- ❑ Hen Harrier *Circus cyaneus*
- ❑ Pallid Harrier *Circus macrourus*
- ❑ Montagu's Harrier *Circus pygargus*
- ❑ Shikra *Accipiter badius*
- ❑ Levant Sparrowhawk *Accipiter brevipes*
- ❑ Eurasian Sparrowhawk *Accipiter nisus*
- ❑ Northern Goshawk *Accipiter gentilis*
- ❑ Steppe Buzzard *Buteo buteo vulpinus*
- ❑ Long-legged Buzzard *Buteo rufinus*
- ❑ Lesser Spotted Eagle *Aquila pomarina*
- ❑ Greater Spotted Eagle *Aquila clanga*
- ❑ Steppe Eagle *Aquila nipalensis*

- ❑ Eastern Imperial Eagle *Aquila heliaca*
- ❑ Golden Eagle *Aquila chrysaetos*
- ❑ Booted Eagle *Aquila pennata*
- ❑ Bonelli's Eagle *Aquila fasciatus*
- ❑ Lesser Kestrel *Falco naumanni*
- ❑ Common Kestrel *Falco tinnunculus*
- ❑ Red-footed Falcon *Falco vespertinus*
- ❑ Amur Falcon *Falco amurensis*
- ❑ Eleonora's Falcon *Falco eleonorae*
- ❑ Sooty Falcon *Falco concolor*
- ❑ Merlin *Falco columbarius*
- ❑ Eurasian Hobby *Falco subbuteo*
- ❑ Lanner Falcon *Falco biarmicus*
- ❑ Saker Falcon *Falco cherrug*
- ❑ Peregrine Falcon *Falco peregrinus*
- ❑ Barbary Falcon *Falco (peregrinus) pelegrinoides*
- ❑ Macqueen's Bustard *Chlamydotis macqueenii*
- ❑ Water Rail *Rallus aquaticus*
- ❑ Corncrake *Crex crex*
- ❑ White-breasted Waterhen *Amaurornis phoenicurus*
- ❑ Little Crake *Porzana parva*
- ❑ Baillon's Crake *Porzana pusilla*
- ❑ Spotted Crake *Porzana porzana*
- ❑ Purple Swamphen *Porphyrio porphyrio*
- ❑ Common Moorhen *Gallinula chloropus*
- ❑ Red-knobbed Coot *Fulica cristata*
- ❑ Eurasian Coot *Fulica atra*
- ❑ Demoiselle Crane *Anthropoides virgo*
- ❑ Common Crane *Grus grus*
- ❑ Eurasian Stone-curlew *Burhinus oedicnemus*
- ❑ Eurasian Oystercatcher *Haematopus ostralegus*
- ❑ Crab-plover *Dromas ardeola*
- ❑ Black-winged Stilt *Himantopus himantopus*
- ❑ Pied Avocet *Recurvirostra avosetta*
- ❑ Northern Lapwing *Vanellus vanellus*
- ❑ Spur-winged Lapwing *Vanellus spinosus*
- ❑ Red-wattled Lapwing *Vanellus indicus*
- ❑ Sociable Lapwing *Vanellus gregarius*
- ❑ White-tailed Lapwing *Vanellus leucurus*
- ❑ Eurasian Golden Plover *Pluvialis apricaria*
- ❑ Pacific Golden Plover *Pluvialis fulva*
- ❑ Grey Plover *Pluvialis squatarola*
- ❑ Common Ringed Plover *Charadrius hiaticula*
- ❑ Little Ringed Plover *Charadrius dubius*
- ❑ Kittlitz's Plover *Charadrius pecuarius*
- ❑ Kentish Plover *Charadrius alexandrinus*
- ❑ Lesser Sand Plover *Charadrius atrifrons*
- ❑ Greater Sand Plover *Charadrius leschenaultii*
- ❑ Caspian Plover *Charadrius asiaticus*

- ❑ Eurasian Dotterel *Charadrius morinellus*
- ❑ Greater Painted Snipe *Rostratula benghalensis*
- ❑ Eurasian Woodcock *Scolopax rusticola*
- ❑ Jack Snipe *Lymnocryptes minimus*
- ❑ Pin-tailed Snipe *Gallinago stenura*
- ❑ Great Snipe *Gallinago media*
- ❑ Common Snipe *Gallinago gallinago*
- ❑ Black-tailed Godwit *Limosa limosa*
- ❑ Bar-tailed Godwit *Limosa lapponica*
- ❑ Whimbrel *Numenius phaeopus*
- ❑ Eurasian Curlew *Numenius arquata*
- ❑ Spotted Redshank *Tringa erythropus*
- ❑ Common Redshank *Tringa totanus*
- ❑ Marsh Sandpiper *Tringa stagnatilis*
- ❑ Common Greenshank *Tringa nebularia*
- ❑ Lesser Yellowlegs *Tringa flavipes*
- ❑ Green Sandpiper *Tringa ochropus*
- ❑ Wood Sandpiper *Tringa glareola*
- ❑ Terek Sandpiper *Xenus cinereus*
- ❑ Common Sandpiper *Actitis hypoleucos*
- ❑ Ruddy Turnstone *Arenaria interpres*
- ❑ Great Knot *Calidris tenuirostris*
- ❑ Red Knot *Calidris canutus*
- ❑ Sanderling *Calidris alba*
- ❑ Red-necked Stint *Calidris ruficollis*
- ❑ Little Stint *Calidris minuta*
- ❑ Temminck's Stint *Calidris temminckii*
- ❑ Long-toed Stint *Calidris subminuta*
- ❑ Pectoral Sandpiper *Calidris melanotos*
- ❑ Curlew Sandpiper *Calidris ferruginea*
- ❑ Dunlin *Calidris alpina*
- ❑ Broad-billed Sandpiper *Limicola falcinellus*
- ❑ Buff-breasted Sandpiper *Tryngites subruficollis*
- ❑ Ruff *Philomachus pugnax*
- ❑ Wilson's Phalarope *Steganopus tricolor*
- ❑ Red-necked Phalarope *Phalaropus lobatus*
- ❑ Grey Phalarope *Phalaropus fulicarius*
- ❑ Cream-coloured Courser *Cursorius cursor*
- ❑ Collared Pratincole *Glareola pratincola*
- ❑ Oriental Pratincole *Glareola maldivarum*
- ❑ Black-winged Pratincole *Glareola nordmanni*
- ❑ Small Pratincole *Glareola lactea*
- ❑ Brown Noddy *Anous stolidus*
- ❑ Lesser Noddy *Anous tenuirostris*
- ❑ Black-legged Kittiwake *Rissa tridactyla*
- ❑ Sabine's Gull *Xema sabini*
- ❑ Slender-billed Gull *Chroicocephalus genei*
- ❑ Brown-headed Gull *Chroicocephalus brunnicephalus*
- ❑ Common Black-headed Gull *Chroicocephalus ridibundus*

- ❑ Little Gull *Hydrocoloeus minutus*
- ❑ Franklin's Gull *Larus pipixcan*
- ❑ Mediterranean Gull *Larus melanocephalus*
- ❑ Great Black-headed Gull *Larus ichthyaetus*
- ❑ White-eyed Gull *Larus leucophthalmus*
- ❑ Sooty Gull *Larus hemprichii*
- ❑ Common Gull *Larus canus*
- ❑ Caspian Gull *Larus cachinnans*
- ❑ Baltic Gull *Larus fuscus fuscus*
- ❑ Heuglin's Gull *Larus heuglini*
- ❑ Steppe Gull *Larus barabensis*
- ❑ Gull-billed Tern *Gelochelidon nilotica*
- ❑ Caspian Tern *Hydroprogne caspia*
- ❑ Swift Tern *Sterna bergii*
- ❑ Lesser Crested Tern *Sterna bengalensis*
- ❑ Sandwich Tern *Sterna sandvicensis*
- ❑ Little Tern *Sternula albifrons*
- ❑ Saunders's Tern *Sternula (albifrons) saundersi*
- ❑ Bridled Tern *Onychoprion anaethetus*
- ❑ Sooty Tern *Onychoprion fuscatus*
- ❑ Roseate Tern *Sterna dougallii*
- ❑ Common Tern *Sterna hirundo*
- ❑ White-cheeked Tern *Sterna repressa*
- ❑ Arctic Tern *Sterna paradisaea*
- ❑ Whiskered Tern *Chlidonias hybrida*
- ❑ White-winged Tern *Chlidonias leucopterus*
- ❑ Black Tern *Chlidonias niger*
- ❑ Brown/South Polar Skua *Stercorarius antarcticus/maccormicki*
- ❑ Pomarine Skua *Stercorarius pomarinus*
- ❑ Arctic Skua *Stercorarius parasiticus*
- ❑ Long-tailed Skua *Stercorarius longicaudus*
- ❑ Pallas's Sandgrouse *Syrrhaptes paradoxus*
- ❑ Pin-tailed Sandgrouse *Pterocles alchata*
- ❑ Chestnut-bellied Sandgrouse *Pterocles exustus*
- ❑ Spotted Sandgrouse *Pterocles senegallus*
- ❑ Black-bellied Sandgrouse *Pterocles orientalis*
- ❑ Lichtenstein's Sandgrouse *Pterocles lichtensteinii*
- ❑ Rock Dove *Columba livia*
- ❑ Feral Pigeon *Columba livia forma domestica*
- ❑ Stock Dove *Columba oenas*
- ❑ Common Woodpigeon *Columba palumbus*
- ❑ European Turtle Dove *Streptopelia turtur*
- ❑ Rufous Turtle Dove *Streptopelia (orientalis) meena*
- ❑ Eurasian Collared Dove *Streptopelia decaocto*
- ❑ Red Turtle Dove *Streptopelia tranquebarica*
- ❑ Laughing Dove *Spilopelia senegalensis*
- ❑ Namaqua Dove *Oena capensis*
- ❑ Alexandrine Parakeet *Psittacula eupatria*
- ❑ Rose-ringed Parakeet *Psittacula krameri*

- ❑ Great Spotted Cuckoo *Clamator glandarius*
- ❑ Pied Cuckoo *Oxylophus jacobinus*
- ❑ Asian Koel *Eudynamys scolopaceus*
- ❑ Common Cuckoo *Cuculus canorus*
- ❑ (Western) Barn Owl *Tyto alba*
- ❑ Pallid Scops Owl *Otus brucei*
- ❑ Eurasian Scops Owl *Otus scops*
- ❑ Pharaoh Eagle Owl *Bubo ascalaphus*
- ❑ Little Owl *Athene noctua*
- ❑ Lilith Owl *Athene (noctua) lilith*
- ❑ Long-eared Owl *Asio otus*
- ❑ Short-eared Owl *Asio flammeus*
- ❑ European Nightjar *Caprimulgus europaeus*
- ❑ Egyptian Nightjar *Caprimulgus aegyptius*
- ❑ Sykes's Nightjar *Caprimulgus mahrattensis*
- ❑ Alpine Swift *Tachymarptis melba*
- ❑ Common Swift *Apus apus*
- ❑ Pallid Swift *Apus pallidus*
- ❑ Fork-tailed Swift *Apus pacificus*
- ❑ Little Swift *Apus affinis*
- ❑ Indian Roller *Coracias benghalensis*
- ❑ European Roller *Coracias garrulus*
- ❑ White-throated Kingfisher *Halcyon smyrnensis*
- ❑ Grey-headed Kingfisher *Halcyon leucocephala*
- ❑ Collared Kingfisher *Todirhamphus chloris*
- ❑ Common Kingfisher *Alcedo atthis*
- ❑ Pied Kingfisher *Ceryle rudis*
- ❑ White-throated Bee-eater *Merops albicollis*
- ❑ Green Bee-eater *Merops orientalis*
- ❑ Blue-cheeked Bee-eater *Merops persicus*
- ❑ European Bee-eater *Merops apiaster*
- ❑ Eurasian Hoopoe *Upupa epops*
- ❑ Eurasian Wryneck *Jynx torquilla*
- ❑ Brown Shrike *Lanius cristatus*
- ❑ Red-backed Shrike *Lanius collurio*
- ❑ Daurian Shrike *Lanius isabellinus*
- ❑ Turkestan Shrike *Lanius (isabellinus) phoenicuroides*
- ❑ Bay-backed Shrike *Lanius vittatus*
- ❑ Long-tailed Shrike *Lanius schach*
- ❑ Lesser Grey Shrike *Lanius minor*
- ❑ Great Grey Shrike *Lanius excubitor*
- ❑ Southern Grey Shrike *Lanius meridionalis*
- ❑ Steppe Grey Shrike *Lanius (meridionalis) pallidirostris*
- ❑ Woodchat Shrike *Lanius senator*
- ❑ Masked Shrike *Lanius nubicus*
- ❑ Eurasian Golden Oriole *Oriolus oriolus*
- ❑ Black Drongo *Dicrurus macrocercus*
- ❑ Ashy Drongo *Dicrurus leucophaeus*
- ❑ House Crow *Corvus splendens*

- ❑ Brown-necked Raven *Corvus ruficollis*
- ❑ Northern Raven *Corvus corax*
- ❑ Fan-tailed Raven *Corvus rhipidurus*
- ❑ Hypocolius *Hypocolius ampelinus*
- ❑ Greater Hoopoe-Lark *Alaemon alaudipes*
- ❑ Calandra Lark *Melanocorypha calandra*
- ❑ Bimaculated Lark *Melanocorypha bimaculata*
- ❑ Bar-tailed Lark *Ammomanes cinctura*
- ❑ Desert Lark *Ammomanes deserti*
- ❑ Greater Short-toed Lark *Calandrella brachydactyla*
- ❑ Lesser Short-toed Lark *Calandrella rufescens*
- ❑ Dunn's Lark *Eremalauda dunni*
- ❑ Crested Lark *Galerida cristata*
- ❑ Eurasian Skylark *Alauda arvensis*
- ❑ Oriental Skylark *Alauda gulgula*
- ❑ Black-crowned Sparrow-Lark *Eremopterix nigriceps*
- ❑ Temminck's Lark *Eremophila bilopha*
- ❑ Red-whiskered Bulbul *Pycnonotus jocosus*
- ❑ White-eared Bulbul *Pycnonotus (leucogenys) leucotis*
- ❑ Red-vented Bulbul *Pycnonotus cafer*
- ❑ White-spectacled Bulbul *Pycnonotus xanthopygos*
- ❑ Brown-throated Martin *Riparia paludicola*
- ❑ Sand Martin *Riparia riparia*
- ❑ Pale Martin *Riparia (riparia) diluta*
- ❑ Barn Swallow *Hirundo rustica*
- ❑ Wire-tailed Swallow *Hirundo smithii*
- ❑ Eurasian Crag Martin *Ptyonoprogne rupestris*
- ❑ Pale Crag Martin *Ptyonoprogne (fuligula) obsoleta*
- ❑ Common House Martin *Delichon urbicum*
- ❑ Asian House Martin *Delichon (urbicum) dasypus*
- ❑ Red-rumped Swallow *Cecropis daurica*
- ❑ Streak-throated Swallow *Petrochelidon fluvicola*
- ❑ Cetti's Warbler *Cettia cetti*
- ❑ Willow Warbler *Phylloscopus trochilus*
- ❑ Common Chiffchaff *Phylloscopus collybita*
- ❑ Siberian Chiffchaff *Phylloscopus (collybita) tristis*
- ❑ Plain Leaf Warbler *Phylloscopus neglectus*
- ❑ Eastern Bonelli's Warbler *Phylloscopus orientalis*
- ❑ Wood Warbler *Phylloscopus sibilatrix*
- ❑ Dusky Warbler *Phylloscopus fuscatus*
- ❑ Radde's Warbler *Phylloscopus schwarzi*
- ❑ Yellow-browed Warbler *Phylloscopus inornatus*
- ❑ Hume's Leaf Warbler *Phylloscopus humei*
- ❑ Arctic Warbler *Phylloscopus borealis*
- ❑ Green Warbler *Phylloscopus nitidus*
- ❑ Basra Reed warbler *Acrocephalus griseldis*
- ❑ Great Reed Warbler *Acrocephalus arundinaceus*
- ❑ Indian Reed Warbler *Acrocephalus (stentoreus) brunnescens*
- ❑ Moustached Warbler *Acrocephalus melanopogon*

- ❑ Sedge Warbler *Acrocephalus schoenobaenus*
- ❑ Paddyfield Warbler *Acrocephalus agricola*
- ❑ Blyth's Reed Warbler *Acrocephalus dumetorum*
- ❑ Caspian Reed Warbler *Acrocephalus (scirpaceus) fuscus*
- ❑ Marsh Warbler *Acrocephalus palustris*
- ❑ Booted Warbler *Iduna caligata*
- ❑ Sykes's Warbler *Iduna rama*
- ❑ Eastern Olivaceous Warbler *Iduna pallida*
- ❑ Upcher's Warbler *Hippolais languida*
- ❑ Icterine Warbler *Hippolais icterina*
- ❑ Common Grasshopper Warbler *Locustella naevia*
- ❑ River Warbler *Locustella fluviatilis*
- ❑ Savi's Warbler *Locustella luscinioides*
- ❑ Scrub Warbler *Scotocerca inquieta*
- ❑ Graceful Prinia *Prinia gracilis*
- ❑ Arabian Babbler *Turdoides squamiceps*
- ❑ Eurasian Blackcap *Sylvia atricapilla*
- ❑ Garden Warbler *Sylvia borin*
- ❑ Barred Warbler *Sylvia nisoria*
- ❑ Lesser Whitethroat *Sylvia curruca*
- ❑ Desert Whitethroat *Sylvia minula*
- ❑ Hume's Whitethroat *Sylvia althaea*
- ❑ Eastern Orphean Warbler *Sylvia crassirostris*
- ❑ Asian Desert Warbler *Sylvia nana*
- ❑ Common Whitethroat *Sylvia communis*
- ❑ Ménétriés's Warbler *Sylvia mystacea*
- ❑ Bank Myna *Acridotheres ginginianus*
- ❑ Common Myna *Acridotheres tristis*
- ❑ Pied Myna *Sturnus contra*
- ❑ Brahminy Starling *Sturnia pagodarum*
- ❑ Rose-coloured Starling *Pastor roseus*
- ❑ Common Starling *Sturnus vulgaris*
- ❑ Wattled Starling *Creatophora cinerea*
- ❑ Violet-backed Starling *Cinnyricinclus leucogaster*
- ❑ Ring Ouzel *Turdus torquatus*
- ❑ Eurasian Blackbird *Turdus merula*
- ❑ Eyebrowed Thrush *Turdus obscurus*
- ❑ Black-throated Thrush *Turdus atrogularis*
- ❑ Dusky Thrush *Turdus eunomus*
- ❑ Fieldfare *Turdus pilaris*
- ❑ Redwing *Turdus iliacus*
- ❑ Song Thrush *Turdus philomelos*
- ❑ Mistle Thrush *Turdus viscivorus*
- ❑ European Robin *Erithacus rubecula*
- ❑ Bluethroat *Luscinia svecica*
- ❑ Thrush Nightingale *Luscinia luscinia*
- ❑ Eastern Nightingale *Luscinia (megarhynchos) golzii*
- ❑ Red-flanked Bluetail *Tarsiger cyanurus*
- ❑ White-throated Robin *Irania gutturalis*

- ❑ Rufous-tailed Scrub Robin *Cercotrichas galactotes*
- ❑ Black Scrub Robin *Cercotrichas podobe*
- ❑ Eversmann's Redstart *Phoenicurus erythronotus*
- ❑ Black Redstart *Phoenicurus ochruros*
- ❑ Common Redstart *Phoenicurus phoenicurus*
- ❑ Whinchat *Saxicola rubetra*
- ❑ European Stonechat *Saxicola rubicola*
- ❑ Siberian Stonechat *Saxicola maurus*
- ❑ Pied Stonechat *Saxicola caprata*
- ❑ Isabelline Wheatear *Oenanthe isabellina*
- ❑ Northern Wheatear *Oenanthe oenanthe*
- ❑ Kurdistan Wheatear *Oenanthe xanthoprymna*
- ❑ Red-tailed Wheatear *Oenanthe chrysopygia*
- ❑ Pied Wheatear *Oenanthe pleschanka*
- ❑ Eastern Black-eared Wheatear *Oenanthe (hispanica) melanoleuca*
- ❑ Desert Wheatear *Oenanthe deserti*
- ❑ Eastern Mourning Wheatear *Oenanthe lugens*
- ❑ Finsch's Wheatear *Oenanthe finschii*
- ❑ Variable Wheatear *Oenanthe picata*
- ❑ Hume's Wheatear *Oenanthe albonigra*
- ❑ White-crowned Wheatear *Oenanthe leucopyga*
- ❑ Hooded Wheatear *Oenanthe monacha*
- ❑ Blackstart *Cercomela melanura*
- ❑ Rufous-tailed Rock Thrush *Monticola saxatilis*
- ❑ Blue Rock Thrush *Monticola solitarius*
- ❑ Spotted Flycatcher *Muscicapa striata*
- ❑ Eurasian Pied Flycatcher *Ficedula hypoleuca*
- ❑ Semi-collared Flycatcher *Ficedula semitorquata*
- ❑ Red-breasted Flycatcher *Ficedula parva*
- ❑ Taiga Flycatcher *Ficedula albicilla*
- ❑ Blue-and-white Flycatcher *Cyanoptila cyanomelana*
- ❑ Purple Sunbird *Cinnyris asiaticus*
- ❑ Indian House Sparrow *Passer (domesticus) indicus*
- ❑ Spanish Sparrow *Passer hispaniolensis*
- ❑ Dead Sea Sparrow *Passer moabiticus*
- ❑ Eurasian Tree Sparrow *Passer montanus*
- ❑ Pale Rockfinch *Carpospiza brachydactyla*
- ❑ Yellow-throated Sparrow *Gymnoris xanthocollis*
- ❑ Golden-backed Weaver *Ploceus jacksoni*
- ❑ Streaked Weaver *Ploceus manyar*
- ❑ Indian Silverbill *Lonchura malabarica*
- ❑ Scaly-breasted Munia *Lonchura punctulata*
- ❑ Forest Wagtail *Dendronanthus indicus*
- ❑ Yellow Wagtail *Motacilla flava*
  - ❑ - Sykes's Wagtail *Motacilla (flava) beema*
  - ❑ - White-headed Wagtail *Motacilla (flava) leucocephala*
  - ❑ - Yellow-headed Wagtail *Motacilla (flava) lutea*
  - ❑ - Grey-headed Wagtail *Motacilla (flava) thunbergi*
- ❑ Black-headed Wagtail *Motacilla (flava) feldegg*

- ❑ Citrine Wagtail *Motacilla citreola*
- ❑ Grey Wagtail *Motacilla cinerea*
- ❑ White Wagtail *Motacilla albas*
- ❑ Masked Wagtail *Motacilla (alba) personata*
- ❑ Richard's Pipit *Anthus (novaeseelandiae) richardi*
- ❑ Blyth's Pipit *Anthus godlewskii*
- ❑ Tawny Pipit *Anthus campestris*
- ❑ Long-billed Pipit *Anthus similis*
- ❑ Meadow Pipit *Anthus pratensis*
- ❑ Tree Pipit *Anthus trivialis*
- ❑ Olive-backed Pipit *Anthus hodgsoni*
- ❑ Red-throated Pipit *Anthus cervinus*
- ❑ Buff-bellied Pipit *Anthus (rubescens) japonicus*
- ❑ Water Pipit *Anthus spinoletta*
- ❑ Common Chaffinch *Fringilla coelebs*
- ❑ Brambling *Fringilla montifringilla*
- ❑ Eurasian Siskin *Carduelis spinus*
- ❑ European Goldfinch *Carduelis carduelis*
- ❑ Common Linnet *Carduelis cannabina*
- ❑ Trumpeter Finch *Bucanetes githagineus*
- ❑ Common Rosefinch *Carpodacus erythrinus*
- ❑ Corn Bunting *Emberiza calandra*
- ❑ Yellowhammer *Emberiza citrinella*
- ❑ Pine Bunting *Emberiza leucocephalos*
- ❑ White-capped Bunting *Emberiza stewarti*
- ❑ Eastern Cinereous Bunting *Emberiza (cineracea) semenowi*
- ❑ Ortolan Bunting *Emberiza hortulana*
- ❑ Striolated Bunting *Emberiza striolata*
- ❑ Little Bunting *Emberiza pusilla*
- ❑ Rustic Bunting *Emberiza rustica*
- ❑ Yellow-breasted Bunting *Emberiza aureola*
- ❑ Black-headed Bunting *Emberiza melanocephala*
- ❑ Red-headed Bunting *Emberiza bruniceps*
- ❑ Common Reed Bunting *Emberiza schoeniclus*

INDEX

T

Printed and bound by CPI Group (UK) Ltd, Croydon, CR0 4YY
26/11/2025
02006096-0001